Nature-Inspired Computation and Swarm Intelligence

Algorithms, Theory and Applications

Nature-Inspired Computation and Swarm Intelligence

Algorithms, Theory and Applications

Edited by

Xin-She Yang

Middlesex University London
School of Science and Technology
London, United Kingdom

ACADEMIC PRESS
An imprint of Elsevier

Library of Congress Cataloging-in-Publication Data
A catalog record for this book is available from the Library of Congress

British Library Cataloguing-in-Publication Data
A catalogue record for this book is available from the British Library

ISBN: 978-0-12-819714-1

For information on all Academic Press publications
visit our website at https://www.elsevier.com/books-and-journals

Publisher: Mara Conner
Acquisitions Editor: Tim Pitts
Editorial Project Manager: Emily Thomson
Production Project Manager: Nirmala Arumugam
Designer: Victoria Pearson

Typeset by VTeX

Contents

PART 1 ALGORITHMS

CHAPTER 1 Nature-inspired computation and swarm intelligence: a state-of-the-art overview **3**

Xin-She Yang, Mehmet Karamanoglu

PART 3 APPLICATIONS

CHAPTER 8 Fine-tuning restricted Boltzmann machines using quaternion-based flower pollination algorithm 111

Leandro Aparecido Passos, Gustavo Henrique de Rosa,
Douglas Rodrigues, João Paulo Papa

CHAPTER 9 Traveling salesman problem: a perspective review of recent research and new results with bio-inspired metaheuristics 135

Eneko Osaba, Xin-She Yang, Javier Del Ser

List of contributors

Utkarsh Agrawal
Maharaja Agrasen Institute of Technology, Delhi, India

Ashik Ahmed
Islamic University of Technology (IUT),
Electrical and Electronic Engineering, Gazipur, Bangladesh

Jatin Arora
Maharaja Agrasen Institute of Technology, Delhi, India

Oscar Castillo
Tijuana Institute of Technology, Division of Graduate Studies and Research,
Tijuana, Mexico

Francesco Contino
Université Catholique de Louvain, Thermodynamics and Fluid Mechanics
Group, Louvain, Belgium

Javier Del Ser
TECNALIA, Basque Research and Technology Alliance (BRTA), Derio, Spain
University of the Basque Country (UPV/EHU), Bilbao, Spain

Floriano De Rango
University of Calabria, DIMES Department, Cosenza, Italy

Gustavo Henrique de Rosa
São Paulo State University, Department of Computing, Bauru, Brazil

Thomas Edward
The University of the West Indies Cave Hill Campus, Department of Computer
Science, Mathematics and Physics, Bridgetown, Barbados

Qin-Wei Fan
Xi'an Polytechnic University, College of Science, Xi'an, China

Andreas Floros
Ionian University, Department of Audio and Vidual Arts, Corfu, Greece

Akemi Gálvez
Toho University, Department of Information Sciences, Faculty of Sciences,
Funabashi, Japan
University of Cantabria, Department of Applied Mathematics and Computational
Sciences, Santander, Spain

Curtis Gittens

The University of the West Indies Cave Hill Campus, Department of Computer Science, Mathematics and Physics, Bridgetown, Barbados

Mechelle Gittens

The University of the West Indies Cave Hill Campus, Department of Computer Science, Mathematics and Physics, Bridgetown, Barbados

Deepak Gupta

Maharaja Agrasen Institute of Technology, Delhi, India

Xing-Shi He

Xi'an Polytechnic University, College of Science, Xi'an, China

Jacob Hunte

Western University, Computer Science Department, Middlesex College, London, Ontario, Canada

Andrés Iglesias

Toho University, Department of Information Sciences, Faculty of Sciences, Funabashi, Japan
University of Cantabria, Department of Applied Mathematics and Computational Sciences, Santander, Spain

Anestis I. Kalfas

Aristotle University, Department of Mechanical Engineering, Thessaloniki, Greece

Maximos Kaliakatsos-Papakostas

Athena Research and Innovation Centre,
Institute for Language and Speech Processing, Athens, Greece

Mehmet Karamanoglu

Middlesex University London, School of Science and Technology, London, United Kingdom

Ashish Khanna

Maharaja Agrasen Institute of Technology, Delhi, India

Kirti

Chitkara University Institute of Engineering and Technology, Chitkara University, Punjab, India

Serdar Kockanat

Sivas Cumhuriyet University, Department of Electrical and Electronics Engineering, Sivas, Turkey

Piotr A. Kowalski

AGH University of Science and Technology,
Faculty of Physics and Applied Computer Science, Kraków, Poland
Systems Research Institute, Polish Academy of Sciences, Warsaw, Poland

Konstantinos Kyprianidis

Mälardalens Högskola, Energy Engineering, Västerås, Sweden

Szymon Łukasik

AGH University of Science and Technology,
Faculty of Physics and Applied Computer Science, Kraków, Poland
Systems Research Institute, Polish Academy of Sciences, Warsaw, Poland

Hanan Lutfiyya

Western University, Computer Science Department, Middlesex College, London,
Ontario, Canada

Patricia Melin

Tijuana Institute of Technology, Division of Graduate Studies and Research,
Tijuana, Mexico

Eneko Osaba

TECNALIA, Basque Research and Technology Alliance (BRTA), Derio, Spain

Nunzia Palmieri

University of Calabria, DIMES Department, Cosenza, Italy

João Paulo Papa

São Paulo State University, Department of Computing, Bauru, Brazil

Leandro Aparecido Passos

São Paulo State University, Department of Computing, Bauru, Brazil

Douglas Rodrigues

São Paulo State University, Department of Computing, Bauru, Brazil

Daniela Sánchez

Tijuana Institute of Technology, Division of Graduate Studies and Research,
Tijuana, Mexico

Anshu Singla

Chitkara University Institute of Engineering and Technology, Chitkara University,
Punjab, India

Patricia Suárez

University of Cantabria, Department of Applied Mathematics and Computational
Sciences, Santander, Spain

Swati Swayamsiddha

Kalinga Institute of Industrial Technology, School of Electronics Engineering, KIIT Deemed to be University, Bhubaneswar, India

Mauro Tropea

University of Calabria, DIMES Department, Cosenza, Italy

Panagiotis Tsirikoglou

Limmat Scientific AG, Zurich, Switzerland

Quazi Nafees Ul Islam

Islamic University of Technology (IUT),
Electrical and Electronic Engineering, Gazipur, Bangladesh

Michael N. Vrahatis

University of Patras, Department of Mathematics, Patras, Greece

Xin-She Yang

Middlesex University London, School of Science and Technology, London, United Kingdom

Yu-Xin Zhao

Harbin Engineering University, College of Automation, Harbin, China

About the editor

Xin-She Yang obtained his PhD in applied mathematics from the University of Oxford. He then worked at Cambridge University and the National Physical Laboratory (UK) as a senior research scientist. Now he is reader at Middlesex University London, and an elected by-fellow at Cambridge University.

He is also the IEEE Computer Intelligence Society (CIS) chair for the Task Force on Business Intelligence and Knowledge Management, director of the International Consortium for Optimization and Modelling in Science and Industry (iCOMSI), and an editor of Springer's book series *Springer Tracts in Nature-Inspired Computing* (STNIC).

With more than 20 years of research and teaching experience, he has authored 12 books and edited more than 15 books. He has published more than 250 research papers in international peer-reviewed journals and conference proceedings. With more than 41,000 citations, he has been on the prestigious lists of Clarivate Analytics and Web of Science Highly Cited Researchers for four consecutive years (2016–2019). He also serves on the editorial boards of many international journals, including the *International Journal of Bio-Inspired Computation*, Elsevier's *Journal of Computational Science* (JoCS), the *International Journal of Parallel, Emergent and Distributed Systems*, and the *International Journal of Computer Mathematics*. He is also the editor-in-chief of the *International Journal of Mathematical Modelling and Numerical Optimisation*.

Preface

Nature-inspired computation and swarm intelligence have become popular and effective tools for solving problems in optimization, computational intelligence, soft computing, and data science, as well as machine learning. Nature-inspired algorithms include bio-inspired algorithms, physics/chemistry-inspired algorithms, and social algorithms. A few good examples are ant colony optimization, the bat algorithm, cuckoo search, particle swarm optimization, the flower pollination algorithm, and genetic algorithms. These algorithms are flexible, versatile, efficient, and yet simple to implement. As global optimizers, these algorithms have been applied in almost every area of science and engineering as well as industrial applications.

In recent years, new algorithms and new applications have emerged, and the literature is rapidly expanding. It is time to provide a comprehensive review of relevant state-of-the-art developments in algorithms, theory, and applications concerning nature-inspired algorithms and swarm intelligence. Thus, this edited book intends to review and document the new developments with a focus on the introduction of nature-inspired algorithms, theoretical analysis of algorithms, and guides for implementations in the context of practical applications.

The book consists of three parts: algorithms, theory, and applications. Part 1 starts with an overview of most widely used nature-inspired algorithms for optimization, followed by the qualitative analysis and the discussions of open problems. Then, some tutorials on the bat algorithm, cuckoo search, the firefly algorithm, and flower pollination algorithms are given with MATLAB® codes. Further introduction of algorithms in the context of wireless communication is also carried out in this part.

Part 2 introduces the fundamentals for mathematical analyses of algorithms, starting with the basic formulations of optimization problems to complexity theory and then focusing on the probability theory and Markov chains. A unified framework is then discussed for analyzing nature-inspired algorithms from various angles and perspectives, including algorithmic complexity, dynamical systems, convergence analysis, no-free-lunch theorems, benchmarking, and others.

Part 3 provides a diverse range of applications and case studies. This part starts with the tuning of restricted Boltzmann machines, followed by a detailed review on the traveling salesman problem and metaheuristic approaches to clustering. Then, classification of blood cells is carried out by the bat algorithm, and modular granular neural networks with firefly algorithm are used for time series predictions. In addition, music generation using artificial intelligence has been reviewed in a comprehensive manner, followed by the controller design for islanded microgrids using the nondominated sorting firefly algorithm and a detailed case study of bat robots as swarm robotics. Further applications include electrical harmonics estimation in power grids, intelligent image segmentation, and optimal order allocation in e-markets. This part concludes with swarm robotics with the emphasis on multirobot coordination, followed by the engineering designs and applications of metaheuristics. It can be expected that all these diverse topics will inspire more research in this active area of research and application.

The book strives to provide a contemporary snapshot for the state-of-the-art developments in nature-inspired computation and swarm intelligence, with a balance of theory and diverse applications so that readers who are interested in either algorithms or theory and applications will all benefit from this book. Therefore, this book can be a useful reference for graduates, lecturers, and researchers in computer science, swarm intelligence, nature-inspired computing, machine learning, engineering optimization, data science, and management science, as well as industries.

Xin-She Yang
London, United Kingdom
7 Nov 2019

Acknowledgments

I would like to thank all the independent reviewers who have reviewed all chapter manuscripts and provided useful comments. I also thank my editors, Tim Pitts, Emily Thomson, and Sheela Bernardine Josy, and staff at Elsevier for their help and professionalism. Last but not least, I thank my family for all the help and support.

Xin-She Yang
7 Nov 2019

Algorithms
Introduction and tutorials

Nature-inspired computation and swarm intelligence: a state-of-the-art overview

Xin-She Yang, Mehmet Karamanoglu

Middlesex University London, School of Science and Technology, London, United Kingdom

CONTENTS

1.1 Introduction

Optimization is very important in many applications, from engineering design and business planning to data mining and machine learning. The purpose of optimization can be anything – to minimize the energy consumption, costs, waste, travel time, and environmental impact, or to maximize the profit, outputs, performance, efficiency, and sustainability. Obviously, the resources, time, and money are always limited in

Nature-Inspired Computation and Swarm Intelligence. https://doi.org/10.1016/B978-0-12-819714-1.00010-5
Copyright © 2020 Elsevier Ltd. All rights reserved.

3

real-world applications, and we have to find optimal solutions to various design problems, subject to a wide range of complex constraints (Boyd and Vandenberghe, 2004; Yang, 2014).

To properly model and formulate optimization problems, we need mathematical optimization or mathematical programming, which can span different subject areas such as operations research, business management, machine learning, data mining, and design engineering, as well as industrial applications. An important part of the tasks is to model and formulate the cost function or objective function so that different design options can be evaluated and compared. The main aim of solving optimization problems is to find the optimal solution or a set of optimal solutions such that the objective function can be minimized or maximized.

To solve optimization problems, optimization algorithms and techniques are needed. As most real-world problems are nonlinear in terms of their objective functions and constraints, it is necessary to use sophisticated optimization tools to deal with such problems. In addition, the evaluations of objective functions can be time consuming, which is especially true for problems related to aerodynamics, protein folding and drug design, large-scale business planning, network flow, scheduling, data mining, and machine learning.

Even with the ever-increasing power of modern computers, simple brute force methods for seeking optimal solutions are not practical and not desirable. Efficient algorithms are crucial to almost all applications. There are a vast array of different algorithms for optimization, including traditional methods such as gradient-based algorithms, the trust-region method and the interior-point method, and evolutionary algorithms such as evolutionary strategy (ES) and genetic algorithm (GA) (Holland, 1975). In recent years, these methods are enriched by a spectrum of new algorithms belonging to nature-inspired algorithms and swarm intelligence, including ant colony optimization (ACO) (Dorigo, 1992), particle swarm optimization (PSO) (Kennedy and Eberhart, 1995), the bat algorithm (BA) (Yang, 2010), the firefly algorithm (FA) (Yang, 2009), cuckoo search (CS) (Yang and Deb, 2009), and others (Yang, 2014).

Significant progress has been made since the 1990s, and thus the main purpose of this chapter is to provide an overview of the most recent developments concerning swarm intelligence and nature-inspired computation, with a focus on optimization algorithms.

1.2 Optimization and optimization algorithms

Many problems can be formulated as optimization problems. For example, design problems, routing problems, and scheduling problems can all be formulated as optimization. Even classification and clustering in data mining, machine learning, and artificial intelligence can all be formulated as or converted into optimization problems. As long as there is something to be maximized or minimized, subject to constraints, a suitable optimization problem can be formulated. However, formulating optimization problems is surely helpful to gain some insight into the problem, but it does not always make it easier to find solutions. In fact, many optimization problems are really hard to solve; a well-known example is the traveling salesman problem (TSP), which concerns the shortest distance to visit each city exactly once. This TSP is nondeterministic polynomial-time (NP) hard, which means that there are no efficient algorithms to solve such problems in general, and approximation and heuristic methods seem to be the main alternatives (Arara and Barak, 2009; Chabert, 1999; Ouaarab et al., 2014; Osaba et al., 2016).

1.2.1 Mathematical formulations

Mathematically speaking, an optimization problem consists of a set of design parameters or decision variables x_i ($i = 1, 2, ..., D$), which spans a decision space or design space by vectors in the form of $x = (x_1, x_2, ..., x_D) \in \mathbb{R}^D$.

The aim is to either minimize or maximize the cost or objective function $f(x)$, subject to some equality and inequality constraints. The optimal solution is a specific set of solutions (often a single solution) that can globally minimize or maximize $f(x)$, and such solutions should satisfy all the constraints (Yang, 2014).

In general, an optimization problem can be formulated as

$$\text{minimize} \quad f(x), \tag{1.1}$$

subject to

$$h_i(x) = 0, \quad (i = 1, 2, ..., M), \tag{1.2}$$

$$g_j(x) \leq 0, \quad (j = 1, 2, ..., N), \tag{1.3}$$

where h_i and g_j are the equality constraints and inequality constraints, respectively. As maximization problems can be converted into minimization by simply multiplying the objective by -1, we can formulate all optimization problems as minimization. In the special cases when all problem functions $f(x)$, h_i, and g_j are linear, it becomes a linear programming (LP) problem. Linear programs can be solved relatively easily by using the simplex method. However, the problem functions $f(x)$, $h_i(x)$, and $g_j(x)$ in most cases are all nonlinear, which can be challenging to solve.

If decision variables x_i can take only discrete values, the optimization problem becomes a discrete optimization problem. In the case x_i can only take integer values, such optimization is called integer programming. If some variables are discrete, while others are continuous, the optimization problem becomes a mixed integer program. Both mixed integer programming and integer programming are discrete or combinatorial optimization problems that are challenging to solve.

Optimization techniques are very diverse, ranging from the simplex method for linear programming to quadratic programming and convex optimization, and from the interior-point method to the trust-region method and many others (Boyd and Vandenberghe, 2004). Since recently, nature-inspired algorithms for optimization are also

widely used (Yang, 2014). For large-scale combinatorial problems, there are no efficient methods, and thus approximation methods and heuristic and metaheuristic
algorithms become useful alternatives.

1.2.2 Gradient-based algorithms

Among traditional optimization methods and techniques, gradient-based methods
are widely used. The essence of such gradient-based methods is to generate a new
solution x_{k+1} from the current solution x_k at iteration k using an increment Δx_k
calculated by the gradient of the objective function $f(x)$. That is,

$$x_{k+1} = x_k + \Delta x_k, \tag{1.4}$$

where

$$\Delta x_k = \alpha g(\nabla f, x^{(k)}). \tag{1.5}$$

Here, $g(\nabla f, x_k)$ is a function of the gradient ∇f and α is the step size that can vary
during iterations.

In general, the increment or modification Δx_k is a vector whose direction is usually along the negative gradient direction (i.e., $-\nabla f$) for minimization problems,
while it is usually in the positive gradient direction for maximization problems. There
are a class of such methods, and different variants or approaches are mainly different
in terms of $g(\nabla f, x^{(k)})$ and variations of step size α.

The well-known Newton method, or Newton–Raphson method, for optimization
was based on Newton's root finding algorithm, because optimality should occur at
$f'(x) = 0$. The main iteration equation in Newton's method is

$$x_{k+1} = x_k - \frac{\nabla f(x_k)}{\nabla^2 f(x_k)} = x_k - H^{-1} \nabla f(x_k), \tag{1.6}$$

which involves the use of both its gradient vector ∇f and its second derivative matrix
$\nabla^2 f$, called Hessian matrix,

$$H(x) \equiv \nabla^2 f(x) \equiv \begin{pmatrix} \frac{\partial^2 f}{\partial x_1^2} & \cdots & \frac{\partial^2 f}{\partial x_1 \partial x_n} \\ \vdots & \ddots & \vdots \\ \frac{\partial^2 f}{\partial x_n \partial x_1} & \cdots & \frac{\partial^2 f}{\partial x_n^2} \end{pmatrix}. \tag{1.7}$$

As the iterative calculations of the Hessian matrix can be computationally expensive, an alternative approach is to approximate it by an identity matrix I with a
scaling factor. That is, $H^{-1} = \alpha I$, leading to a quasi-Newton method,

$$x_{k+1} = x_k - \alpha I \nabla f(x_k) = x_k - \alpha \nabla f(x_k). \tag{1.8}$$

The parameter α is usually called the learning rate in the machine learning and optimization literature.

For large-scale problems, even the calculations of gradients can become expensive, which is especially true in deep learning (LeCun et al., 2015). In this case, the approximation or partial evaluation of the gradient vector can be desirable. One such method is the stochastic gradient descent (SGD) method, which only evaluates the gradients sparsely over a fraction of data, and the choice of such data may involve some randomness. In fact, the use of randomization can be truly advantageous in nature-inspired optimization algorithms to be discussed later where gradient information is not needed and randomization is mainly used to increase the diversity of the solution set.

1.2.3 Gradient-free algorithms

As we discussed earlier, the calculations of the derivatives can be expensive for large-scale problems, and thus it may be advantageous not to use such information. In addition, for problems with discontinuities, the objective function can be nonsmooth, and thus a gradient may not exist. In this case, gradient-free methods should be used.

Apart from the pattern search, one of the classical gradient-free methods is the Nelder–Mead simplex method (Nelder and Mead, 1965), which uses a simplex to guide search moves and only the nodal objective values of the simplex are compared as fitness. No gradient information is needed, and consequently the method is versatile and flexible to deal with both smooth and nonsmooth problems. In fact, most evolutionary algorithms are gradient-free methods; examples include ES and GA. In addition, almost all nature-inspired algorithms are also gradient-free algorithms, including simulated annealing, PSO, FA, CS, flower pollination algorithms (FPAs), and many others (Yang, 2014).

1.3 Nature-inspired algorithms for optimization

The main initiation of heuristic and metaheuristic algorithms was probably in the 1940s by the pioneer Alan Turing (Turing, 1948; Copeland, 2004). Then, GA and ES as well as evolutionary programming were developed in the 1960s (Holland, 1975; Fogel et al., 1966; Goldberg, 1989). In the 1980s, the simulated annealing appeared first (Kirkpatrik et al., 1983), followed by the seminal work on tabu search and metaheuristics by Fred Glover (Glover, 1986) and heuristics (Judea, 1984).

The 1990s were a golden decade for evolutionary computation and metaheuristics. First, ACO was developed by Marco Dorigo (Dorigo, 1992), followed by the development of PSO by James Kennedy and Russell C. Eberhart in 1995 (Kennedy and Eberhart, 1995). Then, in 1997, differential evolution (DE) was developed by R. Storn and K. Price (Storn and Price, 1997). At about the same time, the profound no-free-lunch theorems were proved by D.H. Wolpert and W.G. Macready (Wolpert and Macready, 1997).

The rapid development continued in the first decade of the 21st century, including the harmony search (Geem et al., 2001), the bee algorithm (Karaboga, 2005),

FA (Yang, 2008, 2009), CS (Yang and Deb, 2009), and BA (Yang, 2010). More algorithms, such as FPA, appeared in the literature (Yang, 2012; Yang et al., 2013a; Yang, 2014), and the literature continues to expand rapidly.

Loosely speaking, evolutionary or nature-inspired algorithms can in general be put into two categories, i.e., heuristic and metaheuristic, though their difference is small. Here, "heuristic" means *to find or to discover by trial and error*, while "meta" means *beyond*. In most studies, we will call nature-inspired algorithms metaheuristics.

One of the main advantages of metaheuristic algorithms is that high-quality solutions, even potentially optimal solutions, can be found quite quickly in practice. The philosophy of these methods is that they should work reasonably well most of the times, but not all the time because there is no guarantee that optimal solutions will be reached. Thus, there are many different ways to design such algorithms. It is estimated that there are over 100 different algorithms and variants (Yang, 2014).

It is not our intention to review all the relevant algorithms. Instead, we highlight a few algorithms with the emphasis on their key ideas and characteristics.

1.3.1 Genetic algorithms

Among the nature-inspired algorithms, GA is probably the most well-known example. GA, which is an evolutionary algorithm, was developed by John Holland (Holland, 1975), inspired by the Darwinian evolution of biological systems in nature.

Solution vectors to an optimization problem are encoded as binary strings of 0s and 1s, called chromosomes. Three genetic operators (crossover, mutation, and selection) are used to modify the strings (De Jong, 1975; Goldberg, 1989; Yang, 2014). Two new (child) solutions are formed via crossover from two parent solutions by swapping one part or multiple parts of their chromosomes. Mutation generates a new solution by mutating one bit or multiple bits of an existing solution or binary string.

Elitism and selection of the best solutions can be carried out, according to their fitness. Such fitness is typically represented as a normalized value associated with the function values of the objective. For example, fitness can be proportional to the objective value for maximization problems. Though the details may depend on the actual implementations, crossover tends to occur more often, with a higher probability of typically 0.6 to 0.95, while mutation tends to be less frequent, with a probability ranging from 0.001 to 0.05. As GAs usually do not have any explicit equations in terms of generating new solutions, it is a detailed algorithmic procedure with many variants (Goldberg, 1989).

1.3.2 Ant colony optimization

ACO, developed by Marco Dorigo in 1992 (Dorigo, 1992), was the first swarm intelligence-based algorithm. In essence, ACO mimics the foraging behavior of social ants in a colony, and pheromone is used for simulating the local interactions and communications among ants. Pheromone is deposited by each ant and it evaporates gradually with time. The exact form of evaporation model may vary, depending on the variant and form of ACO used in implementations. Both incremental deposition and exponential decay of pheromone are widely used.

ACO is particularly suitable for discrete optimization problems. For example, for routing problems, a route or path is encoded as a solution. When ants explore different paths, explored routes are marked with deposited pheromone that evaporates over time. The fitness or quality of a path (a solution) is related to the concentration of pheromone on the path. Routes with higher pheromone concentrations will be preferred or be chosen with a higher probability at a junction. Similar to GA, ACO is a mixed procedure with many variants and applications (Dorigo, 1992).

1.3.3 Differential evolution

Though DE, developed by R. Storn and K. Price (Storn and Price, 1997), is not a nature-inspired algorithm, it is still a representative in ways of formulating search equations. Thus, we will briefly discuss it.

Its main mutation equation is to generate a donor or mutation vector

$$v_i^{t+1} = x_p^t + F(x_q^t - x_r^t), \tag{1.9}$$

where x_p^t, x_q^t, and x_r^t are three distinct solution vectors at iteration t. Here, $F \in [0, 2]$ is a parameter controlling the strength of the mutation. Whether this mutation is accepted or not is dependent on the crossover, controlled by a crossover probability C_r. The actual manipulation is done by either a binomial or an exponential scheme, though we will use the binomial scheme here by drawing a random number $r \in [0, 1]$ for each component of x_i. That is,

$$x_{i,j}^{t+1} = \begin{cases} v_{i,j}^{t+1}, & \text{if } r \le C_r, \\ x_{i,j}^t, & \text{otherwise.} \end{cases} \tag{1.10}$$

Here $x_{i,j}^t$ is the jth component of the D-dimensional vector x_i, and $j = 1, 2, ..., d$.

The selection is carried out so that x_i^{t+1} is accepted if $f(x_i^{t+1}) \le f(x_i^t)$ for minimization, otherwise x_i^t remains unchanged. Clearly, from the above scheme, four distinct vectors are required in DE. Obviously, mutation can involve more vectors and the best solution vector found so far, which gives various options to design different mutation schemes. In fact, more than a dozen variants exist in the literature (Price et al., 2005).

1.3.4 Particle swarm optimization

Another popular swarm intelligence-based algorithm is PSO, which was developed by Kennedy and Eberhart in 1995 (Kennedy and Eberhart, 1995). The PSO system uses a set of multiple agents (or n particles) to simulate the swarming behavior of fish and birds such as starlings.

If we interpret a solution vector x_i as the position of a particle (say, particle i), we can also associate the motion of this particle with a velocity v_i. Thus, for each particle, its position and velocity at any iteration or pseudotime t can be updated iteratively using

$$v_i^{t+1} = v_i^t + \alpha \epsilon_1 [g^* - x_i^t] + \beta \epsilon_2 [x_i^* - x_i^t], \qquad (1.11)$$

$$x_i^{t+1} = x_i^t + v_i^{t+1} \Delta t, \qquad (1.12)$$

where ϵ_1 and ϵ_2 are two uniformly distributed random numbers in [0,1] and $\Delta t = 1$ is the time increment. Since all iterative systems can be considered as time-discrete with unit time increments, we can set $\Delta t = 1$. In addition, g^* is the best solution of the population at iteration t, while x_i^* is the individual best solution for particle i among its search history up to iteration t.

In addition, the learning parameters α and β usually take values in the range of [0,2]. The importance of these parameter values was analyzed and they can affect the stability of the algorithm (Clerc and Kennedy, 2002). So, it is no surprise that there are many variants (Kennedy et al., 2001).

1.3.5 Firefly algorithm

FA was developed by Xin-She Yang in 2008. It is based on the swarming and light flashing behavior of tropical fireflies (Yang, 2008, 2009). In FA, a solution vector to an optimization problem is represented as the position of a firefly. The position vector x_i of firefly i at iteration t is updated by

$$x_i^{t+1} = x_i^t + \beta_0 e^{-\gamma r_{ij}^2} (x_j^t - x_i^t) + \alpha\, \epsilon_i^t, \qquad (1.13)$$

where $\beta_0 > 0$ is the attractiveness at zero distance, that is, $r_{ij} = 0$. Though there is no best solution explicitly expressed in the equation, the fittest solution is selected from the population of n solutions at each iteration. In addition, γ is a scale-dependent parameter that controls the visibility of fireflies, while α controls the strength of the randomization term.

The strong nonlinearity in FA can usually result in subdivision of the whole swarm into multiple subgroups or subswarms. Thus, FA is naturally suitable for multimodal optimization problems, and studies have confirmed this (Yang, 2013; Fister et al., 2013; Marichelvam et al., 2014; Osaba et al., 2017; Rango et al., 2018).

1.3.6 Cuckoo search

The CS algorithm, developed by Xin-She Yang and Suash Deb in 2009, is another nonlinear algorithm (Yang and Deb, 2009). CS is based on the aggressive brooding parasitism of some cuckoo species and their egg laying strategy (Davies, 2011). CS can show some scale-free characteristics in its search moves.

In the real cuckoo–host species system, the eggs laid by cuckoos are sufficiently similar to the eggs of the host species in terms of texture, size, and color. The cuckoo–host species form an arms race system where cuckoos' eggs can be discovered and abandoned with a probability p_a.

Loosely speaking, we can consider this system with n eggs, and we can encode the position of an egg as a solution vector x_i to an optimization problem. The similarity of two eggs (solutions x_i and x_j) can be roughly measured by their difference ($x_j - x_i$). Thus, the position at iteration t can be updated (Yang and Deb, 2014) by

$$x_i^{t+1} = x_i^t + \alpha s \otimes H(p_a - \epsilon) \otimes (x_j^t - x_k^t), \qquad (1.14)$$

where we have used a Heaviside step function H to simulate the discovery probability, in combination with the use of a random number ϵ in [0,1]. The step size s is scaled by a parameter α so as to limit its strength.

Compared with other algorithms, Lévy flights are used in the search mechanism, and this feature comes from that fact that the host birds may abandon their nests and fly away if they suspect their eggs were replaced or contaminated. This can be mimicked as a step size s by Lévy flights (Reynods and Frye, 2007; Pavlyukevich, 2007):

$$x_i^{t+1} = x_i^t + \alpha L(s, \lambda), \qquad (1.15)$$

where the Lévy flights are random walks with steps being drawn from

$$L(s, \lambda) \sim \frac{\lambda \Gamma(\lambda) \sin(\pi \lambda / 2)}{\pi} \frac{1}{s^{1+\lambda}}, \qquad (s \gg 0). \qquad (1.16)$$

Here, $\alpha > 0$ is the step size scaling factor.

Mathematically speaking, the Lévy distribution is heavy-tailed with some long jumps, in comparison with Gaussian random walks (Pavlyukevich, 2007). Consequently, CS can be more effective in exploration (Yang and Deb, 2010; Yang, 2013). CS has also been extended to solve multiobjective and combinatorial optimization problems (Yang and Deb, 2013; Ouaarab et al., 2014).

1.3.7 Bat algorithm

The well-known echolocation of bats, especially microbats, is the main inspiration for BA, which was developed by Xin-She Yang in 2010 (Yang, 2010). BA uses frequency tuning in a range from $f_{\min}$ to $f_{\max}$, in combination with varying pulse emission rate r and loudness A (Altringham, 1996; Colin, 2000).

The position of a bat is used to represent a solution vector to an optimization problem, and a set of n solutions form the population. The iterative update of the position vectors is carried out by

$$f_i = f_{\min} + (f_{\max} - f_{\min})\beta, \qquad (1.17)$$

$$v_i^t = v_i^{t-1} + (x_i^{t-1} - x_*) f_i, \qquad (1.18)$$

$$x_i^t = x_i^{t-1} + v_i^t \Delta x_i^t, \qquad (1.19)$$

where x_* is the best solution among the population and β is a random number in [0,1]. The time-discrete increment $\Delta x_i^t = 1$ can be used.

Though the actual echolocation and variations of pulse emission and loudness can be very complicated, we use a simple monotonic form for both r and A. We have

$$A_i^{t+1} = \alpha A_i^t, \quad r_i^{t+1} = r_i^0 (1 - e^{-\gamma t}), \tag{1.20}$$

where the parameters ($0 < \alpha < 1$ and $\gamma > 0$) control the variation properties.

Recent studies have suggested that BA can have a faster convergence rate with guarantees convergence within proper ranges of parameters (Chen et al., 2018). BA has been extended in different ways with many variants, including multiobjective BA (MOBA) and chaotic BA (Yang, 2011; Gandomi and Yang, 2014; Yang and He, 2013; Osaba et al., 2019). A new directional BA has also been developed by Chakri et al. (Chakri et al., 2017).

1.3.8 Flower pollination algorithm

FPA is also a nature-inspired algorithm, though it is no swarm intelligence-based algorithm. FPA was based on the pollination processes and characteristics of flowering plants (Yang, 2012; Yang et al., 2013b), including biotic and abiotic pollination as well as flower constancy (Waser, 1986).

In FPA, the position of a pollen particle is represented as a solution vector x_i, and the long-distance pollination process can be simulated by

$$x_i^{t+1} = x_i^t + \gamma L(\lambda)(g_* - x_i^t), \tag{1.21}$$

where γ is a scaling parameter and g_* is the best solution found so far at iteration t. In the above equation, $L(\lambda)$ can be considered as a random number vector to be drawn from a Lévy distribution with an exponent of λ (Pavlyukevich, 2007).

Other pollination features such as flower constancy can be simulated by

$$x_i^{t+1} = x_i^t + U(x_j^t - x_k^t), \tag{1.22}$$

where U is a random number uniformly distributed in [0,1].

Applications of FPA are quite diverse (Alam et al., 2015; Yang et al., 2013b) and there are also many variants. For a more detailed review, please refer to Alyasseri et al. (2018).

1.3.9 Other algorithms

The above algorithms are among the most widely used algorithms, which may highlight the main characteristics of swarm intelligence (Fisher, 2009; Yang et al., 2013a). As mentioned before, there are more than 100 different algorithms and variants in the current literature, including the artificial immune system, the artificial bee colony, gravitational search, the krill herd algorithm, and others. Interested readers can refer to more specialized literature such as journal articles and books (Yang, 2013, 2014).

Now we will look at nature-inspired algorithms in general and highlight their main characteristics qualitatively.

1.4 **Algorithms and self-organization**

There are different ways of analyzing algorithms, depending on the actual perspectives and points of view. In this chapter, we will mainly focus on the algorithmic characteristics and self-organization. We will develop more in-depth analysis in the coming chapters.

1.4.1 **Algorithmic characteristics**

By looking at the characteristics of nature-inspired algorithms, we can focus on their main steps, search moves and search mechanisms, exploration and exploitation, and the updating equations or updating dynamics (Blum and Roli, 2003; Yang and He, 2019). We can also look at them as self-organizing systems (Ashby, 1962; Keller, 2009). Briefly, we can summarize all algorithms as follows:

* All algorithms are population-based, consisting of a population of multiple agents (e.g., ants, bees, bats, cuckoos, fireflies, particles, etc.). The position of each agent in the search space is considered as a solution vector to the optimization problem under consideration. These agents/particles will interact and coevolve according to certain rules, often local rules.
* The iterative evolution of the population is carried out by modifying the existing population in terms of some operators (e.g., mutation, crossover). Modifications and perturbations of existing solutions in the population are achieved by randomization with selection, biased towards solutions with higher fitness. All algorithms can have certain forms of both local exploitation moves and global exploration moves, or exploitation and exploration.
* Selection of best solutions is carried out, according to the "survival of the fittest," in combination with different forms of elitism. The main idea is to allow fitter solutions to pass onto the next generation more frequently.
* The diversity of the population tends to be higher at the early stage of iterations with higher randomness. As iterations continue, diversity will gradually reduce. When all solutions become sufficiently similar, the system may lead to some self-organized state or converged states. A converged population typically has some strong similarity and structure, but their diversity has been reduced to minimum.

1.4.2 **Comparison with traditional algorithms**

Traditional algorithms such as gradient-based methods tend to use a single solution path, while nature-inspired algorithms usually use multiple paths. Based on the analysis on traditional optimization algorithms and nature-inspired algorithms (Boyd and Vandenberghe, 2004; Yang, 2014), the comparison of their main features and characteristics can be summarized in Table 1.1.

Table 1.1 Comparison of key features between nature-inspired algorithms and traditional optimization algorithms.

Feature	Traditional algorithms	Nature-inspired algorithms
Agent/population size	A single agent	Multiple agents (population-based)
Gradient	Widely used	Rarely used
Hessian	Frequently used	None
Iteration	Fewer iterations	Much more iterations
Randomness	Almost no use	Widely used
Selection	Always	Always/elitism
Hyperparameter	Fewer or mostly one	A few or even many

1.4.3 Self-organized systems

Another insightful angle is to look at an algorithm as a self-organized system. There are some strong similarities between the characteristics of algorithms and the behavior of self-organization (Ashby, 1962; Keller, 2009; Yang and He, 2019). The comparison of algorithmic components and self-organization systems as well as their role or properties is summarized in Table 1.2.

Table 1.2 Role or properties of components or features in nature-inspired algorithms.

Components or features	Role or properties
Population	Diversity, parallel sampling
Derivatives	Exploitation or intensification
Randomization	State perturbation, exploration, escaping local optima
Selection	Driving force, leading to convergence
Elitism	Biased or fitness-proportional selection
Mutation	Modification of solutions for diversity and exploration
Crossover	Mixing of solutions, diversity
Iteration	Evolution of the population

Obviously, there are some significant differences as well. For example, the route to self-organization may not be always clear, and timing is not that important. But for algorithms, the updating equations and evolution of populations tend to be clearer, and time is a crucial factor because the rate of convergence is practically important.

It is worth pointing out that there is no free lunch in algorithms (Wolpert and Macready, 1997), and there is no single best algorithm that is able to solve all the problems. Thus, one of the main tasks of research is to identify what types of problems an algorithm can solve and what algorithm(s) should be used for a given type of problems.

1.5 **Open problems for future research**

Though nature-inspired algorithms are becoming popular, they still lack in-depth theoretical understanding and a rigorous mathematical framework. There are still some significant gaps between theory and practice. For example, a vast number of studies have shown in different ways that nature-inspired algorithms can solve challenging optimization problems in practice; however, we still lack in-depth understanding of why and exactly how they work. For a given type of problems, especially new problems, we cannot predict if these algorithms can work well until we actually try them. In addition, the rate of convergence seems to be quite a mystery. For some problems, some algorithms converge quickly, while for other problems, the same algorithms can converge very slowly.

To compile all open problems can be an impossible task. Here, we will highlight a few open problems so as to inspire more research.

1. Theoretical framework: A mathematical framework will allow rigorous analysis of different algorithms so as to gain deeper understanding of their properties and search mechanisms to use these algorithms more effectively and even potentially design more efficient algorithms. Theoretical analyses can be in terms of Markov chains, dynamical systems, and others. It can be expected that a multidisciplinary approach would be more fruitful.
2. Convergence: The study of the convergence of algorithms and their exact rate of convergence will be very useful. This allows researchers to identify under what conditions and parameter settings a given algorithm can lead to the best convergence, without causing premature convergence. Algorithm-dependent parameters can be tuned so as to maximize the rate of convergence, though how an algorithm can be tuned most effectively is still unknown.
3. Stability and robustness: Even though an algorithm may be efficient, its stability is also very important. The stability and robustness of an algorithm can be controlled by its parameter settings, the type of problem to be solved, and the noise of data used. Unstable algorithms are not useful because they cannot obtain sensible, robust solutions. In many cases, dynamical system theory can be useful to gain some understanding of algorithmic stability; however, many algorithms, such as FA, can be highly nonlinear, which means nonlinear stability should be considered.
4. Fair comparison: There many different algorithms in the literature. If different variants are also considered, there are more than a hundred different algorithms and variants. Comparison studies should be fair, though the fairness depends on the performance measures used for the comparison. Biased performance measures may lead to biased results. It is not clear what measures are fair. Currently, the accuracy, success rates, number of function evaluations, computing time, tolerance, means, best values, and standard derivations are all used for comparison. It is difficult to say if these are fair performance measures. More detailed studies with rigorous mathematics are highly needed.

5. Large-scale problems: Whatever the algorithm may be, the ultimate aim is to solve real-world problems with thousands or even millions of parameters effectively. Currently, almost all algorithms solve small-scale toy problems, benchmark problems of moderate sizes. It is not clear if these algorithms can be scaled up to solve large-scale real-world problems. For example, it is usually the case that simple parallelization may not be enough to scale up. This still remains a major challenge for any algorithm to actually solve large-scale problems efficiently.

Obviously, there are other challenges as well. It is our hope that this chapter and this book can inspire more research in this area. In the next few chapters, we will introduce more algorithms in greater details and then introduce mathematical foundations for analyzing algorithms for optimization. A wide spectrum of applications and case studies will form the topics of other chapters in this book.

References

Alam, D.F., Yousri, D.A., Eteiba, M.B., 2015. Flower pollination algorithm based solar PV parameter estimation. Energy Conversion and Management 101 (2), 410–422.

Altringham, J.D., 1996. Bats: Biology and Behaviour. Oxford University Press, Oxford, UK.

Alyasseri, Z.A.A., Khader, A.T., Al-Betar, M.A., Awadallah, M.A., Yang, X.S., 2018. Variants of the flower pollination algorithm: a review. In: Yang, X.S. (Ed.), Nature-Inspired Algorithms and Applied Optimization. Springer, Cham, pp. 91–118.

Arara, S., Barak, B., 2009. Computational Complexity: A Modern Approach. Cambridge University Press, Cambridge, UK.

Ashby, W.A., 1962. Principles of the self-organizing systems. In: von Foerster, H., Zopf Jr., G.W. (Eds.), Principles of Self-Organization: Transactions of the University of Illinois Symposium. Pergamon Press, London, pp. 255–278.

Blum, C., Roli, A., 2003. Metaheuristics in combinatorial optimization: overview and conceptual comparison. ACM Computing Surveys 25 (2), 268–308.

Boyd, S.P., Vandenberghe, L., 2004. Convex Optimization. Cambridge University Press, Cambridge, UK.

Chabert, J.L., 1999. A History of Algorithms: From the Pebble to the Microchips. Springer-Verlag, Heidelberg.

Chakri, A., Khelif, R., Benouaret, M., Yang, X.S., 2017. New directional bat algorithm for continuous optimization problems. Expert Systems with Applications 69 (1), 159–175.

Chen, S., Peng, G.H., Xing-Shi Yang, X.S., 2018. Global convergence analysis of the bat algorithm using a markovian framework and dynamic system theory. Expert Systems with Applications 114 (1), 173–182.

Clerc, M., Kennedy, J., 2002. The particle swarm: explosion, stability, and convergence in a multidimensional complex space. IEEE Transactions on Evolutionary Computation 6 (1), 58–73.

Colin, T., 2000. The Variety of Life. Oxford University Press, Oxford, UK.

Copeland, B.J., 2004. The Essential Turing. Oxford University Press, Oxford, UK.

Davies, N.B., 2011. Cuckoo adaptations: trickery and tuning. Journal of Zoology 284 (1), 1–14.

De Jong, K., 1975. Analysis of the Behaviour of a Class of Genetic Adaptive Systems. Ph.D. Thesis. University of Michigan, Ann Arbor, MI, USA.

Dorigo, M., 1992. Optimization, Learning, and Natural Algorithms. Ph.D. Thesis. Politecnico di Milano, Milan, Italy.

Fisher, L., 2009. The Perfect Swarm: The Science of Complexity in Everyday Life. Basic Books, New York.

Fister, I., Fister Jr, I., Brest, J., Yang, X.S., 2013. A comprehensive review of firefly algorithms. Swarm and Evolutionary Computation 13 (1), 34–46.

Fogel, L.J., Owens, A.J., Walsh, M.J., 1966. Artificial Intelligence Through Simulated Evolution. Wiley, New York, USA.

Gandomi, A.H., Yang, X.S., 2014. Chaotic bat algorithm. Journal of Computational Science 5 (2), 224–232.

Geem, Z.W., Kim, J.H., Loganathan, G.V., 2001. A new heuristic optimization algorithm: harmony search. Simulation 76 (2), 60–68.

Glover, F., 1986. Future paths for integer programming and links to artificial intelligence. Computers & Operations Research 13 (5), 533–549.

Goldberg, D.E., 1989. Genetic Algorithms in Search, Optimization and Machine Learning. Addison-Wesley, Reading, MA, USA.

Holland, J., 1975. Adaptation in Nature and Artificial Systems. University of Michigan Press, Ann Arbor, MI, USA.

Judea, P., 1984. Heuristics. Addison-Wesley, New York, USA.

Karaboga, D., 2005. An idea based on honeybee swarm for numerical optimization. Technical report. Eriyes University, Turkey.

Keller, E.F., 2009. Organisms, machines, and thunderstorms: a history of self-organization, part ii. Complexity, emergence, and stable attractors. Historical Studies in the Natural Sciences 39 (1), 1–31.

Kennedy, J., Eberhart, R., 1995. Particle swarm optimization. In: Proceedings of the IEEE International Conference on Neural Networks. IEEE, Piscataway, NJ, USA, pp. 1942–1948.

Kennedy, J., Eberhart, R.C., Shi, Y., 2001. Swarm Intelligence. Academic Press, London, UK.

Kirkpatrik, S., Gellat, C.D., Vecchi, M.P., 1983. Optimization by simulated annealing. Science 220 (4598), 671–680.

LeCun, Y., Bengio, Y., Hinton, G.E., 2015. Deep learning. Nature 521 (7553), 436–444.

Marichelvam, M.K., Prabaharan, T., Yang, X.S., 2014. A discrete firefly algorithm for the multi-objective hybrid flowshop scheduling problems. IEEE Transactions on Evolutionary Computation 18 (2), 301–305.

Nelder, J.A., Mead, R., 1965. A simplex method for function optimization. Computer Journal 7 (4), 308–313.

Osaba, E., Yang, X.S., Diaz, F., Lopez-Garcia, P., Carballedo, R., 2016. An improved discrete bat algorithm for symmetric and asymmetric traveling salesman problems. Engineering Applications of Artificial Intelligence 48 (1), 59–71.

Osaba, E., Yang, X.S., Diaz, F., Onieva, E., Masegosa, A., Perallos, A., 2017. A discrete firefly algorithm to solve a rich vehicle routing problem modelling a newspaper distribution system with recycling policy. Soft Computing 21 (18), 5295–5308.

Osaba, E., Yang, X.S., Fister Jr., I., Lopez-Garcia, P., Vazquez-Paravila, A., 2019. A discrete and improved bat algorithm for solving a medical goods distribution problem with pharmacological waste collection. Swarm and Evolutionary Computation 44 (1), 273–286.

Ouaarab, A., Ahiod, B., Yang, X.S., 2014. Discrete cuckoo search algorithm for the travelling salesman problem. Neural Computing and Applications 24 (7–8), 1659–1669.

Pavlyukevich, I., 2007. Lévy flights, non-local search and simulated annealing. Journal of Computational Physics 226 (2), 1830–1844.

Price, K., Storn, R., Lampinen, J., 2005. Differential Evolution: A Practical Approach to Global Optimization. Springer, Berlin, Germany.

Rango, F.D., Palmieri, N., Yang, X.S., Marano, S., 2018. Swarm robotics in wireless distributed protocol design for coordinating robots involved in cooperative tasks. Soft Computing 22 (13), 4251–4266.

Reynods, A.M., Frye, M.A., 2007. Free-flight odor tracking in drosophila is consistent with an optimal intermittent scale-free search. PLoS One 2 (4), e354–e363.

Storn, R., Price, K., 1997. Differential evolution: a simple and efficient heuristic for global optimization. Journal of Global Optimization 11 (4), 341–359.

Turing, A.M., 1948. Intelligent machinery. Technical report. National Physical Laboratory, London, UK.

Waser, N.M., 1986. Flower constancy: definition, cause and measurement. American Naturalist 127 (5), 596–603.

Wolpert, D.H., Macready, W.G., 1997. No free lunch theorems for optimization. IEEE Transactions on Evolutionary Computation 1 (1), 67–82.

Yang, X.S., 2008. Nature-Inspired Metaheurisic Algorithms. Luniver Press, Bristol, UK.

Yang, X.S., 2009. Firefly algorithms for multimodal optimization. In: Watanabe, O., Zeugmann, T. (Eds.), Proceedings of Fifth Symposium on Stochastic Algorithms, Foundations and Applications. In: Lecture Notes in Computer Science, vol. 5792. Springer, pp. 169–178.

Yang, X.S., 2010. A new metaheuristic bat-inspired algorithm. In: Cruz, C., González, J.R., Pelta, D.A., Terrazas, G. (Eds.), Nature Inspired Cooperative Strategies for Optimization (NISCO 2010). In: Studies in Computational Intelligence, vol. 284. Springer, Berlin, Germany, pp. 65–74.

Yang, X.S., 2011. Bat algorithm for multi-objective optimisation. International Journal of Bio-Inspired Computation 3 (5), 267–274.

Yang, X.S., 2012. Flower pollination algorithm for global optimization. In: Durand-Lose, J., Jonoska, N. (Eds.), Unconventional Computation and Natural Computation (UCNC 2012), vol. 7445. Springer, Berlin, Heidelberg, Germany, pp. 240–249.

Yang, X.S., 2013. Cuckoo Search and Firefly Algorithm: Theory and Applications. Studies in Computational Intelligence, vol. 516. Springer, Heidelberg, Germany.

Yang, X.S., 2014. Nature-Inspired Optimization Algorithms. Elsevier Insight, London.

Yang, X.S., Deb, S., 2009. Cuckoo search via Lévy flights. In: Proceedings of World Congress on Nature & Biologically Inspired Computing (NaBIC 2009). IEEE Publications, USA, pp. 210–214.

Yang, X.S., Deb, S., 2010. Engineering optimisation by cuckoo search. International Journal of Mathematical Modelling and Numerical Optimisation 1 (4), 330–343.

Yang, X.S., Deb, S., 2013. Multiobjective cuckoo search for design optimization. Computers & Operations Research 40 (6), 1616–1624.

Yang, X.S., Deb, S., 2014. Cuckoo search: recent advances and applications. Neural Computing and Applications 24 (1), 169–174.

Yang, X.S., He, X.S., 2013. Bat algorithm: literature review and applications. International Journal of Bio-Inspired Computation 5 (3), 141–149.

Yang, X.S., He, X.S., 2019. Mathematical Foundations of Nature-Inspired Algorithms. Springer Briefs in Optimization. Springer, Cham, Switzerland.

Yang, X.S., Cui, Z.H., Xiao, R.B., Gandomi, A.H., Karamanoglu, M., 2013a. Swarm Intelligence and Bio-Inspired Computation: Theory and Applications. Elsevier, London, UK.

Yang, X.S., Karamanoglu, M., He, X.S., 2013b. Multi-objective flower algorithm for optimization. Procedia Computer Science 18 (1), 861–868.

Bat algorithm and cuckoo search algorithm

Xin-She Yang[a], Xing-Shi He[b]
[a]*Middlesex University London, School of Science and Technology, London, United Kingdom*
[b]*Xi'an Polytechnic University, College of Science, Xi'an, China*

CONTENTS

2.1 Introduction

There are many different nature-inspired algorithms for solving optimization problems. In order to see how algorithms work and their main procedure of implementations, we now focus on two nature-inspired algorithms: the bat algorithm (BA) and cuckoo search (CS). We will introduce each algorithm with the basic ideas, the main steps, pseudocode, implementation details, and an example. In addition, we also provide the MATLAB® codes so that readers can understand how the implementations are carried out and how the algorithms actually work.

2.2 Bat algorithm

Bats, especially microbats, use echolocation for navigation and foraging. These bats emit a series of short, ultrasonic bursts in the frequency range of 25 kHz to about 150 kHz, and such short pulses typically last a few milliseconds. The loudness of such bursts and the rate of such pulse emission vary during hunting, especially when homing for prey (Altringham, 1996; Colin, 2000). The increase of frequency will

Nature-Inspired Computation and Swarm Intelligence. https://doi.org/10.1016/B978-0-12-819714-1.00011-7

reduce the wavelength of the ultrasound pulses, and thus increase the resolution and accuracy of the prey detection.

Such well-known echolocation characteristics of bats can be simulated in BA, which was developed by Xin-She Yang in 2010 (Yang, 2010). BA uses frequency tuning as a randomized driving force, in combination with the use of varying pulse emission rate r and loudness A for the bat population.

2.2.1 Algorithmic equations of BA

In BA, there are n bats that form a population for iterative evolution. The location of each bat is denoted by x_i ($i = 1, 2, ..., n$), which can be considered as the solution vector to an optimization problem under consideration. As bats are flying in the search space, each bat is also associated with a velocity vector v_i.

Mathematically speaking, a position vector is considered as a solution vector to an optimization problem in a D-dimensional search space with D independent design variables,

$$x = [x_1, x_2, x_3, ..., x_D]. \tag{2.1}$$

Here, we have used the row vector, though most mathematical formulations use column vectors. This choice is purely for the convenience of array manipulation in the MATLAB codes to be presented later. Obviously, a row vector can be converted easily into a corresponding column vector by a simple transpose. Since the positions will vary and evolve with iterations (a pseudotime counter t), we can use x_i^t to denote the position vector of bat i at iteration t. Similarly, its corresponding velocity at t can be written as v_i^t.

During the iterations, each bat can vary its pulse emission rate r_i, its loudness A_i, and its frequency f_i. Frequency variations or tuning can be carried out by

$$f_i = f_{\min} + \beta(f_{\max} - f_{\min}), \tag{2.2}$$

where $f_{\min}$ and $f_{\max}$ are the minimum and maximum ranges, respectively, of the frequency f_i for each bat i. Though the variations of frequencies are nonuniform in reality, we will use uniform variations for simplicity. Thus, we have used a random number β, which is a uniformly distributed number drawn from $[0, 1]$. In most applications, we can use $f_i = O(1)$. For example, in our implementations here, we use $f_{\min} = 0$ and $f_{\max} = 2$.

The variations of frequencies are then used for modifying the velocities of the bats in the population such that

$$v_i^{t+1} = v_i^t + (x_i^t - x_*)f_i, \tag{2.3}$$

where x_* is the best solution obtained by the population at iteration t.

Once the velocity of a bat is updated, its position (or solution vector) x_i can be updated by

$$x_i^{t+1} = x_i^t + (\Delta t)v_i^{t+1}, \tag{2.4}$$

where Δt is the iteration or time increment. It is worth pointing out that all iterative algorithms are updated in a discrete manner, which means that we can set $\Delta t = 1$. Thus, we can simply consider the vectors without any physical units, and then write the update equation as

$$x_i^{t+1} = x_i^t + v_i^{t+1}. \tag{2.5}$$

For the local modification around the best solution, we can use

$$x_{new} = x_{old} + \sigma \epsilon_t A^{(t)}, \tag{2.6}$$

where ϵ_t is a random number drawn from a normal distribution N(0,1) and σ is the standard deviation acting as a scaling factor. Here, $A^{(t)}$ is the average loudness at iteration t. For simplicity, we can use $\sigma = 0.1$ in our later implementation.

2.2.2 Pulse emission and loudness

The variations of the loudness and the rate of pulse emission can be included in the BA so that they can influence the way of exploration and exploitation. In the real world, their variants are very complicated, depending on the bat species. However, we use a monotonic variation here. The pulse emission rate r_i can monotonically increase from a lower value $r_i^{(0)}$, while the loudness can decrease from a higher value (such as $A^{(0)} = 1$) to a much lower value. So we have

$$A_i^{t+1} = \alpha A_i^t, \tag{2.7}$$

$$r_i^{t+1} = r_i^{(0)}[1 - \exp(-\gamma t)], \tag{2.8}$$

where $0 < \alpha < 1$ and $\gamma > 0$ are constants.

Based on the above formulas, we can see that when t is large enough ($t \to \infty$), we get $A_i^t \to 0$ and $r_i^t \to r_i^{(0)}$. For simplicity, we can use $\alpha = 0.97$ and $\gamma = 0.1$ in our simple implementations here. In addition, we will use the same pulse emission rate and loudness for all the bats, which makes the implementations much simpler for the demo codes presented in this chapter.

2.2.3 Pseudocode and parameters

Based on the above algorithmic equations and the variations of the pulse emission rate and loudness, the main steps of the BA can be realized with a single loop over the whole population, and the modifications are done, depending on the actual switching conditions during iterations. So the main steps of the bat algorithm can be summarized in the pseudocode shown in Algorithm 1.

There are quite a few parameters in BA. Apart from the population size n and the ranges of frequency f_{min} and f_{max}, there are two parameters (α, γ) and two initial values ($A_0^{(0)}$ and $r_i^{(0)}$). Based on the preliminary parameter studies (Yang, 2010, 2011), we will use the following values in our implementation: $\alpha = 0.97$, $\gamma = 0.1$, $f_{min} = 0$,

---

Algorithm 1 The BA pseudocode.

---

input   : Population size n, parameters such as α and γ
output   : Global best position
auxiliaries: Define function $f(x)$ (objective)

1 Generate initial population x_i and v_i $(i = 1, 2, ..., n)$;
  Set frequencies f_i, pulse rates r_i and loudness A_i;
  Initialize $t = 0$ (iteration counter);
  while $(t < t_{max})$ **do**
2 | Vary r_i and A_i;
  | Adjust frequencies;
  | Generate new solutions (velocities and locations) via Eqs. (2.3) and (2.5);
  | **if** *rand* $> r_i$ **then**
3 | | Select a solution among the best solutions;
  | | Modify locally around the selected best solution;
  |
4 | Fly randomly to generate a new solution;
  | **if** *rand* $> A_i$ *and* $f(x_i) < f(x_j)$ *(for minimization)* **then**
5 | | Accept the new solution;
6 | Rank the current population of bats and find the best solution x_*;
7 Display/output the best solution

$f_{max} = 2$, and $A_i^{(0)} = r_i^{(0)} = 1$ for all bats. For the population size n, we use $n = 10$ in the demo implementation, even though n should vary for different applications.

2.2.4 Demo implementation

The above steps and the pseudocode can be realized in any programming language. Here, we have implemented them as a demo code in MATLAB. This example intends to find the minimum of the multivariate function $f(x)$,

$$\text{minimize } f(x) = (x_1 - 2)^2 + (x_2 - 2)^2 + ... + (x_D - 2)^2, \quad x_i \in \mathbb{R}, \quad (2.9)$$

which has a global minimum of $x_* = (2, 2, ..., 2)$. Though the domain of this function is in the whole domain $\mathbb{R}^D$, it is more convenient to impose some simple bounds in practice. So we use the following lower bound (Lb) and upper bound (Ub):

$$Lb = [-10, -10, ..., -10], \quad Ub = [+10, +10, ..., +10]. \quad (2.10)$$

Though the MATLAB code should be a single file that is executed in a sequential manner, for the ease of description, we now split the whole code into four parts: initialization, initial population, the main loop, and the objective function. The simple bounds should be checked at each iteration when new solutions are generated. However, for simplicity, the implementation here does not carry out this check, which means that care should be taken when extending the simple code to more sophisticated cases. Detailed constraints and limits should be handled carefully with relevant methods such as the penalty method.

The initialization part consists of setting parameter values such as α and γ, as well as the generation of the initial population of $n = 10$ bats. Most of these lines are self-explanatory.

```
function [best,fmin]=bat_algorithm
% Default parameters
n=10;            % Population size, typically 10 to 25
A=1;             % Loudness  (constant or monotonic decreasing)
r=1;             % Pulse rate (constant or monotonic decreasing)
alpha=0.97;      % Parameter alpha
gamma=0.1;       % Parameter gamma
% Frequency range
Freq_min=0;      % Minimum frequency
Freq_max=2;      % Maximum frequency
% Max number of iterations
t_max=1000;
t=0;             % Initialize iteration counter
% Dimension of the search variables
d=5;
% Initialization of arrays
Freq=zeros(n,1);    % Initial frequency array
v=zeros(n,d);       % Initial velocities/array
Lb=-10*ones(1,d);   % Lower bounds
Ub=10*ones(1,d);    % Upper bounds
```

Once the parameters are initialized, an initial population and its solutions should be generated. This leads to a set of initial solution vectors and the initial best solution with the minimum objective value in the population. Here, the fitness means the objective values, so higher fitness for minimization problems means lower objective values.

```
% Initialize the population of n solutions
for i=1:n,
  Sol(i,:)=Lb+(Ub-Lb).*rand(1,d);
  Fitness(i)=Fun(Sol(i,:));
end
% Find the best solution among the initial population
[fmin,I]=min(Fitness);
best=Sol(I,:);
```

The next part is the main part, which includes a major loop over the whole population at each iteration. In addition, the pulse emission rate and loudness are varied first, and then new solutions are generated and evaluated. Though each bat can have its own pulse emission rate r_i and loudness A_i, again for simplicity, we have used the same values of r and A for all bats. This means that the average of A_i is now simply A itself. The solutions of the new population are checked and the current best solution is found and updated.

```
% Start the iterations -- Bat Algorithm
while (t<t_max)
    % Varying parameters
    r=r*(1-exp(-gamma*t));
    A=alpha*A;
    % Loop over all bats/solutions
    for i=1:n,
    Freq(i)=Freq_min+(Freq_max-Freq_min)*rand;
    v(i,:)=v(i,:)+(Sol(i,:)-best)*Freq(i);
    S(i,:)=Sol(i,:)+v(i,:);
    % Check a switching condition
    if rand>r,
    S(i,:)=best+0.1*randn(1,d)*A;
    end

    % Evaluate new solutions
    Fnew=Fun(S(i,:));
    % If the solution improves or not too loudness
     if (Fnew<=Fitness(i)) & (rand>A) ,
       Sol(i,:)=S(i,:);
       Fitness(i)=Fnew;
     end

      % Update the current best solution in the population
     if Fnew<=fmin,
       best=S(i,:);
       fmin=Fnew;
     end
   end % end of for
  t=t+1;  % Update iteration counter
end
% Output the best solution
disp(['Best =',num2str(best),' fmin=',num2str(fmin)]);
```

The final part is the objective function, which is problem-dependent. The fitness of a new solution is evaluated by calling the objective or cost function. It is worth pointing out that this part should also include the checking of simple bounds and constraints, though we have not implemented this here.

```
%% Define the objective function
function z=cost(x)
z=sum((x-2).^2); % Solutions should be (2,2,...,2)
```

This code is relatively simple and can be executed quickly on a computer. With a maximum number of 1000 iterations, we can get the best minimum value as $f_{best} = 2.71 \times 10^{-29}$. Obviously, due to the random numbers used, these results are not exactly repeatable, but the order of the magnitude is easily reachable in the simulation. However, multiple runs should be carried out to get meaningful results.

BA can usually have a fast convergence rate, and guaranteed convergence can be achieved under certain conditions (Chen et al., 2018). BA has been extended to multiobjective optimization and different variants with many applications (Yang, 2011; Yang and Gandomi, 2012; Yang and He, 2013; Chakri et al., 2017; Osaba et al., 2019).

2.3 Cuckoo search algorithm

Extensive studies have shown that many cuckoo species employ aggressive reproduction strategies, and it is estimated that 59 cuckoo of the 141 known cuckoo species are so-called obligate reproduction parasites, i.e., they lay eggs in the nests of host species so as to let host species raise their young chicks (Davies and Brooke, 1991; Davies, 2011). The eggs laid by cuckoos can be sufficiently similar to the eggs of hosts, such as warblers, in terms of egg color, texture, and shape. From an evolutionary point of view, this strategy can maximize the reproductivity probability of the cuckoo species with the minimum investment. However, host species can also fight back by either getting rid of cuckoos' eggs or abandoning such eggs and nests with contaminated alien eggs. This forms an ongoing arms race between cuckoo species and host species (Davies, 2011).

Based on these characteristics, the CS algorithm was developed by Xin-She Yang and Suash Deb in 2009 (Yang and Deb, 2009).

2.3.1 Cuckoo search

For simplicity of describing the cuckoo search algorithm, we now first idealize the main characteristics of the cuckoo–host system as a population of n cuckoos with n nests. In real-world cuckoo–host systems, each nest of a host bird typically has 3 or 4 or more eggs, and each cuckoo can attack quite a few nests by laying its eggs in such nests. For simplicity, we assume that each cuckoo can only lay one egg and affect one host nest at a time, which means that the number of eggs is equal to the number of nests and cuckoos. Thus, we can encode the location of an egg as a solution vector x to an optimization problem. This way, there is no need to distinguish eggs, cuckoos, and nests. Thus, we essentially have the following equality: "egg = cuckoo = nest."

- Each cuckoo lays one egg and dumps it into a randomly chosen host nest. For simplicity, the number of eggs, the number of nests, and the number of cuckoos are all the same.
- Each egg laid by a cuckoo may be discovered and abandoned with a probability p_a. This is equivalent to a fraction p_a of the total population being modified at each iteration t.
- The fitness of an egg or solution is determined by its objective value. The best solutions with the lowest objective values (for minimization) will be passed on to the next generation. This means that the best solutions are retained.

In the standard CS, there are two different mechanisms to modify the solutions. One mechanism simulates the egg laying strategy that eggs should be sufficiently similar, and their similarity is measured by their vector difference $(x_j - x_k)$ at time t. We have

$$x_i^{t+1} = x_i^t + \alpha s \otimes H(p_a - \epsilon) \otimes (x_j^t - x_k^t), \tag{2.11}$$

where x_j^t and x_k^t are two different solution vectors, randomly selected from the population for $i = 1, 2, ..., n$. The discovery of a cuckoo's egg with a probability p_a is realized by a Heaviside function (or a step function) by comparing p_a with a random number ϵ drawn from a uniform distribution in [0,1]. In addition, s is a step size with the random permutation of solution differences, and this is scaled by a scaling factor α. Furthermore, the product notation $\otimes$ of two vectors means entry-wise multiplications.

The other mechanism is the protection mechanism by the host species by flying far away to build a nest so as to avoid egg contamination. This is realized by a nonlocal random walk using Lévy flights,

$$x_i^{t+1} = x_i^t + \alpha L(s, \beta). \tag{2.12}$$

This Lévy step size should be drawn from the Lévy distribution. However, the definition of a Lévy distribution and its pseudonumber generations can be tricky (Pavlyukevich, 2007). Thus, we use a power-law distribution,

$$L(s, \beta) \sim \frac{\beta \Gamma(\beta) \sin(\pi \beta / 2)}{\pi} \frac{1}{s^{1+\beta}}, \tag{2.13}$$

to approximate a Lévy probability distribution with an exponent $0 \leq \beta \leq 2$. Here, the gamma function is defined by

$$\Gamma(q) = \int_0^\infty z^{q-1} e^{-u} du. \tag{2.14}$$

It is worth pointing out that the notation "$\sim$" highlights the fact that they are pseudorandom numbers to be drawn from the probability distribution on the right-hand side of the above equation. The realization of Lévy flights is carried out by using Mantegna's algorithm in the implementation below.

2.3.2 Pseudocode and parameters

CS and its main steps can be summarized as the pseudocode shown in Algorithm 2. This can be easily realized in any programming language.

Algorithm 2 The CS pseudocode.

---

input   : Population size n and all parameters
output   : Global best solution
auxiliaries: Define function $f(x)$ (objective)

1 Initialize a population of n nests x_i $(i = 1, 2, ..., n)$
   for *t=1:MaxGeneration* **do**
2 | Choose a cuckoo/solution randomly among the population;
   | Generate a new solution by Lévy flights via Eq. (2.13);
   | Evaluate the fitness of the new solution $f_i(x_i)$;
   | Select a nest (say, j) among n nests randomly;
   | **if** $f_i < f_j$ *(for minimization)* **then**
3 | | Replace x_j by x_i
4 | Abandon a fraction (p_a) of the worse nests;
   | Generate new nests/solutions by Eq. (2.11);
   | Evaluate the new solution and update the best solution;
5 Display and postprocess the results

All nature-inspired algorithms have algorithm-dependent parameters. Compared with other algorithms, CS has a relatively low number of parameters, which makes it easier to tune the algorithm. Apart from the common population size n, there are essentially only two parameters: the switching probability p_a and the Lévy exponent β. We will use $n = 25$, $p_a = 0.25$, and $\beta = 1.5$ in our implementation below.

2.3.3 Demo implementation

Based on the above pseudocode, it is straightforward to implement the CS using MATLAB. The only issue is to ensure that the Lévy random numbers should be actually drawn from a Lévy distribution. For ease of understanding, we will use the following function $f(x)$:

$$\text{minimize } f(x) = (x_1 - 2)^2 + (x_2 - 2)^2 + ... + (x_D - 2)^2, \quad x_i \in \mathbb{R}, \quad (2.15)$$

which has a global minimum of $x_* = (2, 2, ..., 2)$. Similarly to what we have done for BA, we also apply the following lower bound (Lb) and upper bound (Ub):

$$Lb = [-10, -10, ..., -10], \quad Ub = [+10, +10, ..., +10]. \quad (2.16)$$

Obviously, the whole code should be a single file, but for simplicity, we now split it into seven parts, each part focusing on a specific task. The seven parts are initialization, initial population, the main loop, Lévy flights, selection of solution, simple bounds, and the objective function.

The first part is mainly the initialization of parameter values such as the population size $n = 25$, the number of iterations, and the switching probability $pa = 0.25$.

```
function [bestnest,fmin]=cuckoo_search_demo
%% Initialization of parameters
n=25;          % Number of nests (or population size)
pa=0.25;       % Discovery probability of alien eggs/solutions
nd=15;         % Number of dimensions
Lb=-5*ones(1,nd);  % Lower bounds
Ub=5*ones(1,nd);   % Upper bounds
N_IterTotal=1000;  % Increase it to get better results
```

Then, the whole population of 25 solutions is initially evaluated so as to obtain the initial best solution.

```
% Random initial solutions for the population with n cuckoos
for i=1:n,
nest(i,:)=Lb+(Ub-Lb).*rand(size(Lb));
end
% Get the best solution among the initial population
fitness=10^10*ones(n,1);
[fmin,bestnest,nest,fitness]=get_best_nest(nest,nest,fitness);
```

The third part is the main part, which loops over the whole cuckoo population by applying two different search mechanisms at each iteration. New solutions (or cuckoos) are generated by get_cuckoos() or empty_nests() subroutines to be explained later.

```
%% Starting iterations
for iter=1:N_IterTotal,
    % Generate new solutions (but keep the current best)
     new_nest=get_cuckoos(nest,bestnest,Lb,Ub);
     [fnew,best,nest,fitness]=get_best_nest(nest,new_nest,fitness);
    % Application of discovery probability and randomization
     new_nest=empty_nests(nest,Lb,Ub,pa) ;

    % Evaluate the fitness of the new set of solutions
     [fnew,best,nest,fitness]=get_best_nest(nest,new_nest,fitness);
    % Find the best objective so far
    if fnew<fmin,
        fmin=fnew;
        bestnest=best;
    end
end %% End of iterations
```

```
%% Display the best solution and its corresponding objective value
disp(strcat('Best solution=', num2str(bestnest)));
disp(strcat('Best objective=',num2str(fmin)));
```

The fourth part concerns the Lévy flights and the selection of the best solution. As explained earlier, the generation of pseudorandom numbers obeying the Lévy distribution is realized by Mantegna's algorithm (Mantegna, 1994). That is, the step size is calculated by transforming from two normal distributions u and v via

$$s = \frac{u}{|v|^{1/\beta}}, \tag{2.17}$$

where

$$u \sim N(0, \sigma_u^2), \quad v \sim N(0, 1). \tag{2.18}$$

One normal distribution has a unit variance, while σ_u^2 is calculated by

$$\sigma_u = \left\{ \frac{\Gamma(1+\beta)\sin(\pi\beta/2)}{\beta\,\Gamma[(1+\beta)/2]\,2^{(\beta-1)/2}} \right\}^{1/\beta}. \tag{2.19}$$

To exactly compute this variance involves the gamma function, but it is not important in practice. So it is easier to use a fixed σ value, together with a scaling factor such as 0.01. In our implementation, we draw the Lévy flight step by

$$u \sim N(0, \sigma^2), \quad v \sim N(0, 1), \tag{2.20}$$

and

$$s = \frac{u}{|v|^{1/\beta}}, \quad \beta = 3/2. \tag{2.21}$$

This essentially leads to step sizes to modify or perturb x_i as

$$\text{stepsize} = 0.01(x_i - x_{\text{best}}) \cdot \frac{u}{|v|^{1/\beta}}. \tag{2.22}$$

These formulas can be realized by the following code:

```
%% Get cuckoos randomly by carrying out a ramdom walk
function nest=get_cuckoos(nest,best,Lb,Ub)
% Carry out Levy flights
n=size(nest,1);
beta=3/2;
sigma=(gamma(1+beta)*sin(pi*beta/2)/(gamma((1+beta)/2)
      *beta*2^((beta-1)/2)))^(1/beta);

for j=1:n,
    s=nest(j,:);
```

```
%% Levy flights by Mantegna's algorithm
u=randn(size(s))*sigma;
v=randn(size(s));
step=u./abs(v).^(1/beta);
stepsize=0.01*step.*(s-best);
s=s+stepsize.*randn(size(s));
%% Apply simple bounds/limits
nest(j,:)=simplebounds(s,Lb,Ub);
end
```

The above code also includes the application of the simple bounds, to be discussed below.

The next part is about the selection of solutions. New solutions should be evaluated and the new best solution should be found.

```
%% Find the current best nest
function [fmin,best,nest,fitness]=get_best_nest(nest,newnest,fitness)
% Evaluating all new solutions
for j=1:size(nest,1),
    fnew=fobj(newnest(j,:));
    if fnew<=fitness(j),
        fitness(j)=fnew;
        nest(j,:)=newnest(j,:);
    end
end
% Sort and find the best solution in the current population
[fmin,K]=min(fitness) ;
best=nest(K,:);
```

From the Lévy flights and the simple bounds, we can now complete the "empty_nests" function as follows:

```
%% Replace some nests/solutions by generating new nests/solutions
function new_nest=empty_nests(nest,Lb,Ub,pa)
% Each solution/nest is discovered with a probability pa
n=size(nest,1);
% Record if discovery is successful or not -- a status vector
K=rand(size(nest))>pa;
%% Generate new solutions by biased/selective random walks
stepsize=rand*(nest(randperm(n),:)-nest(randperm(n),:));
new_nest=nest+stepsize.*K;
for j=1:size(new_nest,1)
    s=new_nest(j,:);
    new_nest(j,:)=simplebounds(s,Lb,Ub);
end
```

The sixth part checks all the solutions to ensure that they are within the simple lower and upper bounds.

```
% Application of simple lower and upper bounds
function s=simplebounds(s,Lb,Ub)
  % Check the lower bound
  nsol_tmp=s;
  I=nsol_tmp<Lb;
  nsol_tmp(I)=Lb(I);

  % Check the upper bounds
  J=nsol_tmp>Ub;
  nsol_tmp(J)=Ub(J);
  % Update this new solution within the bounds
  s=nsol_tmp;
```

The final part is the simple calculation of the objective function.

```
%% Cost or objective function
function z=cost(x)
z=sum((x-2).^2); % Solutions should be (2,2,...,2)
```

If we put all the above lines of code into a single file, we can run it. For a maximum number of 1000 iterations, we can usually obtain the best minimum value as $f_{best} = 1.17 \times 10^{-12}$ for 15 decision variables. For a lower-dimensional problem with $D = 5$ (five variables), we can easily get about 3.5×10^{-31}. As pointed out in the previous section, multiple runs are needed to ensure that the meaningful results and statistical measures can be calculated from the results.

Various studies have shown that CS can be very effective (Yang and Deb, 2010; Yang, 2013), and we will discuss this in greater detail in later chapters, where we will carry out some theoretical analysis. Briefly speaking, there is a fraction of long jumps in step sizes generated by Lévy flights, in comparison with the steps generated by isotropic Gaussian random walks (Pavlyukevich, 2007). Thus, the exploration capability of CS is enhanced significantly, which has been confirmed by various studies and applications (Yang and Deb, 2013; Ouaarab et al., 2014).

2.4 Discretization and solution representations

The two algorithms we have discussed in this chapter were initially designed for solving problems with continuous variables. They can also be used to solve discrete optimization problems with some modifications in terms of solutions representations. Therefore, it is useful to discuss how we may convert discrete algorithms for continuous optimization to their corresponding discrete versions for solving discrete and combinatorial optimization problems.

Loosely speaking, there is no need to change the algorithmic structures, and the main change is the way of representing solutions or solution vectors. A major step is to convert or discretize continuous variables into their corresponding discrete variables. One simple way to discretize a continuous variable x is to use the sigmoidal or logistic function

$$S(x) = \frac{1}{1 + e^{-x}},$$

(2.23)

which is an S-shaped function. In theory, this converts a continuous variable x into a binary variable S when $|x|$ is large. Clearly, we have $S \to +1$ when $x \to +\infty$, and $S \to 0$ when $x \to -\infty$. This function has a special property for its derivative. Its derivative can be computed by multiplication, i.e.,

$$\frac{dS}{dt} = -\frac{1}{(1 + e^{-x})^2}(-e^{-x}) = \frac{1}{1 + e^{-x}} \cdot \frac{e^{-x}}{1 + e^{-x}}$$

$$= \frac{1}{1 + e^{-x}} \cdot [1 - \frac{1}{1 + e^{-x}}] = S(1 - S).$$

(2.24)

Since large values are needed, this is not easy to realize in practice. A slight modification is to use a random number $r \in [0, 1]$, together with a conditional switch. So we can set $u = 1$ if $S(x) > r$; otherwise, $u = 0$. This results in a binary variable $u \in \{0, 1\}$. Obviously, once we have a binary variable $u \in \{0, 1\}$, we can convert it to binary variables with other discrete values. For example, we can use $y = 2u - 1 \in \{-1, +1\}$.

An alternative to discretize a continuous variable is to use mod and round-up operations. For example, we can use the round-up

$$y = \lfloor x \rfloor$$

(2.25)

to convert x to integer y. Similarly, we can convert a continuous variable into m discrete integers by using a mod function

$$y = \lfloor x + k \rfloor \quad \mod m,$$

(2.26)

where k and $m > 0$ are integers.

Another class of transformation and discretization is the use of the so-called random keys, which maps one set of random keys to a corresponding set of random integers. For example, in the well-known traveling salesman problem (TSP), a path can be encoded into a set of integers and each integer corresponds to the nodal number of a city. Similarly, in the job-shop scheduling problems, the job numbers are also integers. In this case, we can generate an array of random numbers (random keys)

$$x = [0.93, \ 1.2, \ 0.11, \ 0.42, \ 0.80, \ -0.27, 0.69],$$

(2.27)

which can be converted to city orders and job numbers

$$J = [6, \ 7, \ 2, \ 3, \ 5, \ 1, \ 4].$$

(2.28)

This is done by first ranking the real number vector x, and then mapping them into labels of ranks.

For the ease of implementations, uniformly distributed random numbers in the range of [0,1] are widely used. For example, we have

$$\begin{pmatrix} \text{Real numbers} & 0.68 & 0.17 & 0.33 & 0.23 & 0.91 \\ \downarrow & \downarrow & \downarrow & \downarrow & \downarrow & \downarrow \\ \text{Random keys} & 4 & 1 & 3 & 2 & 5 \end{pmatrix}. \quad (2.29)$$

Obviously, there are other ways to map continuous ways to discrete variables. Whatever the method may be, the essence is to represent the solutions in a simple way so that they can be easily and effectively incorporated in the algorithms. In the current literature, random keys-based approaches have been applied in many applications, such as the TSP and vehicle routing problems (Ouaarab et al., 2014; Osaba et al., 2019).

We have provided a tutorial style of introduction to two algorithms in this chapter. It is worth pointing out that we have not implemented any nonlinear constraints because both examples we have discussed are essentially unconstrained function optimization problems. In reality, nonlinear constrained optimization problems are more challenging in implementations, and one key part of any implementation is the proper handling of nonlinear constraints. Interested readers can refer to more specialized literature on this topic.

References

Altringham, J.D., 1996. Bats: Biology and Behaviour. Oxford University Press, Oxford, UK.

Chakri, A., Khelif, R., Benouaret, M., Yang, X.S., 2017. New directional bat algorithm for continuous optimization problems. Expert Systems with Applications 69 (1), 159–175.

Chen, S., Peng, G.H., Xing-Shi Yang, X.S., 2018. Global convergence analysis of the bat algorithm using a markovian framework and dynamic system theory. Expert Systems with Applications 114 (1), 173–182.

Colin, T., 2000. The Variety of Life. Oxford University Press, Oxford, UK.

Davies, N.B., 2011. Cuckoo adaptations: trickery and tuning. Journal of Zoology 284 (1), 1–14.

Davies, N.B., Brooke, M.L., 1991. Co-evolution of the cuckoo and its hosts. Scientific American 264 (1), 92–98.

Mantegna, R.N., 1994. Fast, accurate algorithm for numerical simulation of Lévy stable stochastic process. Physical Review E 49 (5), 4677–4683.

Osaba, E., Yang, X.S., Fister Jr., I., Lopez-Garcia, P., Vazquez-Paravila, A., 2019. A discrete and improved bat algorithm for solving a medical goods distribution problem with pharmacological waste collection. Swarm and Evolutionary Computation 44 (1), 273–286.

Ouaarab, A., Ahiod, B., Yang, X.S., 2014. Discrete cuckoo search algorithm for the travelling salesman problem. Neural Computing and Applications 24 (7–8), 1659–1669.

Pavlyukevich, I., 2007. Lévy flights, non-local search and simulated annealing. Journal of Computational Physics 226 (2), 1830–1844.

Yang, X.S., 2010. A new metaheuristic bat-inspired algorithm. In: Cruz, C., González, J.R., Pelta, D.A., Terrazas, G. (Eds.), Nature Inspired Cooperative Strategies for Optimization (NISCO 2010). In: Studies in Computational Intelligence, vol. 284. Springer, Berlin, Germany, pp. 65–74.

Yang, X.S., 2011. Bat algorithm for multi-objective optimisation. International Journal of Bio-Inspired Computation 3 (5), 267–274.

Yang, X.S., 2013. Cuckoo Search and Firefly Algorithm: Theory and Applications. Studies in Computational Intelligence, vol. 516. Springer, Heidelberg, Germany.

Yang, X.S., Deb, S., 2009. Cuckoo search via Lévy flights. In: Proceedings of World Congress on Nature & Biologically Inspired Computing (NaBIC 2009). IEEE Publications, USA, pp. 210–214.

Yang, X.S., Deb, S., 2010. Engineering optimisation by cuckoo search. International Journal of Mathematical Modelling and Numerical Optimisation 1 (4), 330–343.

Yang, X.S., Deb, S., 2013. Multiobjective cuckoo search for design optimization. Computers & Operations Research 40 (6), 1616–1624.

Yang, X.S., Gandomi, A.H., 2012. Bat algorithm: a novel approach for global engineering optimization. Engineering Computations 29 (5), 464–483.

Yang, X.S., He, X.S., 2013. Bat algorithm: literature review and applications. International Journal of Bio-Inspired Computation 5 (3), 141–149.

Firefly algorithm and flower pollination algorithm

3

Xin-She Yang[a], Yu-Xin Zhao[b]

[a]*Middlesex University London, School of Science and Technology, London, United Kingdom*
[b]*Harbin Engineering University, College of Automation, Harbin, China*

CONTENTS

3.1 Introduction

Nature-inspired algorithms for optimization are very diverse, and they have been designed by drawing different inspiration from nature. Some algorithms are mostly based on the swarming behavior, and a primary example is the particle swarm optimization (Kennedy and Eberhart, 1995), while others are not explicitly based on swarm intelligence but can work equally. Examples for the latter case are differential evolution and the flower pollination algorithm (FPA). This chapter will introduce two nature-inspired algorithms: the firefly algorithm (FA) and FPA.

We will introduce each algorithm with the basic ideas, the main steps, pseudocode, implementation details, and an example. The MATLAB® codes are presented and explained in detail so that readers can gain deeper understanding of how these algorithms work.

Nature-Inspired Computation and Swarm Intelligence. https://doi.org/10.1016/B978-0-12-819714-1.00012-9

3.2 The firefly algorithm

Fireflies flash and swarm with some amazing biological purpose, and flashing with different patterns is part of their sophisticated signaling system (Lewis and Cratsley, 2008). Based on the flashing characteristics of tropic fireflies, Xin-She Yang developed FA in late 2007 and early 2008 (Yang, 2008, 2009).

3.2.1 Algorithmic equations in FA

In FA, the position vector x_i of a firefly, also as a solution vector to an optimization problem, is updated according to the behavior of attraction, random search movement, and variation of light intensity. At iteration t, the distance r_{ij} between two fireflies (i and j) at two positions x_i^t and x_j^t, respectively, is the Cartesian distance

$$r_{ij} = ||x_j - x_i|| = \sqrt{\sum_{k=1}^{D} (x_{j,k} - x_{i,k})^2}, \tag{3.1}$$

where $k = 1, 2, ..., D$ corresponds to the kth component of x_j or x_i in the D-dimensional search space.

The movement of a firefly is updated according to the following algorithmic equation:

$$x_i^{t+1} = x_i^t + \beta_0 e^{-\gamma r_{ij}^2} (x_j^t - x_i^t) + \alpha \epsilon_i^t, \tag{3.2}$$

where β_0 is the attractiveness of a firefly at zero distance, that is, $r_{ij} = 0$. This basic attractiveness can be set as a constant $\beta_0 = 1$ for all fireflies. In addition, α is a parameter controlling the strength of the randomization term with ϵ_i^t being drawn from a normal distribution N(0,1).

In contrast to algorithms such as particle swarm optimization, FA is a nonlinear system because the nonlinear variation of light intensity has been used. Light intensity I varies with the distance r from the source

$$I(r) = \frac{I_0}{r^2}, \tag{3.3}$$

where I_0 is the light intensity at the source with $r = 0$. In addition, light absorption in the air also makes the light intensity decrease in an exponential manner, i.e.,

$$I(r) = I_0 e^{-\gamma r}, \tag{3.4}$$

where γ is the absorption coefficient. Though not physically exact, we can loosely combine these two effects as a single approximate equation,

$$I(r) = I_0 e^{-\gamma r^2}. \tag{3.5}$$

This approximation makes later discussions easier. In addition, we can associate the light intensity with the attractiveness β of a firefly, and thus write the above equation as the attractiveness

$$\beta = \beta_0 e^{-\gamma r^2}, \tag{3.6}$$

where β_0 is the attractiveness at $r = 0$. In general, fireflies tend to be attracted towards the more attractive fireflies in the neighborhood. As the distance is between any two fireflies, the attractiveness is relative, which means that the beauty is indeed in the eye of the beholder.

3.2.2 FA pseudocode

From the discussion of FA, we can see that pair-wise comparison is needed to evaluate the relative attractiveness in the eye of the beholder. This means that two loops are usually needed to loop over all possible pairs. Consequently, the updates of positions or solutions are carried out iteratively, and new solutions are evaluated within the loops. The pseudocode of the FA is summarized in Algorithm 1.

Algorithm 1 The FA pseudocode.
---

input   : Population size n, α_0, θ, β_0, and γ
output   : Global best solution
auxiliaries: Define function $f(x)$ (objective)

1 Generate n fireflies as the initial population;
   Set the iteration counter $t = 0$;
   while $(t < t_{\max})$ **do**
2   **for** $i = 1 : n$ **do**
3   **for** $j = 1 : i$ **do**
4   Compute the distance $r_{ij} = ||x_i - x_j||$;
            Draw a random vector $\epsilon \sim N(0, 1)$;
            Update the positions/solutions of fireflies i and j;
            Evaluate $f(x_i)$ and $f(x_j)$;
            if $f(x_i) < f(x_j)$ *(for minimization)* **then**
5   Move firefly j towards i via Eq. (3.2);
6   Rank all fireflies and update the current best solution g_*;
7 Display the best solution and its objective value

3.2.3 Scalings and parameters

Before we start to implement the FA using Algorithm 1, let us discuss the scalings and parameters in the firefly algorithm.

The main term of influencing the scaling is the nonlinear attractiveness strength β between two fireflies,

$$\beta = \beta_0 \exp[-\gamma r_{ij}^2]. \tag{3.7}$$

As all the terms in Eq. (3.2) should be in the same order, we can set $\beta_0 = O(1)$ in all applications, and in most cases we can use $\beta_0 = 1$.

The value of γ can affect this term quite dramatically. On the one hand, if γ is too small (i.e., $\gamma \to 0$), we get $\beta \to \beta_0$. This means that the light intensity or attractiveness remains close to a constant β_0, which corresponds to the high-visibility case where a flashing firefly can be seen by all the other fireflies in the search space. The whole population is visible to each other, and the information is essentially shared by the whole population.

On the other hand, we have $\beta \to 0$ if $\gamma \to +\infty$, which means that the visibility is very low. This case corresponds to a scenario like in a dense fog where one firefly cannot see other fireflies. Each firefly essentially acts individually and performs local moves randomly.

The former case becomes a linear system (when $\gamma \to +\infty$) with a higher exploitation ability, while the latter case becomes a highly nonlinear system (when $\gamma = O(1)$) with a higher exploration ability. In general, FA should work between these two extremes. Thus, the setting of γ is important. Based on the empirical and mathematical analysis, a simple rule of setting γ is to allow $\gamma r_{ij}^2 = 1$. That is,

$$\gamma = \frac{1}{L^2}, \tag{3.8}$$

where L is the average length scale of the optimization problem. For example, if a variable varies from -10 to $+10$, its scale is $L = 10$, so we can use $\gamma = 1/10^2 = 0.01$, which is the value to be used in our implementation. If there is no prior knowledge of the problem modality, $\gamma = 0.01$ to 1 can be used first.

In FA, there are three parameters, i.e., α, β_0, and γ. We have discussed β_0 and γ, and now we focus on α. The factor α controls the step size of the movements of the fireflies. On the one hand, if α is large, the steps are large, which gives large steps, but these steps may be too far away. This may lead to strong exploration and potentially waste some search moves due to their infeasibility. On the other hand, steps can be too small if α is too small. In this case, the search moves are primarily local and thus may limit the exploration ability. Thus, a fine balance with the right value of α is important. Furthermore, α is also associated with the randomization term and can be considered as the strength of the randomization because there is essentially no randomization if $\alpha = 0$. As the iterations continue, it makes sense to reduce such randomness gradually. So, α should be gradually reduced as the iterations proceed. There can be different ways to reduce α; however, we will use

$$\alpha = \alpha_0 \theta^t, \tag{3.9}$$

where $0 < \theta < 1$ is a constant and α_0 is the initial value of α. Based on preliminary parameter studies, we can use $\theta = 0.9$ to 0.99 and $\alpha_0 = 1$ in most cases. We will use these values in our demo implementation.

3.2.4 Demo implementation

From the descriptions of the main steps of FA, it straightforward to implement it in any programming language as this algorithm has only a single updating equation for the positions of fireflies. In this chapter, we use a simple demo implementation using MATLAB to solve a function optimization problem,

$$\text{minimize } f(x) = x_1^2 + x_2^2 + ... + x_D^2 = \sum_{i=1}^{D} x_i^2, \quad x_i \in \mathbb{R}, \tag{3.10}$$

which has a global minimum of $x_* = (0, 0, ..., 0)$. In theory, the whole domain is spanned by all the real numbers in the D-dimensional space. It is useful to use some simple lower bound (Lb) and upper bound (Ub),

$$Lb = [-10, -10, ..., -10], \quad Ub = [+10, +10, ..., +10]. \tag{3.11}$$

Though the demo MATLAB code is a single file, we split it into five parts so as to ease their descriptions. These five parts are initialization of parameters, initialization of the population, the main loops, application of the limits, and the objective function.

The first part is mainly initialization of parameter values such as α_0, β_0, and γ, as well as θ and the maximum number of iterations.

```
function fa_demo  % Start the FA demo
n=20;             % Population size (number of fireflies)
alpha=1.0;        % Randomness strength 0--1 (highly random)
beta0=1.0;        % Attractiveness constant
gamma=0.01;       % Absorption coefficient
theta=0.95;       % Randomness reduction factor for alpha
d=10;             % Number of dimensions
tMax=500;         % Maximum number of iterations
Lb=-10*ones(1,d); % Lower bounds/limits
Ub=10*ones(1,d);  % Upper bounds/limits
```

Then, a population of $n = 20$ fireflies is generated and evaluated so as to find the initial best solution (or the brightest or most attractive firefly).

```
% Generating the initial locations of n fireflies
    for i=1:n,
    ns(i,:)=Lb+(Ub-Lb).*rand(1,d);
    Lightn(i)=cost(ns(i,:));  % Evaluate objectives
    end
```

The next part is the main part with two loops. Within the loops, the positions of the fireflies are updated by Eq. (3.2).

```
for k=1:tMax,           %%%%% start the iterations (main loop) %%%%%
 alpha=alpha*theta;    % Reduce alpha by a factor theta
 scale=abs(Ub-Lb);     % Scale of the optimization problem
% Two loops over all the n fireflies
for i=1:n,
  for j=1:n,
      % Evaluate the objective values of current solutions
      Lightn(i)=cost(ns(i,:));      % Call the objective
      % Update moves
      if Lightn(i)>Lightn(j),       % Brighter/more attractive
      r=sqrt(sum((ns(i,:)-ns(j,:)).^2));
      beta=beta0*exp(-gamma*r.^2);  % Attractiveness
      steps=alpha.*(rand(1,d)-0.5).*scale;
      % The FA equation for updating position vectors
      ns(i,:)=ns(i,:)+beta*(ns(j,:)-ns(i,:))+steps;
      end
   end % end for j
end % end for i
```

Once the new solutions are generated, they are first checked to see if they remain in the bounds of the domain. Then they are evaluated by calling the objective function and their fitness is ranked.

```
% Check if the new solutions/locations are within limits/bounds
ns=findlimits(n,ns,Lb,Ub);
%% Rank fireflies by their light intensity/objectives
[Lightn,Index]=sort(Lightn);
nsol_tmp=ns;
for i=1:n,
 ns(i,:)=nsol_tmp(Index(i),:);
end
%% Find the current best solution and display outputs
fbest=Lightn(1), nbest=ns(1,:)
end % End of t loop (up to tMax)
```

Here, the simple bounds are imposed by a simple dimension-wise comparison with the limits.

```
% Make sure that new fireflies are within the bounds/limits
function [ns]=findlimits(n,ns,Lb,Ub)
for i=1:n,
    nsol_tmp=ns(i,:);
  % Apply the lower bound
```

```
    I=nsol_tmp<Lb;  nsol_tmp(I)=Lb(I);
    % Apply the upper bounds
    J=nsol_tmp>Ub;  nsol_tmp(J)=Ub(J);
    % Update this new move
    ns(i,:)=nsol_tmp;
end
```

The final part is the objective function, which is very straightforward to write.

```
%% Define the objective or cost function
function z=cost(x)
z=sum(x.^2); % Solutions should be (0,0,...,0)
```

If we put all the above lines of codes into a file, we can run it. For example, one run gives $f_{\min} = 6.1 \times 10^{-21}$ for $n = 20$ after 500 iterations. Due to the random nature of all nature-inspired algorithms, multiple runs should be carried out so as to obtain meaningful statistical results.

FA has been applied in many applications and has also been extended to multiobjective optimization (Yang, 2013b).

3.2.5 Multiobjective FA

Though FA can be effective in solving multimodal problems (Yang, 2009, 2013a), the standard FA is mainly for single-objective optimization. To solve multiobjective optimization problems, some modifications are needed. In fact, multiobjective optimization itself is a very active research area, and we will not be able to introduce this topic in much detail in this chapter. Instead, we briefly introduce a simple way to rewrite a multiobjective problem into a single-objective problem and then use the above algorithm to solve it.

In order to show the main idea, we now use an example with two objectives,

$$f_1(x) = x^2, \quad f_2(x) = (x-2)^2, \quad -1000 \le x \le 1000, \tag{3.12}$$

whose nondominated solutions form a Pareto front. The weighted sum approach is to combine two objectives into one using an additional parameter (either a or b here) in the following form:

$$f(x) = af_1 + bf_2, \quad a, b \in [0, 1], \tag{3.13}$$

where $a + b = 1$. In this case, the optimal solution of $f(x)$ will clearly depend on a (or $b = 1 - a$). From basic calculus, it is straightforward to obtain

$$f'(x) = [ax^2 + (1-a)(x-2)^2] = 2ax + 2(1-a)(x-2) = 0, \tag{3.14}$$

which gives

$$x_* = 2(1-a). \tag{3.15}$$

Clearly, this solution x_* depends on a. As a varies from 0 to 1 (thus b from 1 to 0), the solutions will trace the Pareto front.

Similarly, we can write an optimization problem with m objective functions,

$$\text{Minimize } f_1(x), f_2(x), ..., f_m(x), \quad x \in \mathbb{R}^D, \tag{3.16}$$

as a single-objective optimization problem,

$$\text{Minimize } F(x) = \sum_{i=1}^{m} w_i f_i, \quad \forall w_i \in [0, 1], \quad x \in \mathbb{R}^D, \tag{3.17}$$

where all the weights should sum to one. That is, $\sum_{i=1}^{m} w_i = 1$. Then, we can in principle solve it using any nature-inspired algorithm (including FA).

However, things are not so simple in practice. There are many additional challenges associated with multiobjective optimization. First, this weighted method is only valid for problems with convex Pareto fronts, and the Pareto fronts are not convex in many cases. Second, selection of the new solutions generated should be carried out so as to find the nondominated solutions. Third, even with nondominated solutions, some sorting is needed so as to produce points that can be reasonably spread out on the Pareto fronts. There are other issues as well, including solution selection measures and computing costs. In the current literature, there are many different approaches for multiobjective optimization, including multiobjective FA (MOFA), multiobjective differential evolution (MODE), and the nondominated sorting genetic algorithm (NSGA-II) (Deb et al., 2002). Interested readers can refer to more specialized literature about multiobjective optimization.

3.3 Flower pollination algorithm

Now let us introduce FPA in the remainder of this chapter. FPA was initially designed for both single-objective optimization and multiobjective optimization (Yang, 2012; Yang et al., 2013). The main idea of FPA was based on the characteristics of flower pollination processes of flowering plants, including both biotic pollination and abiotic pollination as well as flower constancy of certain flower species and their pollinators (Waser, 1986).

Similar to the notation of FA we discussed earlier, the position of a pollen particle is represented as a solution vector. There are two main algorithmic equations in FPA, taking from different inspirations. One equation simulates the long-distance pollination mechanism by

$$x_i^{t+1} = x_i^t + \gamma L(\lambda)(g_* - x_i^t), \tag{3.18}$$

where g_* is the best solution found so far at iteration t, the parameter γ is a scaling parameter, and the random step size $L(\lambda)$ should be drawn from a Lévy distribution, characterized by an exponent λ (Pavlyukevich, 2007). Lévy pseudorandom numbers can be generated by Mantegna's algorithm (Mantegna, 1994).

The other search mechanism, inspired by local pollination characteristics and flower constancy, can be simulated by

$$x_i^{t+1} = x_i^t + U(x_j^t - x_k^t), \tag{3.19}$$

where U is a uniformly distributed random number and x_j^t and x_k^t represent two different pollen particles/positions.

As there are two different search mechanisms represented by two equations, a switching probability p is used for switching or selecting each mechanism. This probability is inspired by the actual abiotic pollination and biotic pollination as well as cross pollination and self-pollination.

Similar to the role of Lévy flights in the cuckoo search algorithm, the use of Lévy flights in the FPA can also enhance the search capability because a fraction of steps generated by Lévy flights are larger than those used in the Gaussian distribution (Pavlyukevich, 2007).

3.3.1 FPA pseudocode and parameters

The main steps of FPA can be summarized as the pseudocode shown in Algorithm 2.

Algorithm 2 FPA pseudocode.

---

input   : Population size n and all parameters
output   : Global best solution
auxiliaries: Function $f(x)$ (objective)

1 Initialize a population of n pollen particles x_i $(i = 1, 2, ..., n)$;
   for *t=1:MaxGeneration* **do**
2   **for** $i = 1 : n$ **do**
3   **if** *rand> p* **then**
4   Draw a step size vector L from a Lévy distribution;
               Update solutions by Eq. (3.18);
5   **else**
6   Draw a random number U from [0,1];
               Update solution by Eq. (3.19);
7   Evaluate new solutions and accept if better;
8   Find the current best solution g_* among the population;
9 Display and postprocess the results

There are three parameters in FPA, i.e., the switching probability p, the population size n, and the scaling parameter γ. Though a native $p = 0.5$ seems to make sense, $p = 0.8$ seems to be more effective and more relevant to real-world pollination processes. Though the population size is huge in real-world settings, empirical observations and numerical simulations suggest that a small population is sufficient. For the scaling parameters, we can in principle combine them with the Lévy step sizes, though $\gamma = 0.01$ to 0.1 can be used. We will use $n = 10$, $p = 0.8$, and $\gamma = 0.1$ in our demo implementation.

3.3.2 Demo implementation

Based on the above pseudocode, it is relatively straightforward to implement it in any programming language. We provide here a simple demo code in MATLAB. The code consists of five parts in the following order: initialization, the main loop, Lévy flights, simple bounds, and the objective function.

The code starts with the initialization of all relevant parameters and the initial population. The maximum number of iterations is set to 1000.

```
function [best,fmin]=fpa_demo
n=10;                  % Population size, typically 10 to 25
p=0.8;                 % Probability for switch
N_iter_max=1000;       % Maximum number of iterations
% Dimension of the search variables
d=5;
Lb=-10*ones(1,d);      % Lower bounds
Ub=10*ones(1,d);       % Upper bounds
% Initialize the population/solutions
for i=1:n,
  Sol(i,:)=Lb+(Ub-Lb).*rand(1,d);
  Fitness(i)=Fun(Sol(i,:));
end
% Find the current best
[fmin,I]=min(Fitness);
best=Sol(I,:);
S=Sol;  % Record the initial solutions
```

The main loop has two branches, switching with a probability p between the first search mechanism and the second search mechanism. New solutions will be updated if better solutions are found.

```
% Start the FPA iterations (main loop)
for t=1:N_iter_max,
        for i=1:n,
        % Realize the first search mechanism
          if rand>p,
            L=Levy(d);  % Carry out Levy flights
            dS=L.*(Sol(i,:)-best);
```

```
S(i,:)=Sol(i,:)+dS;
% Check if the simple limits/bounds are satisfied
S(i,:)=simplebounds(S(i,:),Lb,Ub);
else
    U=rand; % Draw a uniform random number
    JK=randperm(n);
    % Second mechanism: x_i^{t+1}=x_i^t+U*(x_j^t-x_k^t)
    S(i,:)=S(i,:)+U*(Sol(JK(1),:)-Sol(JK(2),:));
    % Check if the simple limits/bounds are met
    S(i,:)=simplebounds(S(i,:),Lb,Ub);
end
% Evaluate new solutions
 Fnew=Fun(S(i,:));
% Update if fitness improves (better solutions found)
  if (Fnew<=Fitness(i)),
      Sol(i,:)=S(i,:);
      Fitness(i)=Fnew;
  end
% Update the current global best (fmin)
if Fnew<=fmin,
      best=S(i,:);
      fmin=Fnew ;
  end
end
end    % End of the main loop
% Display/output the best solution and its objective value
disp(['Best solution=',num2str(best)]);
disp(['fmin=',num2str(fmin)]);
```

The full Lévy flights usually need a sophisticated algorithm to realize. Here, we have used a simplified version of Mantegna's algorithm by fixing $\sigma = 0.7$, based on the values called from a formula with an exponent $\beta = 3/2$.

```
% Draw a vector of n Levy flight samples (combining with gamma)
function L=Levy(d)
gamma=0.1;  % A scaling parameter
beta=3/2;   % Levy exponent
sigma=0.7;  % Calculated from beta=3/2
    u=randn(1,d)*sigma;
    v=randn(1,d);
    step=u./abs(v).^(1/beta);
L=gamma*step; % The equivalent Levy steps
```

The next part is to verify if the simple bounds are satisfied for all new solutions generated.

```
% Application of simple bounds
function s=simplebounds(s,Lb,Ub)
  % Apply the lower bound
  nsol_tmp=s;
  I=nsol_tmp<Lb;  nsol_tmp(I)=Lb(I);
  % Apply the upper bounds
  J=nsol_tmp>Ub;  nsol_tmp(J)=Ub(J);
  % Update this new solution if within bounds
  s=nsol_tmp;
```

Obviously, the optimization problem needs an objective function and here we have a simple multivariate function.

```
% Definite the objective or cost function
function z=Fun(x)
z=sum(x.^2);
```

By putting together all the lines of the codes discussed above, we have the MAT-LAB demo code for the FPA. For this simple function optimization problem, we can easily get $f_{min} = 7.1 \times 10^{-16}$ for $n = 10$ after 1000 iterations. Again, it is worth pointing out that multiple runs are needed to ensure meaningful statistics.

3.4 Constraint handling

In the above implementations, we have only considered the simple limits or bounds because the examples we used are mainly unconstrained optimization problems. For constrained optimization problems with complex equality and inequality constraints, proper constraint handling techniques should be used to ensure the solutions obtained are feasible, satisfying all the constraints.

There are many constraint handling techniques, including penalty methods, dynamic penalty, evolutionary methods, multiobjective approaches, ϵ-constraint approaches, and others (Boyd and Vandenberghe, 2004; Yang, 2014). Interested readers can refer to more specialized literature.

3.5 Applications

In this chapter, we have introduced both FA and FPA. Each of these two algorithms has already a wide range of applications and different variants (Gandomi et al., 2013; Alyasseri et al., 2018).

FA has been applied to mixed-variable structure optimization (Gandomi et al., 2011), clustering (Senthilnath et al., 2011), nonconvex economic dispatch problems (Yang et al., 2012), multiobjective hybrid flowshop scheduling problems (Marichelvam et al., 2014), and others (Fister et al., 2013; Yang, 2013a). FPA has been applied to solve parameter estimation problems (Alam et al., 2015), multiobjective optimization (Yang et al., 2013), and many others (Abdel-Basset and Shawky, 2018) with various new variants (Alyasseri et al., 2018). Other applications will be presented in later chapters of this book.

References

Abdel-Basset, M., Shawky, L.A., 2018. Flower pollination algorithm: a comprehensive review. Artificial Intelligence Review. https://doi.org/10.1007/s10462--018--9624--4.

Alam, D.F., Yousri, D.A., Eteiba, M.B., 2015. Flower pollination algorithm based solar PV parameter estimation. Energy Conversion and Management 101 (2), 410–422.

Alyasseri, Z.A.A., Khader, A.T., Al-Betar, M.A., Awadallah, M.A., Yang, X.S., 2018. Variants of the flower pollination algorithm: a review. In: Yang, X.S. (Ed.), Nature-Inspired Algorithms and Applied Optimization. Springer, Cham, pp. 91–118.

Boyd, S.P., Vandenberghe, L., 2004. Convex Optimization. Cambridge University Press, Cambridge, UK.

Deb, K., Pratap, A., Agarwal, S., Mayarivan, T., 2002. A fast and elitist multiobjective algorithm: Nsga-ii. IEEE Transactions on Evolutionary Computation 6 (2), 182–197.

Fister, I., Fister Jr, I., Brest, J., Yang, X.S., 2013. A comprehensive review of firefly algorithms. Swarm and Evolutionary Computation 13 (1), 34–46.

Gandomi, A.H., Yang, X.S., Alavi, A.H., 2011. Mixed variable structural optimization using firefly algorithm. Computers & Structures 89 (23–24), 2325–2336.

Gandomi, A.H., Yang, X.S., Talatahari, S., Alavi, A.H., 2013. Firefly algorithm with chaos. Communications in Nonlinear Science and Numerical Simulation 18 (1), 89–98.

Kennedy, J., Eberhart, R., 1995. Particle swarm optimization. In: Proceedings of the IEEE International Conference on Neural Networks. IEEE, Piscataway, NJ, USA, pp. 1942–1948.

Lewis, S.M., Cratsley, C.K., 2008. Flash signal evolution, mate choice and predation in fireflies. Annual Review of Entomology 53 (2), 293–321.

Mantegna, R.N., 1994. Fast, accurate algorithm for numerical simulation of Lévy stable stochastic process. Physical Review E 49 (5), 4677–4683.

Marichelvam, M.K., Prabaharan, T., Yang, X.S., 2014. A discrete firefly algorithm for the multi-objective hybrid flowshop scheduling problems. IEEE Transactions on Evolutionary Computation 18 (2), 301–305.

Pavlyukevich, I., 2007. Lévy flights, non-local search and simulated annealing. Journal of Computational Physics 226 (2), 1830–1844.

Senthilnath, J., Omkar, S.N., Mani, V., 2011. Clustering using firefly algorithm: performance study. Swarm and Evolutionary Computation 1 (3), 164–171.

Waser, N.M., 1986. Flower constancy: definition, cause and measurement. American Naturalist 127 (5), 596–603.

Yang, X.S., 2008. Nature-Inspired Metaheurisic Algorithms. Luniver Press, Bristol, UK.

Yang, X.S., 2009. Firefly algorithms for multimodal optimization. In: Watanabe, O., Zeugmann, T. (Eds.), Proceedings of Fifth Symposium on Stochastic Algorithms, Foundations and Applications. In: Lecture Notes in Computer Science, vol. 5792. Springer, pp. 169–178.

Yang, X.S., 2012. Flower pollination algorithm for global optimization. In: Durand-Lose, J., Jonoska, N. (Eds.), Unconventional Computation and Natural Computation (UCNC) 2012, vol. 7445. Springer, Berlin, Heidelberg, Germany, pp. 240–249.

Yang, X.S., 2013a. Cuckoo Search and Firefly Algorithm: Theory and Applications. Studies in Computational Intelligence, vol. 516. Springer, Heidelberg, Germany.

Yang, X.S., 2013b. Multiobjective firefly algorithm for continuous optimization. Engineering with Computers 29 (2), 175–184.

Yang, X.S., 2014. Nature-Inspired Optimization Algorithms. Elsevier Insight, London.

Yang, X.S., Hosseini, S.S.S., Gandomi, A.H., 2012. Firefly algorithm for solving non-convex economic dispatch problems with valve loading effect. Applied Soft Computing 12 (3), 1180–1186.

Yang, X.S., Karamanoglu, M., He, X.S., 2013. Multi-objective flower algorithm for optimization. Procedia Computer Science 18 (1), 861–868.

Bio-inspired algorithms: principles, implementation, and applications to wireless communication

4

Swati Swayamsiddha

Kalinga Institute of Industrial Technology, School of Electronics Engineering, KIIT Deemed to be University, Bhubaneswar, India

CONTENTS

4.1 Introduction

Bio-inspired algorithms have evolved to a new dimension in solving engineering optimization problems. They have recently captured the attention of researchers owing to their simple implementation and superior performance. Since they are inspired by the processes of nature they are adaptive and robust (Del Ser et al., 2019). These algorithms are basically used as global optimization tools and they have been implemented in various optimization engineering problems. Selected algorithms are discussed for different strategic applications in an effective way. The main focus of this work is to introduce the important bio-inspired techniques available in the liter-

Nature-Inspired Computation and Swarm Intelligence. https://doi.org/10.1016/B978-0-12-819714-1.00013-0

ature. The method of development, implementation and the corresponding step-wise description is highlighted along with their applications to the wireless communication domain.

Bio-inspired algorithms have gained relevance, particularly for solving complex optimization problems in the engineering domain. These are stochastic search techniques which are developed to achieve near-optimal solutions to large-scale optimization problems (Yang, 2012). The traditional mathematical optimization techniques may often fail (being trapped in a local optimum) in solving NP-hard problems having a large number of variables and nonlinear objective functions, which led to the development of alternative solutions. In recent times, the bio-inspired algorithms have been explored for various telecommunication applications, including routing in sensor networks (Chaudhry et al., 2019), electromagnetic antenna design (Yang et al., 2002; Mohammed et al., 2016), mobility management (Swayamsiddha et al., 2018, 2019; Parija et al., 2017), filter design (Storn, 1996), home automation networks (Wang et al., 2015), spectrum sensing in cognitive radio (Azmat et al., 2015), and Internet of Things (IoT) (Deng et al., 2018; Li and Chen, 2016), and their performance is as good as or better than the conventional techniques. These algorithms are inspired by the natural biological evolution and/or social behavior of species (Elbeltagi et al., 2005). These techniques are based on biological structures and behaviors that have self-adaptation and self-organization potential (Floreano and Mattiussi, 2008) and are used as powerful optimization tools. In the course of time, the area of bio-inspired computing is getting highlighted. Because of the growing complexity of the optimization problems whose exact solution is not feasible, the soft computing techniques based on bio-inspired algorithms prove beneficial. Among the metaheuristics, the bio-inspired algorithms are gaining prominence which can adapt, adjust, and accommodate just like the biological entities. Metaheuristics is defined as heuristics at a higher level of frameworks. "Heuristics" means to find and "meta" means higher-level. So, metaheuristics are problem-independent strategies, whereas heuristics are problem-specific. The bio-inspired algorithms are broadly classified as evolutionary algorithms, which are based on Darwin's theory of survival of the fittest and natural selection, and swarm intelligence algorithms, which are based on behavioral models of social creatures such as ants, honey bees, fireflies, fish, birds, etc. The performance analysis of bio-inspired computing-based algorithms is studied for various applications, grouped under direct modeling, inverse modeling, and mobile location management in cellular networks. A well-organized study of said areas is presented in four sections, beginning with the current section. The remainder of the chapter is organized as follows: Section 4.2 presents the step-wise implementation details of the selected bio-inspired algorithms. Section 4.3 describes the application of bio-inspired techniques in various areas in the wireless communication domain. Section 4.4 gives the conclusion of the chapter.

4.2 Selected bio-inspired techniques: principles and implementation

This section presents the underlying principles and step-wise implementation details of bio-inspired computing techniques such as genetic algorithm (GA), differential evolution (DE), particle swarm optimization (PSO), and bacterial foraging optimization (BFO).

4.2.1 Genetic algorithm

GA was proposed by J.H. Holland of the University of Michigan in 1965, though his major work was published in 1975, where the mathematical foundation of GA was established using the schema theorem or building block hypothesis (Holland, 1975). Later substantial work on GA was carried out by Goldberg. The different variants of GA available in the literature are binary-coded GA (Goldberg and Holland, 1988), real-coded GA (Goldberg, 1991; Eshelman and Schaffer, 1993), micro-GA (Krishnakumar, 1990), messy GA (Goldberg et al., 1989), etc. This population-based evolutionary computing algorithm is based on Darwin's principle of natural selection and follows the process of "survival of the fittest." GA is a popular probabilistic, iterative controlled search technique used for global optimization. The working cycle of GA has four basic steps, namely, initialization, selection, crossover, and mutation. The crucial parameters involved are population size, the probability of crossover, the probability of mutation, the selection method, and the number of generations. The flow diagram of GA is given in Fig. 4.1, and the steps of binary GA are discussed below.

Step 1: Initialization of population

The initial population of GA consists of randomly generated binary numbers termed chromosomes or GA strings, consisting of bits called genes. These are the probable solutions to the optimization problem. The number of bits n_b assigned to represent a variable in the chromosome depends on the precision ϵ and the range of the variable $[x_{\min}, x_{\max}]$, and is given by

$$n_b = \log_2 \left(\frac{x_{\max} - x_{\min}}{\epsilon} \right) \qquad (4.1)$$

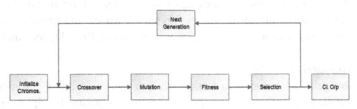

FIGURE 4.1

Flow diagram of GA.

Step 2: Fitness evaluation

The fitness value of each of the GA strings is evaluated by first determining the decoded values of the variables D, and next the corresponding real values are obtained using the following linear mapping rule:

$$x = x_{min} + \frac{x_{max} - x_{min}}{2^{n_b} - 1} \times D. \tag{4.2}$$

Then the fitness function values are calculated knowing the real values of design variables.

Step 3: Reproduction/selection

After the fitness values are evaluated, this step selects the chromosomes with better fitness values to participate in the crossover. Various selection modes have been proposed in the literature, namely, roulette wheel selection or proportionate selection, rank-based selection, and tournament selection (Thiele and Blickle, 1995). In proportionate selection, the probability of a chromosome to be selected in the mating pool is directly proportional to its fitness value. So, the chromosome having better fitness value has a higher expected count in the mating pool. This may lead to premature convergence of the solution as there is a chance of losing the diversity. The rank-based selection, where the chromosomes are ranked in accordance to their fitness values and then the probability of selection into the mating pool is evaluated, is expected to perform better than the proportionate selection. The advantage of tournament selection lies in its computational efficiency as this scheme is faster compared with the other two selection methods. In this scheme, n chromosomes are randomly picked from the population of solutions, where n represents the tournament size. The chromosome having the best fitness value is copied to the mating pool and all the n GA strings are returned to the population. This process is repeated for obtaining all the individuals of the mating pool.

Step 4: Crossover

In this step, the genes are exchanged between two parent chromosomes, giving rise to new set of solutions, termed children. The crossover operation represents the selection pressure or exploitation of good chromosomes for even better solutions. The crossover probability P_c determines the number of individuals participating in the crossover operation, and this control parameter value is optimally chosen as nearly equal to 1.0. In the literature, various schemes of crossover are found, like single-point crossover, two-point crossover, multipoint crossover, and uniform crossover, and their comparative study also exists (Spears and Anand, 1991; De Jong and Spears, 1992).

Step 5: Mutation

The change of a bit from 0 to 1 and from 1 to 0 in the solution chromosome is termed mutation, and it is used for exploration of new solutions. It helps to come out of the local basin and search for the global solution. The mutation probability P_m, which determines the number of mutations, is generally kept very low, as the good

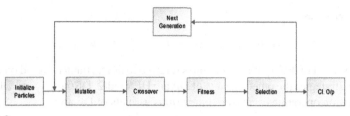

FIGURE 4.2

Flow diagram of DE.

solutions may be lost if its value is high. The range of P_m is given as

$$\frac{0.1}{l} \leqslant P_m \leqslant \frac{1}{l}, \tag{4.3}$$

where l represents the length of the GA string. Steps 2, 3, 4, and 5 are repeated until the termination criterion (maximum number of generations or desired precision of solution) is met.

4.2.2 Differential evolution

The DE algorithm (Storn and Price, 1997) is similar to GA, but the differential mutation is performed prior to the crossover operation. This algorithm is much faster compared with GA and its working cycle involves four basic operations, namely, initialization, differential mutation, crossover, and selection, as shown in Fig. 4.2 The crucial parameters are population size, scaling factor, and crossover probability. The steps of DE are discussed as follows (Swayamsiddha and Thethi, 2018).

Step 1: Initialization

The initialization step involves generation of NP initial parameter vector solutions. Each vector has p parameters, and the lower and upper bounds of each parameter are fixed. Each of the parameters is a random number within the specified range. The initial ith vector for the jth parameter in generation g is denoted as $v_{i,j}(g)$. It is given as follows:

$$v_{i,j}(g) = rand_j(0, 1).(p_U - p_L), \tag{4.4}$$

where p_U and p_L are upper and lower bounds.

Step 2: Differential mutation

Let us consider the first vector of population as target vector. With respect to this target vector three random vectors ($v_{r1,j}$, $v_{r2,j}$, $v_{r3,j}$) are chosen. Then the difference between the corresponding elements of the last two vectors is taken, and each element of the difference vector is multiplied by the scaling factor F. The resultant vector becomes the mutant vector of the first target vector. This process is continued until the last population number. Thus, for each target vector number the corresponding mutant vectors $m_{i,j}(g + 1)$ are generated. The equation used to generate the mutant vector is given as

$$m_{i,j}(g+1) = v_{r1,j}(g) + F.(v_{r2,j}(g) - v_{r3,j}(g)). \tag{4.5}$$

This variant of differential mutation is referred to as DE/rand/1. Based on different mutation strategies the other variants of DE are DE/rand/2, DE/best/1, and DE/best/2. The mutation operations carried out in various variants of DE are

$$m_{i,j}(g+1) = v_{r5,j}(g) + F.(v_{r1,j}(g) + v_{r2,j}(g) - v_{r3,j}(g) - v_{r4,j}(g)), \tag{4.6}$$

$$m_{i,j}(g+1) = v_{best,j}(g) + F.(v_{r2,j}(g) - v_{r3,j}(g)), \tag{4.7}$$

$$m_{i,j}(g+1) = v_{best,j}(g) + F.(v_{r1,j}(g) + v_{r2,j}(g) - v_{r3,j}(g) - v_{r4,j}(g)), \tag{4.8}$$

where F is a real constant $\in [0, 2]$. This control parameter amplifies the differential variation; $v_{best,j}$ is the best member of the population.

Step 3: Crossover

The crossover operation involves the exchange of parameters between the initially chosen target vector and the mutant vector based on the probability of the crossover ratio CR. The resultant vector called the trial vector $u_{i,j}(g+1)$ is obtained as

$$u_{i,j}(g+1) = \begin{cases} m_{i,j}(g+1), & \text{if} \quad rand(0,1) \leqslant CR \quad \text{or} \quad j = j_{rand}, \\ v_{i,j}(g), & \text{if} \quad rand(0,1) > CR \quad \text{or} \quad j \neq j_{rand}, \end{cases} \tag{4.9}$$

where CR lies between 0 and 1 and decides the probability of parameters from the mutant vector that are to be copied to the trial vector. A random number $rand(0,1)$ is generated and if its value is less than or equal to CR, then the parameter from the mutant vector is inherited to the trial vector; otherwise, the trial vector takes the parameter from the target vector. This process is repeated for all pairs of target and mutant vectors.

Step 4: Selection

The cost function is evaluated for the resultant trial vector $u_{i,j}(g)$. If the cost of the trial vector is better compared with that of the target vector, then the trial vector survives and replaces the target vector in the next generation; otherwise, the target vector is retained for another generation. Mathematically, the expression for the process of selection is

$$v_{i,j}(g+1) = \begin{cases} u_{i,j}(g), & \text{if} \quad f(u_{i,j}) \leqslant f(v_{i,j}), \\ v_{i,j}(g), & \text{otherwise}, \end{cases} \tag{4.10}$$

where $f(u_{i,j})$ and $f(v_{i,j})$ represent the cost of trial and target vectors, respectively. Steps 2, 3, and 4 are repeated until the termination criterion is met.

4.2.3 Particle swarm optimization

PSO is a swarm intelligence-based algorithm proposed by Kennedy and Eberhart in 1995. This population-based algorithm derives its inspiration from the flocking of

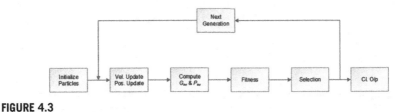

FIGURE 4.3

Flow diagram of PSO.

birds, fish schooling, etc. The controlled random search operation is analogous to
the combined movement of birds or fish in search of food considering the individual
effort as well as the collaborative effort. In PSO the global and local search operations
are carried out simultaneously. The population of potential solutions is termed as the
swarm and the individual solutions are called particles. The main advantage of the
PSO search technique is that the particles have a memory of where their previous
best positions are stored, termed as personal best P_{best}. The particle having the best
fitness value is termed as global best, P_g. The flow diagram of PSO is presented in
Fig. 4.3 and steps of PSO are discussed below (Shi and Eberhart, 1999; Padhy and
Simon, 2015; Shi and Eberhart, 1998).

Step 1: Initialization

The swarm of potential solutions is generated with random position and velocities.
The ith particle in d-dimensional space may be denoted as $X_i = (x_{i1}, x_{i2}, ..., x_{id})$ and
$i = 1, 2...., N$, where N denotes the size of the swarm.

Step 2: Fitness evaluation

The corresponding fitness values of the particles are evaluated.

Step 3: Determination of personal and global best

The personal best, P_{best}, of the individual particle is stored, the particle having the best fitness value is determined for the current generation, and the global
best value P_g is updated. The P_{best} of the ith particle is given by $P_{best,i} =
(p_{i1}, p_{i2}, ..., p_{id})$ and the global best P_g is given as $P_g = (p_{g1}, p_{g2}, ..., p_{gd})$.

Step 4: Velocity and position update

The velocity and position of the ith particle are updated according to the following
equations:

$$v_{id}^{t+1} = w v_{id}^t + c_1 rand()^t (p_{id}^t - x_{id}^t) + c_2 Rand()^t (p_{gd}^t - x_{id}^t), \qquad (4.11)$$

$$x_{id}^{t+1} = x_{id}^t + v_{id}^{t+1}, \qquad (4.12)$$

where $V_{id} = \left(v_{i1}, \quad v_{i2}, \quad . \quad ., \quad v_{id}\right)^T$ is the velocity (position change), p_{gd} and
p_{id} represent the global best and personal best solution, the superscript t denotes
the iteration number, c_1 is a cognitive parameter (acceleration constant) representing
personal thinking, c_2 is a social parameter (acceleration constant) representing the
social component which pulls the particle towards the globally best particle found
so far, $rand()$ and $Rand()$ are random numbers in the range of [0, 1], and w is the

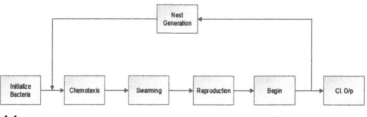

FIGURE 4.4

Flow diagram of BFO.

inertia weight which provides momentum. Steps 2, 3, and 4 are repeated until the stopping criterion is met.

4.2.4 Bacterial foraging optimization

BFO, proposed by K.M. Passino, is based on the food gathering process of the bacterium *E. coli* (Passino, 2002). Foraging means locating, handling, and ingesting food. The objective of this algorithm is to maximize the energy per unit time spent on foraging. The movement of the bacteria in search of food is categorized as run (swim) and tumble. The counterclockwise rotation of whip-like structures called flagella on the body of the bacterium enables it to swim, and the clockwise rotation of flagella causes tumble movement. In other words, the covering unit walk in the same direction is called swim and covering unit walk in the random direction is called tumble. The steps of this algorithm consist of operations like chemotaxis, swarming, reproduction, elimination, and dispersal, as shown in Fig. 4.4 (Das et al., 2009).

Step 1: Chemotaxis

"Chemo" means chemical and "taxis" means movement, and in this process, the bacterium swims or tumbles. The position of the bacterium is updated by

$$\theta^i(j+1,k,l) = \theta^i(j,k,l) + c(i)\phi(j), \qquad (4.13)$$

where $\theta_i(j,k,l)$ represents the ith bacterium at the jth chemotactic, kth reproductive, and lth elimination and dispersal step, $c(i)$ is the step size, and $\phi(j)$ is the direction angle of the jth step (when the bacterium swims, $\phi(j)$ is the same as $\phi(j-1)$; otherwise, it takes a random angle in the range $[0 \quad 2\pi]$). If the fitness function value is better at $\theta^i(j+1,k,l)$, then the bacterium proceeds in the same direction with step size $c(i)$. As long as the fitness function value is better than the previous value, the swim operation continues but to a maximum of N_s steps.

Step 2: Swarming

Under stress conditions the bacteria attract each other, and in order to maintain the minimum distance from each other, they release repellents. This interbacterial communication process may be represented mathematically by

$$J_{cc}\left(\theta, P\left(j,k,l\right)\right) = \sum_{i=1}^{S} J_{cc}\left(\theta, \theta^{i}\left(j,k,l\right)\right),$$

$$J_{cc}\left(\theta, P\left(j,k,l\right)\right) = \sum_{i=1}^{S}\left[-d_{a}exp\left(-w_{a}\sum_{m=1}^{P}\left(\theta_{m}-\theta_{m}^{i}\right)^{2}\right)\right]$$

$$+ \sum_{i=1}^{S}\left[-h_{r}exp\left(-w_{r}\sum_{m=1}^{P}\left(\theta_{m}-\theta_{m}^{i}\right)^{2}\right)\right], \qquad (4.14)$$

where $J_{cc}\left(\theta, P\left(j,k,l\right)\right)$ represents the fitness function value to be added to the actual fitness function, S is the total number of bacteria, p represents the number of variables to be optimized, $\theta = \begin{bmatrix} \theta_1 & \theta_2 & . & . & \theta_p \end{bmatrix}^{T}$ is a point in the p-dimensional search space, d_a and w_a are the depth and width of the attractant, respectively, and h_r and w_r are the height and width of the repellent, respectively.

Step 3: Reproduction

The reproduction step is carried out after N_c chemotactic steps. In this step, $S_r = S/2$, where the S_r least healthy bacteria die and S_r healthy bacteria split into two identical ones, keeping the population size constant.

Step 4: Elimination and dispersal

This step is carried out to prevent the solution to be trapped in local optima where the bacteria are chosen according to a preset probability P_{ed} to be dispersed and moved to another position within the environment. The above steps are repeated until the termination criterion is reached.

4.3 Application of bio-inspired optimization techniques in wireless communication

The wireless communication engineering domain involves real-time optimization scenarios vital for enhancing the performance of communication systems. The bio-inspired optimization algorithms are preferred over the traditional methods for solving complex problems in various areas of wireless communication. In this chapter, the performance of these bio-inspired optimization techniques is analyzed for different applications such as direct modeling, inverse modeling, and mobile location management in cellular networks. These applications are viewed as optimization problems, where the bio-inspired techniques are used for obtaining global optimal solutions (Kar, 2016; Nanda and Panda, 2014; Swayamsiddha et al., 2017b). Currently, one of the most challenging issues in telecommunication problems is the selection of a suitable strategy and choosing appropriate intelligent algorithms from a range of bio-inspired techniques for the optimization problem. The developing stages of bio-inspired computing techniques are introduced from an application perspective to wireless communication by describing the upgrades of their importance in direct modeling, inverse modeling, and mobility management in cellular networks, and the relationship between these applications and bio-inspired optimization. Usually for good channel estimation we require better channel modeling and channel equalization, which in turn improves the signal quality to support mobility management in wireless environments. So, this chapter proceeds towards mobility management from channel modeling and equalization. Nonlinear dynamic system modeling and identification is an interface between the real-world applications and mathematical computations and can be developed as an optimization problem. It not only proves to be useful in telecommunication for estimating parameters and modeling of the communication channels, but also has relevance in wider fields of control theory, model abstractions, and geophysical explorations. Channel equalization is an important aspect in a digital communication scenario, which is useful to nullify the intersymbol interference (ISI). The data transmitted through a band-limited communication channel suffer from linear, nonlinear, and additive distortions. Equalization compensates for this ISI caused by multipath within time-dispersive channels. Adaptive equalization involves training and tracking of the time-varying characteristics of the mobile channel. Equalization is basically an iterative process of minimization of the mean square error, which can be viewed as an optimization problem. In recent times, one of the research focus points in the wireless communication domain lies in the design of a computationally efficient next-generation mobility management system based on bio-inspired algorithms which can provide optimal network configuration by reducing the spectrum utilization as well as the overhead location management cost. Due to the rapid progress in technology with multiple applications and the exponential rise in the number of mobile users, there is an urgent need to track the mobility pattern of the mobile terminals by the networks. The current location update is required so that an incoming call can be routed to the particular mobile terminal without much delay. Mobile location management (MLM) involves two problems, i.e., location update and paging, and the ongoing research focuses on a trade-off between the two such that the MLM cost is optimized. With the tremendous increase in the number of wireless devices in recent times, the spectrum scarcity problem has emerged. The percentile dwell time factor integrated with bio-inspired techniques is an innovative way to reduce the signaling cost in mobility management.

4.3.1 Bio-inspired techniques for direct modeling application

This section starts with the study of bio-inspired techniques and their applications for direct modeling of nonlinear dynamic systems keeping in mind the nature of wireless channel environments, which are highly dynamic, chaotic, and nonlinear (Charalambous et al., 2008). To model, the behavior of the system which can be predicted under different operating conditions is the purpose of direct modeling. The direct modeling of the real-world systems is a complex problem because most of them are nonlinear and dynamic in nature. A literature survey reveals that direct modeling finds relevance in various fields of research such as communication engineering, control theory and prediction, signal processing, geophysical exploration, and much more (Huijberts et al., 2000). Many works on channel identification, modeling of wireless networks, and parameter estimation of the channels have been reported in the recent past (Ma et al., 2011; George and Panda, 2013; Swayamsiddha and Thethi, 2013). The MIMO channel identification, with blind MIMO identification approaches which do not require a full rank channel convolution matrix, has been studied (Ding and Qiu, 2003). An approach to identify the parameters of nonlinear MIMO systems using DE and performance comparison of its variants are explored in Swayamsiddha et al. (2015). Direct modeling of nonlinear dynamic systems using computationally efficient Chebyshev functional link artificial neural networks (FLANNs) has been proposed in Patra and Kot (2002). In Swayamsiddha et al. (2013), hybrid Chebyshev Functional Linked Artificial Neural Network-Differential Evolution (CFLANN-DE) is proposed for direct modeling of nonlinear dynamic systems and its performance is compared with that of hybrid Chebyshev Functional Linked Artificial Neural Network-Genetic Algorithm (CFLANN-GA) and conventionally used hybrid Chebyshev Functional Linked Artificial Neural Network-Back Propagation(CFLANN-BP). A literature survey reveals that the performance comparison of evolutionary computation algorithms has been carried out earlier for various numerical benchmark problems (Vesterstrom and Thomsen, 2004).

4.3.2 Bio-inspired techniques for inverse modeling application

Inverse modeling is an application of adaptive filtering which has a greater significance in wireless communication. Adaptive channel equalization is required as the wireless communication channels are unknown, nonstationary, and time-varying channels. Since the adaptive channel equalizer compensates for the effects of the nonlinear time-varying channel, a suitable adaptive optimization algorithm is to be applied for updating the equalizer coefficients and thus tracking the variations of the channel. In the recent past the adaptive channel equalization was developed using soft computing approaches such as evolutionary and swarm intelligence algorithms. The artificial neural network is employed for adaptive nonlinear channel equalization (Patra et al., 1999), where the computationally efficient, single-layered FLANN is proposed and compared with multilayer perceptron (MLP) and polynomial perceptron networks (PPNs). The DE algorithm is compared with gradient-based algorithms in Wu et al. (2008) and Thethi and Swayamsiddha (2013), where the performance of DE is shown to be superior in terms of convergence rate, quality of solution, and the bit error rate (BER). A modified BFO called self-adaptation BFO (SA-BFO) was proposed for the design of adaptive channel equalizers (Su et al., 2010). The SA-BFO algorithm strikes a balance between exploitation and exploration by adaptively changing the size of run length, hence giving good results.

4.3.3 Bio-inspired techniques for mobility management in cellular networks

The literature suggests the application of bio-inspired algorithms for cost optimization in mobile location management (MLM) (Chaurasia and Singh, 2015). The MLM is formulated as an optimization problem where the total cost comprised of location update cost and paging cost is to be minimized using bio-inspired optimization techniques (Swayamsiddha et al., 2017a). The binary PSO (BPSO), binary DE (BDE) and binary artificial bat algorithm (BABA) have been proposed for cost optimization in the reporting cell planning strategy of MLM. The proposed techniques for MLM in cellular networks are validated for test networks and reference networks, and have been implemented for realistic networks. Also, the comparative performance analysis of various metaheuristic algorithms is carried out in the perspective of MLM application (Almeida-Luz et al., 2011). The bio-inspired technique-based cost optimization for MLM using the percentile dwell time factor is discussed in (Swayamsiddha et al., 2017c). This work proposes normalized percentile dwell time distribution to be incorporated into various bio-inspired algorithms, such as GA and BPSO, for optimizing the total cost of location management.

4.3.4 Bio-inspired techniques for cognitive radio-based Internet of Things

Many recent works on cognitive radio networks and Internet of Things (IoT) have included the application of bio-inspired techniques. Nondominated sorting genetic algorithm-II (NSGA-II) has been applied to solve the multiobjective spectrum allocation problem in cognitive radio-based IoT (Han et al., 2018). The coverage maximization problem in sensor network-based IoT is dealt with using resample PSO, which enhances the performance to a great extent (Wang et al., 2018). The GA-based big sensor data in IoT are yet another application area of bio-inspired techniques in the wireless domain (Deng et al., 2018).

4.4 Conclusion

The growing complexity of real-world problems has motivated the research world to opt for alternative efficient problem solving algorithms which are inspired by biological or natural processes. These bio-inspired metaheuristics have the capability to provide low-cost, fast, and robust solutions to complex optimization problems which are difficult to model mathematically. Moreover, these are global optimization techniques, which alleviates the chance of the solution to be trapped in local optima, and which are suitable for optimization of multimodal and discontinuous objective functions. A literature survey reveals that there has been extensive research in this domain but there is scope for performance improvement using the algorithms which are still unexplored for this application. Though there are different strategies for optimization in engineering problems, the bio-inspired computing techniques being global optimization tools have potential utility in this field. Thus, this chapter provides the details of bio-inspired computing techniques from the literature, which forms the foundation on which the performance of bio-inspired computing techniques for various real-world applications can be based.

References

Almeida-Luz, S.M., Vega-Rodríguez, M.A., Gómez-Púlido, J.A., Sánchez-Pérez, J.M., 2011. Differential evolution for solving the mobile location management. Applied Soft Computing 11 (1), 410–427.

Azmat, F., Chen, Y., Stocks, N., 2015. Bio-inspired collaborative spectrum sensing and allocation for cognitive radios. IET Communications 9 (16), 1949–1959.

Charalambous, C.D., Bultitude, R.J., Li, X., Zhan, J., 2008. Modeling wireless fading channels via stochastic differential equations: identification and estimation based on noisy measurements. IEEE Transactions on Wireless Communications 7 (2), 434–439.

Chaudhry, R., Tapaswi, S., Kumar, N., 2019. A green multicast routing algorithm for smart sensor networks in disaster management. IEEE Transactions on Green Communications and Networking 3 (1), 215–226.

Chaurasia, S.N., Singh, A., 2015. A hybrid swarm intelligence approach to the registration area planning problem. Information Sciences 302, 50–69.

Das, S., Biswas, A., Dasgupta, S., Abraham, A., 2009. Bacterial Foraging Optimization Algorithm: Theoretical Foundations, Analysis, and Applications. Foundations of Computational Intelligence, vol. 3. Springer, pp. 23–55.

De Jong, K.A., Spears, W.M., 1992. A formal analysis of the role of multi-point crossover in genetic algorithms. Annals of Mathematics and Artificial Intelligence 5 (1), 1–26.

Del Ser, J., Osaba, E., Molina, D., Yang, X.S., Salcedo-Sanz, S., Camacho, D., Das, S., Suganthan, P.N., Coello, C.A.C., Herrera, F., 2019. Bio-inspired computation: where we stand and what's next. Swarm and Evolutionary Computation 48, 220–250.

Deng, X., Jiang, P., Peng, X., Mi, C., 2018. An intelligent outlier detection method with one class support Tucker machine and genetic algorithm toward big sensor data in Internet of Things. IEEE Transactions on Industrial Electronics 66 (6), 4672–4683.

Ding, Z., Qiu, L., 2003. Blind MIMO channel identification from second order statistics using rank deficient channel convolution matrix. IEEE Transactions on Signal Processing 51 (2), 535–544.

Elbeltagi, E., Hegazy, T., Grierson, D., 2005. Comparison among five evolutionary-based optimization algorithms. Advanced Engineering Informatics 19 (1), 43–53.

Eshelman, L.J., Schaffer, J.D., 1993. Real-Coded Genetic Algorithms and Interval-Schemata, Foundations of Genetic Algorithms, Vol. 2. Elsevier, pp. 187–202.

Floreano, D., Mattiussi, C., 2008. Bio-Inspired Artificial Intelligence: Theories, Methods, and Technologies. MIT Press.

George, N.V., Panda, G., 2013. Advances in active noise control: a survey, with emphasis on recent nonlinear techniques. Signal Processing 93 (2), 363–377.

Goldberg, D.E., 1991. Real-coded genetic algorithms, virtual alphabets, and blocking. Complex Systems 5 (2), 139–167.

Goldberg, D.E., Holland, J.H., 1988. Genetic algorithms and machine learning. Machine Learning 3 (2), 95–99.

Goldberg, D.E., Korb, B., Deb, K., et al., 1989. Messy genetic algorithms: motivation, analysis, and first results. Complex Systems 3 (5), 493–530.

Han, R., Gao, Y., Wu, C., Lu, D., 2018. An effective multi-objective optimization algorithm for spectrum allocations in the cognitive-radio-based Internet of Things. IEEE Access 6, 12858–12867.

Holland, J., 1975. Adaptation in Natural and Artificial Systems: an Introductory Analysis with Application to Biology, Control and Artificial Intelligence.

Huijberts, H., Nijmeijer, H., Willems, R., 2000. System identification in communication with chaotic systems. IEEE Transactions on Circuits and Systems. I, Fundamental Theory and Applications 47 (6), 800–808.

Kar, A.K., 2016. Bio inspired computing–a review of algorithms and scope of applications. Expert Systems with Applications 59, 20–32.

Krishnakumar, K., 1990. Micro-genetic algorithms for stationary and non-stationary function optimization. Intelligent Control and Adaptive Systems 1196, 289–296.

Li, J., Chen, M., 2016. Multiobjective topology optimization based on mapping matrix and NSGA-II for switched industrial Internet of Things. IEEE Internet of Things Journal 3 (6), 1235–1245.

Ma, X., Olama, M.M., Djouadi, S.M., Charalambous, C.D., 2011. Estimation and identification of time-varying long-term fading channels via the particle filter and the EM algorithm. In: 2011 IEEE Radio and Wireless Symposium, pp. 13–16.

Mohammed, H.J., Abdullah, A.S., Ali, R.S., Abd-Alhameed, R.A., Abdulraheem, Y.I., Noras, J.M., 2016. Design of a uniplanar printed triple band-rejected ultra-wideband antenna using particle swarm optimisation and the firefly algorithm. IET Microwaves, Antennas & Propagation 10 (1), 31–37.

Nanda, S.J., Panda, G., 2014. A survey on nature inspired metaheuristic algorithms for partitional clustering. Swarm and Evolutionary Computation 16, 1–18.

Padhy, N., Simon, S.P., 2015. Soft Computing: with MATLAB Programming. Oxford University Press, Inc..

Parija, S., Singh, S.S., Swayamsiddha, S., 2017. Particle swarm optimization for cost reduction in mobile location management using reporting cell planning approach. In: Recent Developments in Intelligent Nature-Inspired Computing. IGI Global, pp. 171–189.

Passino, K.M., 2002. Biomimicry of bacterial foraging for distributed optimization and control. IEEE Control Systems Magazine 22 (3), 52–67.

Patra, J.C., Kot, A.C., 2002. Nonlinear dynamic system identification using Chebyshev functional link artificial neural networks. IEEE Transactions on Systems, Man and Cybernetics. Part B. Cybernetics 32 (4), 505–511.

Patra, J.C., Pal, R.N., Baliarsingh, R., Panda, G., 1999. Nonlinear channel equalization for QAM signal constellation using artificial neural networks. IEEE Transactions on Systems, Man and Cybernetics. Part B. Cybernetics 29 (2), 262–271.

Shi, Y., Eberhart, R.C., 1998. Parameter selection in particle swarm optimization. In: International Conference on Evolutionary Programming, pp. 591–600.

Shi, Y., Eberhart, R.C., 1999. Empirical study of particle swarm optimization. In: Proceedings of the 1999 Congress on Evolutionary Computation-CEC99. Cat. No. 99TH8406, vol. 3, pp. 1945–1950.

Spears, W.M., Anand, V., 1991. A study of crossover operators in genetic programming. In: International Symposium on Methodologies for Intelligent Systems, pp. 409–418.

Storn, R., 1996. Differential evolution design of an IIR-filter. In: Proceedings of IEEE International Conference on Evolutionary Computation, pp. 268–273.

Storn, R., Price, K., 1997. Differential evolution–a simple and efficient heuristic for global optimization over continuous spaces. Journal of Global Optimization 11 (4), 341–359.

Su, T.J., Cheng, J.C., Yu, C.J., 2010. An adaptive channel equalizer using self-adaptation bacterial foraging optimization. Optics Communications 283 (20), 3911–3916.

Swayamsiddha, S., Thethi, H.P., 2013. Nonlinear system identification using evolutionary computing based training schemes. International Journal of Computer Applications 975, 8887.

Swayamsiddha, S., Thethi, H.P., 2018. Performance comparison of adaptive channel equalizers using different variants of differential evolution. Journal of Engineering Science and Technology 13 (8), 2271–2286.

Swayamsiddha, S., Mondal, S., Thethi, H.P., 2013. Identification of nonlinear dynamic systems using differential evolution based update algorithms and Chebyshev functional link artificial neural network. In: IET International Conference on Computational Intelligence and Information Technology, pp. 508–513.

Swayamsiddha, S., Behera, S., Thethi, H.P., 2015. Blind identification of nonlinear MIMO system using differential evolution techniques and performance analysis of its variants. In: 2015 International Conference on Computational Intelligence and Networks, pp. 63–67.

Swayamsiddha, S., Parija, S., Sahu, P.K., Singh, S.S., 2017a. Optimal reporting cell planning with binary differential evolution algorithm for location management problem. International Journal of Intelligent Systems and Applications 9 (4), 23–31.

Swayamsiddha, S., Parija, S., Singh, S.S., Sahu, P.K., 2017b. Bio-inspired algorithms for mobile location management—a new paradigm. In: Proceedings of the 5th International Conference on Frontiers in Intelligent Computing: Theory and Applications, pp. 35–44.

Swayamsiddha, S., Singh, S.S., et al., 2017c. Location management cost optimization using normalized percentile dwell time. Journal of Engineering Science and Technology Review 10 (5).

Swayamsiddha, S., Singhal, C., Roy, R., 2018. Nature-inspired-algorithms-based cellular location management: scope and applications. In: Handbook of Research on Modeling, Analysis, and Application of Nature-Inspired Metaheuristic Algorithms. IGI Global, pp. 346–362.

Swayamsiddha, S., Singh, S.S., Parija, S., Pratihar, D.K., et al., 2019. Reporting cell planning-based cellular mobility management using a binary artificial bat algorithm. Heliyon 5 (3), e01276.

Thethi, H.P., Swayamsiddha, S., 2013. Performance analysis of adaptive channel equalizer using population based update algorithms. International Journal of Computer Applications 74 (12).

Thiele, L., Blickle, T., 1995. A comparison of selection schemes used in genetic algorithms. TIK Report. Swiss Federal Institute of Technology.

Vesterstrom, J., Thomsen, R., 2004. A comparative study of differential evolution, particle swarm optimization, and evolutionary algorithms on numerical benchmark problems. In: Proceedings of the 2004 Congress on Evolutionary Computation. IEEE Cat. No. 04TH8753, vol. 2, pp. 1980–1987.

Wang, J., Cao, J., Li, B., Lee, S., Sherratt, R.S., 2015. Bio-inspired ant colony optimization based clustering algorithm with mobile sinks for applications in consumer home automation networks. IEEE Transactions on Consumer Electronics 61 (4), 438–444.

Wang, X., Zhang, H., Fan, S., Gu, H., 2018. Coverage control of sensor networks in IoT based on RPSO. IEEE Internet of Things Journal 5 (5), 3521–3532.

Wu, Z., Huang, H., Zhang, X., Yang, B., Dong, H., 2008. Adaptive equalization using differential evolution. In: 2008 IEEE Congress on Evolutionary Computation (IEEE World Congress on Computational Intelligence), pp. 1962–1967.

Yang, X.S., 2012. Nature-inspired metaheuristic algorithms: success and new challenges. Preprint. arXiv: 1211.6658.

Yang, S., Gan, Y.B., Qing, A., 2002. Sideband suppression in time-modulated linear arrays by the differential evolution algorithm. IEEE Antennas and Wireless Propagation Letters 1, 173–175.

Theory
Analysis and framework

Mathematical foundations for algorithm analysis

5

Xin-She Yang

Middlesex University London, School of Science and Technology, London, United Kingdom

CONTENTS

5.1 Introduction

Algorithm analysis tends to start with the iterative characteristics, which often leads to some aspects of the fixed-point theory and stability. This kind of methods can work well for deterministic algorithms and many classical optimization algorithms (Boyd and Vandenberghe, 2004). For algorithms with many parameters and stochastic components, modern analyses tend to use Markov chains, dynamical systems, and others (Yang and He, 2019). We will introduce some of these methods in detail in the next two chapters. Here we will provide a brief introduction of relevant mathematical concepts so as to lay some foundations for algorithm analysis. However, we will assume that the readers have already mastered basic calculus, and experienced researchers can skip this chapter.

5.2 Optimization and optimality

For an optimization problem with $n \geq 1$ independent, real decision variables $(x_1, x_2, ..., x_n)$, we usually use an n-dimensional vector

$$\boldsymbol{x} = (x_1, x_2, ..., x_n)^T \in \mathbb{R}^n \tag{5.1}$$

to represent them. Here we have treated this vector as a column by using the transpose (T), though row vectors are also used in many software packages.

Nature-Inspired Computation and Swarm Intelligence. https://doi.org/10.1016/B978-0-12-819714-1.00015-4

In principle, we can always write an optimization problem with various nonlinear constraints as

$$\text{maximize/minimize } f(x), \tag{5.2}$$

subject to

$$h_i(x) = 0, \quad (i = 1, 2, ..., M), \tag{5.3}$$

$$g_j(x) \leq 0, \quad (j = 1, ..., N), \tag{5.4}$$

where all functions are scalar functions of x, and $f(x)$ is the objective or cost function. Here, equations $h_i(x) = 0$ are equality constraints, and $g_j(x) \leq 0$ are inequality constraints. It is worth pointing out that we have used ≤ 0, which is consistent with many textbooks. Obviously, other forms such as ≥ 0 are also commonly used. Since a simple multiplication by (-1) is sufficient to convert one form into the other, we will use the form stated above. In addition, some textbooks use maximization, while other use minimization. In this chapter, we will mainly use minimization (Gill et al., 1982; Boyd and Vandenberghe, 2004).

In the special case of $M = 0$ and $N = 0$, the constrained optimization problem becomes an unconstrained optimization. In most cases, unconstrained optimization problems are easier to be implemented in practice, as there is no need to worry how to handle the constraints properly.

Both feasibility and optimality are important concepts.

- Feasible solution: a point x (a vector) that satisfies all the constraints is referred to as a feasible point, which is also called a feasible solution to the optimization problem under consideration. All the feasible points form a feasible set, corresponding to a feasible region (or regions).
- A solution x_* that corresponds to a local maximum of the objective function is called a solution of a strong local maximum if it satisfies the condition that $f(x_*) > f(x)$ for $\forall x \in N(x_*, \delta)$ where $\delta > 0$ and $x \neq x_*$.

In the case it is not a strong local maximum, the condition $f(x_*) \geq f(x)$ without equality for $\forall x \in N(x_*, \delta)$ leads to a weak local maximum at x_*. Similarly, if we reverse the inequality signs to $<$ and $\leq$, we have weak and strong local minima, respectively.

From basic calculus, we know that optimal solutions (points) usually occur at $f'(x) = 0$ for a univariate, real-valued function in $x \in \mathbb{R}$. The solution corresponds to a local maximum if $f''(x) < 0$ and to a local minimum if $f''(x) > 0$. Both its first derivative and second derivative are important to determine the solution characteristics.

The above optimal condition can be extended to optimality conditions concerning multivariate objective functions $f(x) = f(x_1, x_2, ..., x_n)$. That is, the first-order partial derivatives should be zero. Thus, this stationary condition requires the gradient vector G to satisfy

$$G = \nabla f = (\frac{\partial f}{\partial x_1}, \frac{\partial f}{\partial x_2}, ..., \frac{\partial f}{\partial x_n})^T = 0. \quad (5.5)$$

However, for a multivariate function, its second derivatives are much more complicated. The local property is determined by its Hessian matrix

$$H \equiv \nabla^2 f(x) \equiv \begin{pmatrix} \frac{\partial^2 f}{\partial x_1^2} & \frac{\partial^2 f}{\partial x_1 \partial x_2} & \cdots & \frac{\partial^2 f}{\partial x_1 \partial x_n} \\ \frac{\partial^2 f}{\partial x_2 \partial x_1} & \frac{\partial^2 f}{\partial x_2^2} & \cdots & \frac{\partial^2 f}{\partial x_2 \partial x_n} \\ \vdots & \vdots & \ddots & \vdots \\ \frac{\partial^2 f}{\partial x_n \partial x_1} & \frac{\partial^2 f}{\partial x_n \partial x_2} & \cdots & \frac{\partial^2 f}{\partial x_n^2} \end{pmatrix}. \quad (5.6)$$

If a point satisfies the stationary condition $G = \nabla f = 0$, it corresponds to a local minimum if H is positive definite, while it corresponds to a local maximum if H is negative definite (Yang, 2018).

The definiteness of a square matrix requires some knowledge of its eigenvalues. Before we introduce eigenvalues and definiteness, let us briefly review the norms of a vector.

5.3 Norms

For an n-dimensional vector $x = (x_1, x_2, ..., x_n)^T$ where $n \geq 1$, its length is its Euclidean norm, or 2-norm,

$$||x|| = \sqrt{x_1^2 + x_2^2 + \cdots + x_n^2}. \quad (5.7)$$

In general, the p-norm or L_p-norm (also L^p-norm) of a vector x can be defined by

$$||x||_p \equiv \left(|x_1|^p + |x_2|^p + \cdots + |x_n|^p \right)^{1/p} = \left(\sum_{i=1}^{n} |x_i|^p \right)^{1/p}, \quad p > 0. \quad (5.8)$$

So the Euclidean distance is an L_2-norm when $p = 2$, also called Cartesian norm.

Different norms can have different properties and potential different applications, and the three most widely used norms are L_1, L_2, and L_∞, where $p = 1, 2$, and ∞, respectively (Yang, 2018).

The L_1-norm of x is defined by

$$||x||_1 = |x_1| + |x_2| + ... + |x_n|, \quad (5.9)$$

which is the sum of its absolute component values.

The L_∞-norm is the largest absolute component, i.e.,

$$||x||_\infty = \max \left\{ |x_1|, |x_2|, ..., |x_n| \right\} = x_{\max}.$$ (5.10)

In general, the norms of any two vectors u and v in the same space satisfy the triangle inequality

$$||u||_p + ||v||_p \ge ||u + v||_p$$ (5.11)

for all $p \ge 0$.

Though norms are defined for vectors, it is possible to extend them for matrices in different ways. For a matrix A

$$A \equiv [a_{ij}] = \begin{pmatrix} a_{11} & a_{12} & \cdots & a_{1n} \\ a_{21} & a_{22} & \cdots & a_{2n} \\ \vdots & \vdots & \ddots & \vdots \\ a_{m1} & a_{m2} & \cdots & a_{mn} \end{pmatrix},$$ (5.12)

its Frobenius norm can be defined by

$$||A||_F = \sqrt{\sum_{i=1}^{m} \sum_{j=1}^{n} |a_{ij}|^2}.$$ (5.13)

This norm can also be linked to the trace of the product of the original matrix and its transpose. We have

$$||A||_F = \sqrt{\operatorname{tr}(A^T A)} = \sqrt{\operatorname{diag}(A^T A)}.$$ (5.14)

The Frobenius norm is essentially a 2-norm. Similarly to those for vectors, we can also define some other norms. For example, the $||A||_1$-norm is the maximum of the absolute column sums, which can be calculated by

$$||A||_1 = \max \sum_{i=1}^{m} |a_{ij}|, \quad \text{for } 1 \le j \le n.$$ (5.15)

Similarly, the $||A||_\infty$-norm is the maximum of the absolute row sums, which is

$$||A||_\infty = \max \sum_{j=1}^{n} |a_{ij}|, \quad \text{for } 1 \le i \le m.$$ (5.16)

There are other norms as well. Norms can be calculated easily, given a vector or matrix.

5.4 Eigenvalues and eigenvectors

For a square matrix A, its main properties in the context of optimization can largely be determined by its eigenvalues λ. In general, the eigenvalues of a square matrix A can be obtained by

$$Au = \lambda u, \tag{5.17}$$

where u is a nonzero eigenvector for a corresponding λ. For an $n \times n$ real matrix, there are at most n different eigenvalues with n corresponding eigenvectors.

The eigenvalues of A can be computed by its characteristic polynomial of order n, which can be written compactly as

$$\det \left| A - \lambda I \right| = 0, \tag{5.18}$$

where I is an $n \times n$ identity matrix.

Based on its eigenvalues, a square symmetric A is called positive definite if all its eigenvalues are strictly positive. That is, $\lambda_i > 0$, where $i = 1, 2, ..., n$. If the equal sign is included (that is, $\lambda_i \geq 0$), it is called positive semidefinite. Conversely, if all its eigenvalues are strictly negative, the matrix is called negative definite. With the inclusion of equal sign, it leads to negative semidefiniteness. In other words, A is called positive semidefinite if $v^T A v \geq 0$ and negative semidefinite if $v^T A v \leq 0$ for all v for any nonzero n-dimensional vector v.

There are many ways of testing the definiteness of a matrix. However, the signs of its elements do not have any direct link with the definiteness. For example, the matrix

$$A = \begin{pmatrix} 7 & 3 \\ 3 & 7 \end{pmatrix} \tag{5.19}$$

is positive definite as its two eigenvalues are 4 and 10, respectively, but

$$B = \begin{pmatrix} 3 & 7 \\ 7 & 3 \end{pmatrix} \tag{5.20}$$

is not since its two eigenvalues are -4 and 10, respectively.

As a simple example, let us study the behavior of the function $f = xy = x_1 x_2$ at the critical point $(0, 0)$ by setting $\partial f / \partial x = 0$ and $\partial f / \partial y = 0$. If we use $x = (x_1, x_2)^T$ as the vector, we can expand the function about $(0, 0)$ and we have

$$f(x) = f(0) + (\nabla f)^T x + \frac{1}{2} x^T H x + \cdots \text{(higher-order terms)}, \tag{5.21}$$

where H is the Hessian matrix, which is given by

$$H = \begin{pmatrix} \frac{\partial^2 f}{\partial x^2} & \frac{\partial^2 f}{\partial x \partial y} \\ \frac{\partial^2 f}{\partial y \partial x} & \frac{\partial^2 f}{\partial y^2} \end{pmatrix} = \begin{pmatrix} 0 & 1 \\ 1 & 0 \end{pmatrix}. \tag{5.22}$$

Since $\nabla f = 0$ at $(0, 0)$, the behavior of the function is controlled by $\boldsymbol{H}$. As the eigenvalues of $\boldsymbol{H}$ are $+1$ and -1, respectively, $\boldsymbol{H}$ is not definite. Thus, the point $(0,0)$ is not a maximum or a minimum. In fact, it is a saddle point.

5.5 Convergence sequences

Almost all algorithms are iterative, and all the solutions obtained during the iterations will form the solution sequences. Even for a simple one-agent algorithm, its solutions at different iterations will form a zig-zag path, leading to a sequence of s_0, s_1, s_2, ..., s_k, ...; that is, s_k $(k = 0, 1, 2, ...)$.

As the iterations continue $(k \to \infty)$, if sequence s_k converges towards a fixed point P or a fixed value, we have

$$\lim_{k \to \infty} s_k = s_\infty = P. \tag{5.23}$$

The order q of the convergence can be defined as

$$\lim_{k \to \infty} \frac{|s_{k+1} - P|}{|s_k - P|^q} = A, \quad A > 0, \tag{5.24}$$

where $q \geq 1$ and A is sometimes called the rate of convergence. Loosely speaking, we can say that the above sequence converges to P with the order of q and the rate of convergence A. If we define the error or difference $E_k = s_k - s_\infty$ from the limiting value s_∞, the ratio of this absolute error to its previous error term can be considered as the rate of convergence,

$$A = \lim_{k \to \infty} \frac{|E_{k+1}|}{|E_k|^q}. \tag{5.25}$$

For a given sequence, this asymptotic error constant A will be a constant, which loosely gives $|E_{k+1}| = A|E_k|^q$. Clearly, the reduction of errors and the convergence should require that $0 < A < 1$, though the special case of $A = 1$ corresponds to sublinear convergence.

More specifically, the convergence is called linear if $q = 1$ and $A < 1$, and superlinear if $q = 1$ but $A = 0$. The convergence is called quadratic if $q = 2$, while such convergence becomes cubic if $q = 3$.

As a simple example, the sequence 1, $1/2$, $1/4$, $1/16$, ... can be written as $s_k = 2^{-k} (k = 0, 1, 2, ...)$ and it converges towards $P = 0$ linearly. This can be verified by showing that $q = 1$ and $A = 1/2$ from

$$\lim_{k \to \infty} \frac{|s_{k+1} - P|}{|s_k - P|^{q=1}} = \lim_{k \to \infty} \frac{|2^{-(k+1)} - 0|}{|2^{-k} - 0|} = \lim_{k \to \infty} \frac{2^{-k}2^{-1}}{2^{-k}} = \frac{1}{2}. \tag{5.26}$$

Another example is that Newton's method for finding roots of a polynomial is typically quadratic under certain conditions (Boyd and Vandenberghe, 2004; Yang and He, 2019).

5.6 **Series**

The sum of an infinite series in general can be written as

$$\sum_{k=1}^{\infty} a_k = a_1 + a_2 + a_3 + \cdots + a_k + \cdots, \tag{5.27}$$

and the sum of its first n terms is denoted by

$$S_n = \sum_{k=1}^{n} a_k. \tag{5.28}$$

This partial sum will form a sequence. In general, if S_n approaches a finite limit S_*, then the original series will converge to the same value, that is,

$$\lim_{n \to \infty} S_n = S_*. \tag{5.29}$$

A well-known example is the geometric series where the ratio r between consecutive terms is a constant. Since $a_k = a_0 r^{k-1}$, we have

$$\sum_{k=1}^{\infty} a_k = a_0 + a_0 r + a_0 r^2 + \cdots = \frac{a_0}{1-r}, \quad |r| < 1. \tag{5.30}$$

Here we have imposed the condition $|r| < 1$; otherwise, the series will not converge. Another implication is that it requires that $a_k \to 0$ when $k \to \infty$. However, the condition $\lim_{k \to \infty} a_k \to 0$ is a necessary condition, but a sufficient condition. For example, the well-known harmonic series

$$\sum_{k=1}^{\infty} \frac{1}{k} = 1 + \frac{1}{2} + \frac{1}{3} + \frac{1}{4} + \frac{1}{5} + \cdots + \frac{1}{k} + \cdots \tag{5.31}$$

is divergent, but $1/k \to 0$ when $k \to \infty$.

A useful test of convergence is the so-called ratio test. For the ratio

$$R = \lim_{k \to \infty} \frac{a_{k+1}}{a_k}, \tag{5.32}$$

the series converges if $|R| < 1$ and diverges if $|R| > 1$. However, the special case of $|R| = 1$ is inconclusive, as it is not possible to determine if the series is convergent or not.

For example, it is straightforward to show that

$$\sum_{k=1}^{\infty} \frac{k^2}{3^k} \tag{5.33}$$

is convergent. From $a_k = k^2/3^k$, we have

$$\lim_{k\to\infty} \frac{a_{k+1}}{a_k} = \lim_{k\to\infty} \frac{(k+1)^2/3^{k+1}}{k^2/3^k} = \lim_{k\to\infty} (1+\frac{1}{k})^2 \cdot \frac{1}{3} = \frac{1}{3}, \tag{5.34}$$

which is less than 1. However, it is worth pointing out that the proof that a series is convergent does not necessarily mean that we can obtain its sum easily. In fact, the calculations of such summation can be tricky, and even impossible in many cases.

There are many other convergence tests, including the term test, root test, integral test, comparison test, limit comparison test, and others. Interested readers can refer to more specialized literature on this topic.

5.7 Computational complexity

Computational complexity is a very important topic, and it concerns the estimation of how hard or easy a problem can be solved computationally. However, the computational complexity will not answer the actual computation time to be taken to solve a particular problem or problem instance because the actual implementations will depend on other factors, such as hardware and software. Therefore, the complexity is an estimate of the order or number of computational operations, rather than time.

Loosely speaking, computational complexity and classes are closely associated with Turing machines. A Turing machine can be considered as an abstract machine which can read an input and manipulate one operation at a time, according to predefined rules. In general, a Turing machine is capable of carrying out the computation of any computable function. Since rules are fixed, one action is carried out at a time, so such a Turing machine becomes deterministic, called a deterministic Turing machine.

On the other hand, if more than one action is allowed and symbol manipulations can have multiple branches, then the transition on the machine may depend on the current states. Multiple decision structures can lead to nondeterministic characteristics, and we often refer to such a Turing machine as a nondeterministic Turing machine.

In the computer science literature, complexity classes are often used for describing different levels of computational complexity (Arara and Barak, 2009; Cook, 1983). Though the rigorous definition requires other mathematical concepts, we can loosely define a complexity class as a set of decision problems that are computable in terms of a Turing machine, subject to the constraint of a fixed resource.

Class P is formed by all the problems or functions that are solvable by a deterministic Turing machine whose computational complexity is a polynomial of problem size n. Here, the computational time is the order of computation, not the actual execution time. Thus, the order notation $O(n^k)$ is widely used. In theory, a problem in Class P can be solved in a polynomial time, but this does not imply that it can be actually solved very quickly in practice. In the case of a large problem size, for example, $n = 100000$ and $k = 5$, this gives $O(n^k) = O(10^{25})$. Thus, for a given problem size, the main aim of algorithm designs is to reduce k. For example, the inverse of a full matrix of size $n \times n$ typically requires $O(n^3)$. Another class of much harder problems is the class of the exponential-time problems, whose complexity is typically $O(2^{n^k})$ in terms of problem size n.

On the other hand, there are many cases where it is much easier to verify a given solution, but it is very hard to guess that solution in the first place. This concerns the nondeterministic Turing machines and the concept of nondeterministic polynomial time (NP). Class NP includes all the problems that can be solved by nondeterministic Turing machines in a polynomial time. If a problem Q_1 can be shown to be reduced to another problem Q_2 in a polynomial time, we can say that Q_1 is reducible to Q_2, and all the techniques to solve Q_2 can be used to solve Q_1 efficiently. The definition of NP-hardness is a bit vague, and a problem is called NP-hard if it is at least as hard as any problem in Class NP. If we can somehow find an algorithm to solve one NP problem in a polynomial time, it is possible to solve all the problems in this class. However, the current belief is that such algorithms do not exist. This is the well-known hypothesis that P $\neq$ NP.

If a problem H in class NP is NP-hard, we say that it is an NP-complete problem if every problem in the NP class can be reduced to this problem in a polynomial time. Consequently, we can say that NP-complete problems are the hardest NP problems (Goldreich, 2008). Many problems belong to this NP-complete class, including the well-known traveling salesman problem, the knapsack problem, the subset sum problem, integer programming, and the graph coloring problem.

The above complexity classes focus on the problems, not the algorithms. Obviously, for the same type of problems, different algorithms can be used to solve them, but different algorithms can have very high complexity. In this case, we are more concerned with the algorithmic complexity. For example, the complexity of the quicksort for sorting n different numbers is $O(n \log n)$. In this context, the number or order of algebraic manipulations is the main concern, not the physical memory or real computational time. Thus, the complexity we discuss in this book is mainly about the algorithmic complexity of different algorithms.

5.8 Convexity

Before we end this chapter, let us discuss convexity, as it is an important concept in optimization. A set $\Omega \in \mathbb{R}^n$ in an n-dimensional, real vector space is a convex set if any two of its elements x and y lead to

$$\alpha x + (1 - \alpha)y \in \Omega, \quad \alpha \in [0, 1], \quad \forall (x, y) \in \Omega. \tag{5.35}$$

Even for a simple univariate function, it is called a convex function if it is defined on a convex set Ω and it satisfies the inequality

$$f(\alpha x + \beta y) \le \alpha f(x) + \beta f(y), \quad \forall x, y \in \Omega, \tag{5.36}$$

where $\alpha, \beta \geq 0$ and $\alpha + \beta = 1$. Based on this definition, it is straightforward to verify that both $f(x) = x^2$ and $g(x) = 2x^2 + 3$ are convex.

Convexity has some interesting properties under some proper conditions. The convexity is preserved under the weighted sum with all nonnegative weights, function composition via affinity, and minimization or maximization of a convex function (Boyd and Vandenberghe, 2004). For example, af is convex if f is convex and $a > 0$. The weighted sum $af + bg$ is convex for $a, b \geq 0$ when both f and g are convex. In addition, if functions f and g are both convex, their composition $h(x) = f(g(x))$ will also be convex, subject to nondecreasing conditions. For example, since $\exp(x)$ is convex, $h(x) = \exp[f(x)]$ will be a convex function if $f(x)$ itself is convex. This can be generalized to a so-called log-sum-exp function,

$$h(x) = \log\left[\sum_{k=1}^{n} e^{x_k}\right], \tag{5.37}$$

which is convex. The convexity of an optimization problem will make it much easier to find its global optimal solution. In fact, there are a wide range of optimization techniques for convex optimization (Boyd and Vandenberghe, 2004; Bertsekas et al., 2003).

We have introduced some fundamentals of mathematical concepts concerning mathematical analysis of algorithms. In order to analyze algorithms from different perspectives so as to gain a full picture, we also need probability theory and statistical measures, which will be the topics of the next chapter.

References

Arara, S., Barak, B., 2009. Computational Complexity: A Modern Approach. Cambridge University Press, Cambridge, UK.

Bertsekas, D.P., Nedic, A., Ozdaglar, A., 2003. Convex Analysis and Optimization, second ed. Athena Scientific, Belmont, MA.

Boyd, S.P., Vandenberghe, L., 2004. Convex Optimization. Cambridge University Press, Cambridge, UK.

Cook, S., 1983. An overview of computational complexity. Commun. ACM 26 (6), 400–408.

Gill, P.E., Murray, W., Wright, M.H., 1982. Practical Optimization. Emerald Publishing, Bingley.

Goldreich, O., 2008. Computational Complexity: A Conceptual Perspective. Cambridge University Press, Cambridge, UK.

Yang, X.S., 2018. Optimization Techniques and Applications with Examples. John Wiley & Sons, Hoboken, NJ, USA.

Yang, X.S., He, X.S., 2019. Mathematical Foundations of Nature-Inspired Algorithms. Springer Briefs in Optimization. Springer, Cham, Switzerland.

Probability theory for analyzing nature-inspired algorithms

6

Xing-Shi He[a], Qin-Wei Fan[a], Xin-She Yang[b]

[a]*Xi'an Polytechnic University, College of Science, Xi'an, China*
[b]*Middlesex University London, School of Science and Technology, London, United Kingdom*

CONTENTS

6.1 Introduction

Randomization components and random numbers are widely used in almost all nature-inspired algorithms and evolutionary computation. Many implementations use some forms of initialization of the initial population in terms of sampling uniformly in the intervals of decision variables. Thus, the analysis of these algorithms and their key components requires statistical measures and probability theory (Grindstead and Snell, 1997).

This chapter will introduce the basics of random variables, commonly used probability distributions, and the concept of Markov chains. In addition, the essential idea of Monte Carlo and the ways of generating random numbers are also briefly introduced.

Nature-Inspired Computation and Swarm Intelligence. https://doi.org/10.1016/B978-0-12-819714-1.00016-6

6.2 Random variables and probability

Randomness is intrinsic in many processes. For example, the noise levels on the street and the random movement of gas molecules are random, and thus they can be modeled by random variables. Probability tools allow us to find any nonrandom regularity by studying random phenomena (Grindstead and Snell, 1997; Bhat and Miller, 2002).

For an even A such as the flipping of a fair coin, we can assign a probability $P(A)$ to show how likely the event occurs when an experiment is performed randomly, and such experiments are usually performed multiple times and independent from each other. Some random variables (e.g., the noise level on the street) are continuous, while other random variables are discrete, including the number of cars on a road, the number of phone calls at a call center, and the number of earthquakes in a particular region.

A random variable X is called discrete if it only takes discrete values. Each discrete value x_i may occur with a corresponding probability $p(x_i)$. The total probabilities of all the discrete values (such as the head and tail of a coin in a coin flipping experiment) must add up to one because one of these values or outcomes must occur. This requires

$$\sum_{i=1}^{n} p(x_i) = 1, \tag{6.1}$$

where n is the number of all possible outcomes or discrete values. This probability function $p(x)$ is often referred to as the probability mass function (PMF).

Sometime, it is necessary to study all the values less than a fixed number or the events over an interval, so we have to use the cumulative probability function (CPF). The CPF of a random variable X is defined by

$$P(X \leq x) = \sum_{x_i < x} p(x_i). \tag{6.2}$$

The above two concepts for discrete variables can be extended to describe continuous random variables. The PMF becomes a probability density function (PDF), which is $p(x)$, defined on an interval $[a, b]$. Either a or b or both can be infinity. The value of a PDF at a particular point is not the probability. However, if we interpret $P(x)$ as the probability of random variable X taking the values in $x < X \leq x + dx$, we have $P(x < X \leq x + dx) = p(x)dx$. Thus, we have

$$\int_{a}^{b} p(x)dx = 1, \tag{6.3}$$

and its corresponding CDF becomes the definite integral

$$\Phi(x) = P(X \leq x) = \int_{a}^{x} p(x)\,dx, \tag{6.4}$$

from the lower limit a to the present value $X = x$.

There are two commonly used statistical measures for random variables, i.e., mean (μ) and variance (σ^2). The mean of a discrete variable is the weighted sum

$$\mathbb{E}[X] = \sum_i x_i p(x_i). \tag{6.5}$$

This becomes an integral for a continuous variable

$$\mu \equiv \mathbb{E}[X] \equiv \int_a^b x p(x) dx. \tag{6.6}$$

The variance ($\text{var}[X] = \sigma^2$) of a discrete variable is the expectation of the derivation squared, that is,

$$\text{var}[X] = \sum_i (x_i - \mu)^2 p(x_i), \tag{6.7}$$

which is equivalent to

$$\text{var}[X] = \mathbb{E}[X^2] - \left(\mathbb{E}[X]\right)^2. \tag{6.8}$$

Clearly, once we know the variance of a random variable, we can calculate its standard deviation σ.

For a continuous variable, the above formulas become

$$\mu \equiv \mathbb{E}[X] \equiv \int_a^b x p(x) dx \tag{6.9}$$

and

$$\text{var}[X] \equiv \mathbb{E}[(X - \mu)^2] = \int_a^b (x - \mu)^2 p(x) dx. \tag{6.10}$$

The mean is the average of the values of the random variable, while the variance measures how spread out these values are. There are many other measures, such as the moments and skewness, as well as entropy. For more details, readers can refer to more advanced literature (Bhat and Miller, 2002; Grindstead and Snell, 1997).

6.3 Probability distributions

There are many different probability density distributions and they can describe different categories of random variables.

6.3.1 Commonly used distributions

The binomial distribution concerns a binary variable that only takes two possible outcomes: success/yes (i.e., 1) with probability p, or failure/no (i.e., 0) with probability

$1 - p$. For n independent trials, the probability of X taking the value of k is

$$B(n, p) = \binom{n}{k} p^k (1 - p)^{n-k}, \quad (k = 0, 1, 2, ..., n), \quad (6.11)$$

where

$$\binom{n}{k} = \frac{n!}{k!(n-k)!}. \quad (6.12)$$

Its mean and variance are $\mu = np$ and $\sigma^2 = np(1 - p)$, respectively.

A very widely used probability density distribution is the Poisson distribution, which can be considered as the limiting case of a binomial distribution for small probability events with a large number of independent trials. This requires that $\lambda = np > 0$ is a finite value with $n \gg 1$ (this $0 < p \ll 1$). The Poisson distribution is given by

$$P(X = n) = \frac{\lambda^n e^{-\lambda}}{n!}, \quad (n = 0, 1, 2, ...), \quad (6.13)$$

where its mean is λ. It is also easy to verify that its variance is also λ.

Probably, the most widely used distribution is the Gaussian distribution or Gaussian normal distribution of continuous random variables. The Gaussian distribution is given by

$$p(x) = \frac{1}{\sigma\sqrt{2\pi}} \exp\left\{ -\frac{(x - \mu)^2}{2\sigma^2} \right\}, \quad (6.14)$$

where μ and σ^2 are the mean and variance, respectively, of the random variable X. As the domain of X is the whole real number, the total probability requires that

$$\int_{-\infty}^{\infty} p(x) dx = 1. \quad (6.15)$$

Its cumulative probability function (CPF) can be obtained by integrating

$$F(x) = P(X < x) = \frac{1}{\sqrt{2\pi\sigma^2}} \int_{-\infty}^{x} e^{-\frac{(u-\mu)^2}{2\sigma^2}} du = \frac{1}{\sqrt{2}}\left[1 + \mathrm{erf}\left(\frac{x - \mu}{\sqrt{2}\sigma}\right)\right], \quad (6.16)$$

where the error function is defined by

$$\mathrm{erf}(x) = \frac{2}{\sqrt{\pi}} \int_0^x e^{-\zeta^2} d\zeta. \quad (6.17)$$

The Gaussian normal distribution is usually denoted by $N(\mu, \sigma^2)$ in the literature. If $\mu = 0$ and $\sigma = 1$, it becomes simply a normal distribution $N(0,1)$.

In the context of nature-inspired algorithms and their initialization, the uniform distribution is commonly used, which is defined by a constant probability p over an interval $[a, b]$,

$$p(x) = \frac{1}{b - a}, \quad x \in [a, b]. \tag{6.18}$$

By simple integration, it is straightforward to show that its mean is $\mathbb{E}[X] = (a+b)/2$ and the variance is $\sigma^2 = (b-a)^2/12$.

Another important distribution is Student's t-distribution,

$$p(t) = A\left(1 + \frac{t^2}{n}\right)^{-(n+1)/2}, \quad A = \frac{\Gamma((n+1)/2)}{\sqrt{n\pi}\,\Gamma(n/2)}, \tag{6.19}$$

where $-\infty < t < +\infty$ and n is the degree of freedom. Here, the special Γ-function is given by

$$\Gamma(\nu) = \int_0^\infty x^{\nu-1} e^{-x} dx, \tag{6.20}$$

which leads to the factorial $\Gamma(n) = (n-1)!$ when $\nu = n$ is a positive integer.

6.3.2 Distributions with long tails

Some probability density distributions can have a long tail, or a heavy tail, when the probability, though small, is still significantly nonzero for sufficiently large x. Such distributions are often called heavy-tailed, fat-tailed, or long-tailed distributions.

The probability density distribution of the exponential distribution can be described by a parameter λ, which can be written as

$$f(x) = \lambda e^{-\lambda x}, \quad x > 0, \tag{6.21}$$

and $f(x) = 0$ for $x \le 0$. It is easy to verify that its mean is $\mu = 1/\lambda$, and its standard deviation is also $\sigma = 1/\lambda$. If λ is very small, this distribution can become long-tailed. In addition, the power-law probability distribution is also long-tailed, and its probability density is

$$p(x) = Ax^{-\alpha}, \quad x \ge x_0 > 0, \quad A = (\alpha - 1)x_0^{\alpha-1}, \tag{6.22}$$

where $\alpha > 1$ is an exponent and $x_0 > 0$ is a threshold parameter. This implicitly assumes that $p(x) = 0$ for $x < x_0$. These two distributions are approximations to heavy-tailed distributions.

A truly heavy-tailed distribution should have an infinite variance. A good example is the Cauchy distribution

$$p(x, \mu, \gamma) = \frac{1}{\pi\gamma}\left[\frac{\gamma^2}{(x-\mu)^2 + \gamma^2}\right], \quad -\infty < x < \infty, \tag{6.23}$$

with two parameters μ and γ. Both its mean and variance are infinite or undefined.

Another truly heavy-tailed distribution is the Pareto distribution

$$p(x) = \begin{cases} \alpha x_0^\alpha x^{-(\alpha+1)}, & \text{if } x \ge x_0, \\ 0, & \text{if } x < x_0, \end{cases} \tag{6.24}$$

where $x_0 > 0$ is the minimum value of x and $\alpha > 0$ is a parameter.

A very important heavy-tailed distribution is the Lévy distribution, which is used in the cuckoo search and flower pollination algorithms. We will introduce this distribution in greater detail later in this chapter.

6.3.3 Entropy and information measures

Though entropy is a concept in physics, it has been extended to information theory. For a known probability function $p(x)$ for a discrete random variable, the Shannon entropy (H) is defined by

$$H(p) = -\sum_i p(x_i) \log_b[p(x_i)], \tag{6.25}$$

where log is usually in the base $b = 2$. As this convention is so widely used, we can drop the base in the writing. In essence, this Shannon entropy $H(p) \ge 0$ is a measure of the average amount of information contained in event outcomes of the discrete variable. If the random variable is continuous in the domain Ω, the sum becomes an integral and its entropy is

$$H(p) = \int_\Omega p(x) \log\left(\frac{1}{p(x)}\right) dx, \tag{6.26}$$

or

$$H(p) = -\int_\Omega p(x) \log\left(p(x)\right) dx, \tag{6.27}$$

which integrates over the whole domain of probability $p(x)$.

Cross entropy is a measure of distance/dissimilarity or similarity between two probability distributions $p(x)$ and $q(x)$, and it is defined by

$$H(p, q) = -\int p(x) \log[q(x)] dx. \tag{6.28}$$

Another important measure of the distance between $p(x)$ and $q(x)$ is the so-called Kullback–Leibler (KL) divergence,

$$D_{KL}(p, q) = \int p(x) \log\left[\frac{p(x)}{q(x)}\right] dx, \tag{6.29}$$

which leads to

$$D_{KL}(p, q) = H(p, q) - H(p). \tag{6.30}$$

This measure represents the difference or distance of $p(x)$ from a given $q(x)$. Clearly, the KL divergence is zero when $p(x) = q(x)$, as it should be.

6.4 Random walks and Lévy flights

Random walks have rigorous mathematical foundations, and they can model many phenomena such as diffusion, percolation, and Brownian motion (Bhat and Miller, 2002).

In general, if each step w_t at time t is drawn from the same probability distribution, the sum of consecutive steps form a random walk, i.e.,

$$S_t = w_1 + w_2 + \cdots + w_t, \tag{6.31}$$

where the sum S_t can be considered as the current state.

This random walk can be considered as a random walk in the D-dimensional space if the steps are considered as random vectors with each of its component being drawn from the same distribution, such as N(0,1).

If the current state S_t is known, the local move step w_{t+1} will give

$$S_{t+1} = S_t + w_{t+1}. \tag{6.32}$$

If w_{t+1} is drawn from the Gaussian distribution

$$w_{t+1} \sim N(0, 1), \tag{6.33}$$

the random walk becomes a Brownian motion. Here we have used the notation "$\sim$" to show that the random steps should be drawn from the probability distribution described by the right-hand side of the equation.

In fact, the diffusion process of pollen particles or ink molecules in water can be modeled by random walks, known as Brownian motion, and the movements obey the standard Gaussian probability distribution. Loosely speaking, the variance of Brownian random walks in a D-dimensional space can be estimated by

$$\sigma^2(t) = (2D\kappa)t, \tag{6.34}$$

where $\kappa = s^2/(2\tau)$ can be effectively considered as a diffusion coefficient. This diffusion coefficient can be estimated by the mean step s (drawn from the Gaussian distribution) as a jump in a short time interval τ. This random walk is isotropic and there is no drifting velocity to a particular direction. As the iteration time is discrete, the pseudotime t can be replaced by the number ($N = t$) of steps. Thus, the average distance d covered by a Brownian random walk is

$$d \propto \sqrt{N}. \tag{6.35}$$

This square root law is a typical feature for many diffusion phenomena.

There is no reason that the steps must be Gaussian, and step lengths can be drawn from any other distributions, including non-Gaussian distributions that correspond to non-Gaussian diffusion processes. In the context of nature-inspired algorithms for optimization, a very special random walk is the Lévy flights, whose steps are drawn from the Lévy distribution.

The rigorous definition of Lévy probability distribution can be tricky. For example, it can be given with a fixed exponent,

$$p(x, \mu, \gamma) = \frac{\sqrt{\frac{\gamma}{2\pi}}}{(x - \mu)^{3/2}} e^{-\frac{\gamma}{2(x-\mu)}}, \quad x \geq \mu, \tag{6.36}$$

where parameter $\mu > 0$ controls its location of this distribution, while γ controls its scale. However, its general definition involves an integral (Gutowski, 2001; Pavlyukevich, 2007), in which case we have

$$p(x) = \frac{1}{\pi} \int_0^\infty \cos(kx) e^{-\alpha|k|^\beta} dk, \quad (0 < \beta \leq 2), \tag{6.37}$$

where $\alpha > 0$. The case of $\beta = 1$ is equivalent to a Cauchy distribution, while $\beta = 2$ leads to a normal distribution.

Though the inverse integral is difficult to calculate, approximations can be obtained when s is sufficiently large. We have

$$L(s) \to \frac{\alpha \beta \Gamma(\beta) \sin(\pi\beta/2)}{\pi |s|^{1+\beta}}, \quad s \gg 0, \tag{6.38}$$

where $\Gamma(\beta)$ is the standard gamma function.

Comparing the variance of the Brownian random walks, the variance of Lévy flights varies nonlinearly (Gutowski, 2001; Pavlyukevich, 2007), i.e.,

$$\sigma^2(t) \sim t^{3-\beta}, \quad 1 \leq \beta \leq 2, \tag{6.39}$$

which means that the variance increases much faster, which is faster than $\sigma^2(t) \sim t$ in a Brownian random walk (Yang, 2014). As a result, the mean distance covered by the Lévy flights after N steps is

$$d \propto N^{(3-\beta)/2}. \tag{6.40}$$

This power-law feature is typical for superdiffusion phenomena.

As we mentioned in earlier chapters, the generation of random numbers that obey a Lévy distribution is quite tricky. From the practical implementation perspective, Mantegna's algorithm is quite effective (Mantegna, 1994). This algorithm was implemented in the cuckoo search algorithm.

6.5 Monte Carlo and pseudorandom numbers

The initialization process of almost all nature-inspired algorithms can be considered as a random sampling procedure and thus is essentially a major step of the Monte Carlo method. Monte Carlo is a class of sampling methods with a wide range of applications (Fishman, 1995). Its main idea is that a system quantity can be expected to be derived by averaging over or evaluating at a finite number of samples.

6.5.1 Monte Carlo

Monte Carlo is widely used for estimating multidimensional integrals over a regular domain. Monte Carlo integration is very efficient, and its error usually decreases with the number (N) of samples in the following manner:

$$E_N \propto 1/\sqrt{N}, \tag{6.41}$$

which is essentially independent of the dimensionality D of the sampling problem. This becomes advantageous for approximating high-dimensional integrals.

Obviously, the sampling of Monte Carlo can be any sampling method in statistics such as importance sampling. Important sampling can focus on a certain region, for example, an integral with an integrand around a sharp peak such as $\exp[-(100x)^2]$. In recent years, the Markov chain Monte Carlo (MCMC) has become a powerful tool for sampling in many applications such as computational biology and mathematical finance.

The rate of convergence or error reduction in the standard Monte Carlo is $O(1/\sqrt{N})$, but other methods can improve this even further. Quasi-Monte Carlo (QMC) has an error reduction of $O\big((\log N)^k/N\big)$, where k is a parameter related to the dimensionality of the problem. However, QMC involves low-discrepancy numbers, which will briefly be introduced later.

6.5.2 Generation of random numbers

In many simulations, especially Monte Carlo simulation, we need random numbers. However, the random numbers used in almost all programming languages are typically generated by an algorithm, and they are thus not truly random. These numbers are pseudorandom numbers. A commonly used generator is the linear congruential generator (Hull and Dobell, 1962),

$$d_i = (ad_{i-1} + c) \mod m, \quad a, c, m \in \mathbb{Z}, \tag{6.42}$$

where integers a and m are relatively prime. For example, in many algorithms, m is set to $m = 2^k$ or $2^k \pm 1$. As the maximum period is m, most implementations usually use $a = 1103515245$, $m = 2^{32}$, and $c = 12345$.

In general, linear congruential generators produce pseudorandom integers, but Marsaglia's generator can generate floating-point numbers by shifting and subtracting (Marsaglia, 1968), i.e.,

$$d_i = d_{i+20} - d_{i+5} - k, \qquad (6.43)$$

where the subscript indices $(i, i + 5, i + 20)$ only take integer values that are residuals after the mod 32. The theoretical period of this generator is about 2^{1492}, but the number generated are multiples of $\epsilon = 2^{-53}$ in the range of $[2^{-53}, 1 - 2^{-53}]$.

Modern algorithms for generating pseudorandom numbers are much more sophisticated and can pass various high-quality statistical tests. For example, the well-known Mersenne twister algorithm can have a very long period of $2^{19937} - 1$.

6.5.3 Quasirandom numbers

Pseudorandom numbers are used for standard Monte Carlo simulations. As we briefly mentioned earlier, QMC can have some advantages, but QMC simulations usually use quasirandom numbers or low-discrepancy numbers. A classic example is the van der Corput sequence, where an integer n is expressed in a prime base b,

$$n = \sum_{j=0}^{m} a_j(n) b^j, \qquad (6.44)$$

where $a_j(n)$ are integer coefficients and can only take values $\{0, 1, 2, ..., b-1\}$. Then, this expression (in base b) is reflected or reversed to give

$$\phi_b(n) = \sum_{j=0}^{m} a_j(n) \frac{1}{b^{j+1}}, \qquad (6.45)$$

where m is the smallest integer giving $a_j(n) = 0$ for all $j > m$. This method can indeed generate a low-discrepancy number whose distribution is uniform in the interval $[0, 1)$. There are many low-discrepancy sequences, including Sobol's quasirandom sequence, which is commonly used in QMC simulations (Sobol, 1994).

The random number sequences we discussed above are main uniform distributions in [0,1]. To generate other distributions, there are a class of techniques, including the inverse transform method, the acceptance–rejection method, and Gibbs sampling. More sophisticated methods are the MCMC based on Markov chains.

6.6 Markov chains

The random walk we discussed earlier is a Markov chain because its next state is essentially dependent only on the current state and transition probability. In other words, S_{t+1} depends only on S_t and transition moves w_{t+1}. More formally, a random variable X is called a Markov process if the transition probability, from state $X_t = S_i$

(at time t) to another state $X_{t+1} = S_j$, can only depend on its current state X_i. This means that

$$P(i, j) \equiv P(X_{t+1} = S_j | X_0 = S_p, ..., X_t = S_i) = P(X_{t+1} = S_j | X_t = S_i). \quad (6.46)$$

This transition probability is often written as

$$P(i, j) \equiv P(i \rightarrow j), \quad \text{or} \quad P_{i \rightarrow j} \equiv P(i, j). \quad (6.47)$$

This means that the transition probability, also called transition kernel, is independent of the states before t. The Markov process generates a random sequence $(X_0, X_1, ..., X_n)$, called a Markov chain.

Following a similar procedure, we can derive the k-step transition probability $P_{ij}^{(k)}$ (from state i to state j)

$$P_{ij}^{(k)} = P(X_{t+k} = S_j | X_t = S_i), \quad (6.48)$$

where $k > 0$ is an integer. The transition matrix $\boldsymbol{P} = [P_{ij}^{(1)}] = [P_{ij}]$ is a stochastic matrix with each row summing to 1. That is,

$$P_{ij} > 0, \quad \sum_{j=1}^{m} P_{ij} = 1, \quad (\text{for } i = 1, 2, ..., m). \quad (6.49)$$

Though we do not intend to provide rigorous definitions here, we still try to define certain terminologies that may be useful to the theoretical analyses in the next chapter. A Markov chain is called regular if its transition matrix $\boldsymbol{P}$ or its power $K > 0$ has only positive entries. In other words, there is an integer $K > 0$ which gives $P_{ij}^{(K)} > 0$ for all i and j. This is equivalent to saying that there is a nonzero probability to go from any state i to any other state j.

Let us use the notation $\pi_i(t)$ to denote the probability of the chain in the state i (or more accurately, S_i) at time t. For a regular Markov chain starting with an initial π_0 with a transition probability matrix $\boldsymbol{P}$, we have after k steps

$$\pi_k = \pi_0 \boldsymbol{P}^k, \quad \text{or} \quad \pi_k = \pi_{k-1} \boldsymbol{P}. \quad (6.50)$$

Markov chain theory shows that when k is sufficiently large, the chain will converge to a unique stationary distribution π^*, defined by

$$\pi^* = \pi^* \boldsymbol{P}. \quad (6.51)$$

This means that the largest eigenvalue of $\boldsymbol{P}$ is 1, and its corresponding eigenvector is π^*.

In nature-inspired computation, different algorithms can use different transition probabilities and different transition matrices will lead to different behavior and characteristics in the random chains used in the algorithms.

The theory of Markov chains is rigorous and its applications are diverse. Many sampling techniques are based on it. The well-known Metropolis–Hasting algorithm, the Gibbs sampling method, and MCMC all belong to this class. Interested readers can refer to more advanced literature (Geyer, 1992; Ghate and Smith, 2008).

References

Bhat, U.N., Miller, G.K., 2002. Elements of Applied Stochastic Processes, third ed. John Wiley & Sons, New York.

Fishman, G.S., 1995. Monte Carlo: Concepts, Algorithms and Applications. Springer, New York, USA.

Geyer, C.J., 1992. Practical Markov chain Monte Marlo. Statistical Science 7 (6), 473–511.

Ghate, A., Smith, R., 2008. Adaptive search with stochastic acceptance probability for global optimization. Operations Research Letters 36 (3), 285–290.

Grindstead, C.M., Snell, J.L., 1997. Introduction to Probability, second ed. American Mathematical Society, Providence, Rhode Island.

Gutowski, M., 2001. Lévy flights as an underlying mechanism for global optimization algorithms. ArXiv Mathematical Physics e-Prints. (Accessed 1 Sept 2019).

Hull, T.E., Dobell, A.R., 1962. Random number generators. SIAM Review 4 (4), 230–254.

Mantegna, R.N., 1994. Fast, accurate algorithm for numerical simulation of Lévy stable stochastic process. Physical Review E 49 (5), 4677–4683.

Marsaglia, G., 1968. Random numbers fall mainly in the planes. Proceedings of the National Academy of Sciences of USA 61 (1), 25–28.

Pavlyukevich, I., 2007. Lévy flights, non-local search and simulated annealing. Journal of Computational Physics 226 (2), 1830–1844.

Sobol, I.M., 1994. A Primer for the Monte Carlo Method. CRC Press, Boca Raton, FL.

Yang, X.S., 2014. Nature-Inspired Optimization Algorithms. Elsevier Insight, London.

Mathematical framework for algorithm analysis

7

Xin-She Yang[a], Xing-Shi He[b], Qin-Wei Fan[b]

[a]*Middlesex University London, School of Science and Technology, London, United Kingdom*
[b]*Xi'an Polytechnic University, College of Science, Xi'an, China*

CONTENTS

7.1 Introduction

For any new algorithms, it would be useful if we can gain some insight into their mechanisms, their iteration characteristics, and their long behavior. More specifically, we wish to see if an algorithm will converge or not, under what conditions, and if it is stable under these conditions. From the implementation perspective, we wish to see how quickly it can converge and if an accurate solution can be obtained.

However, it is often a challenging task to answer any of these questions, unless the algorithm under consideration is sufficiently simple or has some desired mathemati-

Nature-Inspired Computation and Swarm Intelligence. https://doi.org/10.1016/B978-0-12-819714-1.00017-8

cal properties. Traditional numerical analysts tend to start with the iterative functions and attempt to figure out if they can satisfy the conditions for fixed-point theorems.

However, things become more difficult, and often intractable, when there are some randomization and stochastic behavior coming into play. To some extent, this is true for most nature-inspired algorithms. This chapter intends to formulate a mathematical framework for analyzing nature-inspired algorithms from different theoretical perspectives so as to gain some insight into the stability, convergence, and rates of convergence of these algorithms. We will discuss theories and techniques concerning fixed-point theorems, Markov chains, dynamical systems, and the no-free-lunch theorem, as well as benchmarking.

7.2 Iterated function and fixed-point theorems

Almost all algorithms for optimization are iterative, and fixed-point theorems are key results concerning traditional iterative algorithms.

For a relatively simple root finding algorithm such as Newton's method, the roots of a polynomial $p(x) = 0$ can be obtained by

$$x_{k+1} = x_k - \frac{p(x_k)}{p'(x_k)}, \tag{7.1}$$

starting with an educated guess x_0. This can be written as an iterative functional as

$$x_{k+1} = f(x_k) = x_k - \frac{p(x_k)}{p'(x_k)}, \tag{7.2}$$

where $p'(x_k) \neq 0$.

From function composite properties, we know that two functions $f(x)$ and $g(x)$ can form a composite as

$$g\big(f(x)\big) = (g \circ f)(x). \tag{7.3}$$

In the case f and g are identical, we have

$$f^n(x) \equiv (f \circ f \circ \dots \circ f)(x), \tag{7.4}$$

which means that

$$f^{n+1} = (f \circ f^n)(x). \tag{7.5}$$

Thus, Newton's method can be written as

$$x_{k+1} = (f \circ f \circ \dots \circ f)(x_0) = f^k(x_0). \tag{7.6}$$

Obviously, if a correct root x_* can be found, it requires that

$$\lim_{k \to \infty} x_{k+1} = \lim_{k \to \infty} f^k(x_0) = x_*, \qquad (7.7)$$

which means that this root is a fixed point of the iteration functional f.

However, there are some strict conditions for the existence of the fixed point. From experience, we know that not all Newton's iterations will converge. There are a few forms of fixed-point theorems that can guarantee if an iterated function can converge towards a fixed point. Similar to the multiple roots of a polynomial, multiple or different fixed points can exist, depending the conditions and starting points.

The well-known Banach fixed-point theorem states that the fixed point x_* can exist for a contraction mapping $f : x \mapsto x$, satisfying $f(x_*) = x_*$, under the condition that

$$d\big(f(x_i), f(x_j)\big) \le \theta\, d(x_i, x_j), \qquad 0 \le \theta < 1, \qquad (7.8)$$

for all x_i and x_j. Here the function $d(.,.)$ is a metric such as the Cartesian distance between x_i and x_j. Essentially, this condition requires the distance metric is shrinking or contracting. This fixed point can be obtained by generating a sequence via $x_{k+1} = f(x_k)$ starting with x_0. However, this condition may not be true, and there may not exist a parameter $\theta \in [0, 1)$ at all. There are other forms of fixed-point theory, including the Brouwer fixed-point theorem, the Lefschetz fixed-point theorem, and the Schauder fixed-point theorem (Granas and Dugundji, 2003; Khamsi and Kirk, 2001).

In the context of nature-inspired computation, an algorithm A tries to generate a new solution x_{k+1} from the current solution x_k and the current best solution x_* found up to the present iteration k. The way of generating new moves will also depend on some parameters $p_1, p_2, ..., p_m$. This means that

$$x_{k+1} = A(x_k, x_*, p_1, ..., p_m), \qquad (7.9)$$

for $k \ge 0$. Empirical observations suggest this will converge under certain conditions as the iterations continue, thus it is possible that

$$\lim_{k \to \infty} x_{k+1} = x_\infty, \qquad (7.10)$$

where the limit x_∞ is a fixed point of this algorithm. In the case x_∞ diverges or becomes unbounded, we can say that the algorithm diverges.

In general, there is no guarantee that the best solution x_* found so far is the same as or close to the final fixed point x_∞. In the case of $x_\infty \ne x_*$, the limiting point is not the same as the intended best solution x_*, which may indicate that the iteration sequence becomes prematurely converged. However, in a special case when it indeed gives $x_\infty = x_*$, we can say that the algorithm has reached the true optimal solution. In reality, some problems, especially multimodal optimization problems, can have multiple optimal solutions. This requires that the algorithm used should be able to general multiple solutions and follow multiple paths, and thus population-based algorithms are necessary in this case. The good news is that almost all nature-inspired algorithms are population-based algorithms.

Since nature-inspired metaheuristics use some randomness (in terms of a random variable ϵ), the solutions during iterations can be considered in the average sense. Thus, the expectation of the iteration sequence becomes

$$\mathbb{E}[\lim_{k \to \infty} x_{k+1}] = < \lim_{k \to \infty} x_{k+1} > = \langle A(x_k, x_*, p_1, p_2, ..., p_m, \epsilon) \rangle = < x_\infty > . \quad (7.11)$$

Ideally, we can somehow do some rigorous mathematical analysis about this iteration system. However, the fixed-point theorems we discussed earlier cannot be directly applied to algorithms with stochastic components, or algorithm procedures without explicit iterated functions. Despite the significant developments concerning nature-inspired algorithms and their applications in the last few decades, not much progress has been made in the area of theoretical analysis of these algorithms. More theoretical studies are highly needed.

7.3 Complexity of algorithms

Computational complexity is a very active area of research in computer science, and it is also useful to estimate the complexities of nature-inspired algorithms.

7.3.1 Nature-inspired algorithms

Nature-inspired algorithms for optimization can take various forms. However, almost all nature-inspired optimization algorithms use a population of n solutions, and each solution is a vector in the D-dimensional space so as to solve an optimization problem with D decision variables. In addition, the number of iterations is usually limited to a fixed maximum number T. With these parameters, it is straightforward to show that the algorithmic complexities of particle swarm optimization and bat algorithm are $O(nT)$. In addition, the complexities of the cuckoo search and flower pollination algorithms are also $O(nT)$ since the generation of Lévy random numbers does not affect their complexities.

Some algorithms such as the firefly algorithm can use more than one loop, and their complexity will increase slightly. In fact, the complexity of the firefly algorithm is $O(n^2 T)$ as it uses two loops over all n solutions. However, the population size (n) is usually small, compared with a relatively large T, the computational costs of using two loops do not increase much in practice.

Overall, the computational complexities of all nature-inspired optimization algorithms are low. However, compared with traditional gradient-based algorithms, the maximum number of iterations is much higher. In most gradient-based algorithms (e.g., Newton's method), it usually requires 25 to 40 iterations, and it rarely uses more than 100 iterations. In addition, such number of iterations will change little even for higher-dimensional problems. This is a major advantage of traditional gradient-based methods. That is also why many machine learning algorithms are mostly gradient-based, including the popular Adam optimizer used in deep learning.

On the other hand, for evolutionary computation and nature-inspired computation, T can be at least several thousand to half a million. In some literature, it is even suggested that $T = 50000D$ or $100000D$, where D is the number of dimensions of the problem. For problems of moderate dimensions such as $D = 100$, $T = 5 \times 10^6$ to 10^7, which may be acceptable for function optimization because the evaluation of an objective function is very quick. However, this can become an important issue in practice, especially for problems where the computational evaluation of a single objective (such as aerodynamical design) may take a long time. In addition, due to the stochastic nature of nature-inspired algorithms, multiple runs are needed to ensure meaningful statistics and results. This again will increase the computational costs in practice.

7.3.2 Relationship to complexity classes

The algorithmic complexities of all nature-inspired algorithms are relatively low, but extensive studies indicate that they are capable to deal with a wide range of problems, including some very difficult NP-hard problems such as the traveling salesman problem (Ouaarab et al., 2014; Osaba et al., 2016) and scheduling problems (Marichelvam et al., 2014). Thus, it seems that the algorithmic complexity does not somehow provide any implications to the problems that an algorithm can solve, and there is no link to the problem classes. It is still a bit of a mystery why such nature-inspired algorithms with low algorithmic complexity can cope with highly complex problems, and they somehow can manage to find good feasible solutions or even optimal solutions in practice. More research is needed to gain more insight in this area.

7.4 Markov chain framework for convergence

When an algorithm can indeed find the true optimality to the problem under consideration, its iterations must have converged. When analyzing the convergence, we wish to understand the conditions the algorithm converge and how quickly it converges. The convergence can be in a deterministic sense such as the case of Newton's method or in a probabilistic sense. The convergence of nature-inspired algorithms is analyzed mainly in terms of probabilistic convergence due to the stochastic nature of their algorithmic components.

7.4.1 Probabilistic convergence

A common way for analyzing the probabilistic convergence is to use Markov chain theory. For example, the convergence of the genetic algorithm has been analyzed by using Markov chains (Suzuki, 1995; Aytug et al., 1996; Greenhalgh and Marshal, 2000), and the same methodology can also be used for analyzing metaheuristics (Gutjahr, 2010). More recently, Markov chain theory has also been used to analyze the cuckoo search algorithm (He et al., 2018) and the bat algorithm (Chen et al., 2018).

It is worth pointing out that these studies often make some additional assumptions or simplifications of an algorithm under analysis so that the simplified algorithm structure is clearer and can satisfy the conditions or criteria for convergence analysis. Now let us first discuss the criteria for convergence.

7.4.2 Convergence criteria

Each algorithm will be different from other algorithms, even though the difference may be insignificant. However, these differences in algorithms make it difficult to provide a unified approach to analyze them. In the framework of Markov chains, there are two commonly used test criteria (Solis and Wets, 1981).

Let us first define some notations. We can use $\langle H, f \rangle$ to denote an optimization problem with an objective function f in a feasible solution space H. Then, an algorithm A (usually a stochastic optimizer, or any nature-inspired algorithm in general) generates a new solution x_{k+1} at iteration k, based on the solution x_k at k. To avoid the need to write variables in bold font, we have used here the simple x to denote a D-dimensional vector. In addition, all the feasible solutions generated by the algorithm during iterations form a solution set S. Now the iterations of algorithm A can be written as

$$x_{k+1} = A(x_k, S). \tag{7.12}$$

To guarantee that the global optimality is achievable, it requires two conditions or criteria:

- Criterion 1: The sequence $\{f(x_k)\}_{k=0}^{\infty}$ generated by an optimization algorithm A should be decreasing. If $f(A(x, S)) \leq f(x)$ (for minimization) with $S \in H$, we have

$$f(A(x, S)) \leq f(S). \tag{7.13}$$

- Criterion 2: For all subsets $\forall B \in H$ with a probability measure $p_k(B)$ at iteration k of algorithm A on B, the following relationship holds:

$$\prod_{k=0}^{\infty}(1 - p_k(B)) = 0, \tag{7.14}$$

subject to a proper measure $v(B) > 0$ defined on B.

If we define the best solution (g) as the solution with the best fitness (smallest for minimization, largest for maximization) in set S, we can say that the algorithm has indeed converged if the best solution in set S is close to or in the neighborhood R_ε of the optimal solution. Here R_ε is the neighborhood of radius $\varepsilon > 0$, which can be infinitely small.

If the above two criteria are met for a given algorithm, the probability of the algorithm converging to R_ε is

$$\lim_{k \to \infty} P(x_k \in R_\varepsilon) = 1. \tag{7.15}$$

This means that the convergence is in the probabilistic sense. In other words, the probability of this convergence is almost surely one, or the solution set S contains the optimal solution with probability one.

7.4.3 Sketch of convergence analysis

The detailed proof of convergence of an algorithm, including the genetic algorithm, cuckoo search and bat algorithm, can be very lengthy. We will not delve into much detail here, and readers who are interested in convergence analysis can refer to more specialized literature (Gutjahr, 2010; He et al., 2018; Chen et al., 2018). Instead, we will focus on the main steps and key ideas of such proofs, so we can put a few algorithms into a unified approach here.

One of the key parts is to construct transition probability, and it is often necessary to simplify the algorithms for analysis. In the case of cuckoo search, there are two branches. However, we only use the main search mechanism for constructing the transition moves (He et al., 2018),

$$x_{k+1} = x_k + \alpha \otimes L(\lambda). \tag{7.16}$$

For the bat algorithm, the local perturbation term is ignored, and only the main frequency tuning part is used for constructing the transition probability (Chen et al., 2018).

The main steps of proofs are as follows:

- Simplify the algorithms with appropriate assumptions and then construct the transition probability.
- Show all the solutions form a nonempty set, and the sequences can be mapped into a chain of solution states (including the best solution g).
- Prove that the constructed chain is a finite, homogeneous Markov chain.
- Show that both convergence criteria are met. The final probabilistic convergence is guaranteed when the number of iterations becomes very large or sufficiently large.

All the above steps can ensure the convergence of the algorithm under discussion. However, there is a key issue here concerning "sufficiently large." Mathematically, it requires $k \to \infty$, but this is impossible to achieve in practice. Thus, how large "sufficiently large" is still a major issue. This is probably one of the reasons why some literature insisted that the maximum number of iterations T should be something like 100,000 or more. Another issue is that even if an algorithm has been proved to converge, it is still commonly observed that some iterations do not yield good solutions.

This is because the mathematical proofs in the literature have used some simplifications or unrealistic assumptions, thus the version with proved convergence was an idealized variant. Another issue is that the parameter setting may not be properly tuned in the algorithm used for simulation, and thus the actual simulation time is much longer in order to produce quality solutions.

These issues mean that we have to be careful when interpreting theoretical results. There are still some significant gaps between theory and practice, even if we have some theoretical analysis and understanding about a certain algorithm. Obviously, the gap is even bigger if we have not had any theoretical analysis at all. This highlights the need for further research in this key area.

7.4.4 **Rate of convergence**

Even if we can prove the convergence of an algorithm in the probabilistic sense, it does not usually provide enough information about how quickly the algorithm actually converges in practice. To get such information, it requires the rate of convergence. However, this rate is much harder to analyze, if not impossible.

As we know that the largest eigenvalue of a proper Markov chain is $\lambda_1 = 1$, it is believed that the second largest eigenvalue $0 < \lambda_2 < 1$ controls the rate of convergence (Grindstead and Snell, 1997; Ghate and Smith, 2008), i.e.,

$$||E|| \leq Q(1 - \lambda_2)^k, \tag{7.17}$$

where E is the error and $Q > 0$ is a constant. It is expected that the chain converges as $k \to \infty$. But the main difficulty is to figure out this eigenvalue λ_2. In most cases concerning nature-inspired algorithms, this proves to be a quite challenging task, as there is little work on this topic.

For many algorithms, there are some extensive studies using numerical simulations to plot out the convergence graphs for different benchmark functions for a given algorithm. Different functions can have different rates of convergence, while different algorithms can also have different rates of convergence even for the same optimization function. This implies that the rate of convergence could depend on both the algorithm and the problem under consideration. More work is highly needed in this area.

7.5 **Dynamical systems and algorithm stability**

The theory of dynamical systems concerns many nonlinear systems and their applications in physics, biology, mathematics, economics, and astronomy (Khalil, 1996; Hasselblatt and Katok, 2003). It is also a powerful mathematical tool for analyzing the characteristics and stability of iterative algorithms.

7.5.1 Dynamical system theory

For a state variable x that varies with time t, a dynamical system can be written as a differential equation

$$\dot{x} = \phi(x, t), \qquad (7.18)$$

where $\phi(x, t)$ is a known function. In the case the differential equation is autonomous, ϕ does not depend on t explicitly, and it becomes a time-invariant dynamical system

$$\dot{x} = \phi(x). \qquad (7.19)$$

The equilibrium point is determined by $\phi(x) = 0$. A general dynamical system can be extended to an n-dimensional system $x = x = (x_1, x_2, ..., x_n)$. For simplicity and consistency with the notations used in the literature, we do not use the bold font in this case. We have

$$\dot{x}_1 = \phi_1(x), \quad \dot{x}_2 = \phi_2(x), \quad \cdots, \quad \dot{x}_n = \phi_n(x). \qquad (7.20)$$

For example, the well-known nonlinear pendulum equation

$$\ddot{\theta} + \omega^2 \sin\theta = 0, \quad \omega = \sqrt{\frac{g}{L}} \qquad (7.21)$$

is a second-order differential equation for angle θ. Here, g is the acceleration due to gravity, while L is the length of the pendulum. Setting $x_1 = \theta$ and $x_2 = \dot{\theta}$, we can rewrite the second-order nonlinear differential equation as a set of two first-order ordinary differential equations:

$$\dot{x} = \frac{d}{dt}\begin{pmatrix} x_1 \\ x_2 \end{pmatrix} = \frac{d}{dt}\begin{pmatrix} \theta \\ \dot{\theta} \end{pmatrix} = \begin{pmatrix} \dot{\theta} \\ -\omega^2 \sin\theta \end{pmatrix} = \begin{pmatrix} x_2 \\ -\omega^2 \sin x_1 \end{pmatrix}. \qquad (7.22)$$

Thus, we can now focus on the general n-dimensional system with an initial condition $x(0) = x_0$,

$$\dot{x} = \phi(x), \quad \text{with} \quad x(0) = x_0, \qquad (7.23)$$

which is a time-continuous dynamical system. Dynamical system theory shows that the above equation can have a solution in the form (Khalil, 1996)

$$x(t) = \psi(x_0), \qquad (7.24)$$

which is often called a vector flow. However, the exact form of $\psi(x_0)$ depends on the dynamical system.

If the dynamical system is a discrete-time system, also called time-discrete dynamical system, we have an iterative difference equation

$$x_{k+1} = \phi(x_k), \quad x(0) = x_0. \qquad (7.25)$$

There are a spectrum of methods for describing and analyzing the characteristics of dynamical systems, including the Poincaré surface of section, Lyapunov stability, and others (Khalil, 1996; Hasselblatt and Katok, 2003). Here we will only discuss the Lyapunov exponent and its link with stability.

For an iterative system (7.25), the Lyapunov exponent can be calculated by

$$\Lambda = \lim_{k \to \infty} \frac{1}{k} \sum_{m=1}^{k-1} \ln \left| (\phi^m)'(x_m) \right|, \tag{7.26}$$

where $(\phi^m)'(x_m) = \phi'(x_{m-1})...\phi'(x_1)\phi'(x_0)$ and ϕ^m is the function composite. That means that

$$\phi^m = (\phi \circ \phi^{m-1})(x_0) = \phi(\phi^{m-1}(x_0)). \tag{7.27}$$

Lyapunov stability requires that the Lyapunov exponent should be negative (Khalil, 1996).

For a time-discrete linear dynamical system that can be written as

$$x_{k+1} = Ax_k, \tag{7.28}$$

its solution can be written as

$$x(k) = A^k x_0. \tag{7.29}$$

Thus, its stability will depend on the eigenvalues of A. The Lyapunov stability requires that all the n eigenvalues λ_i of A must satisfy

$$|\lambda_i| \leq 1. \tag{7.30}$$

In the case of $|\lambda_i| < 1$ (without equality), it becomes globally asymptotically stable.

Now let us use the above framework of dynamical system theory to analyze the stability of particle swarm optimization and the bat algorithm under their usual parameter settings.

7.5.2 Stability of particle swarm optimization

In the particle swarm optimization (Kennedy and Eberhart, 1995), the main iterative equations for position x_i and velocity v_i for particle i can be written as

$$v_i^{t+1} = v_i^t + \alpha \epsilon_1 (g^* - x_i^t) + \beta \epsilon_2 (x_i^* - x_i^t) \tag{7.31}$$

and

$$x_i^{t+1} = x_i^t + v_i^t, \tag{7.32}$$

where parameters α and β are called learning parameters, while ϵ_1 and ϵ_2 are two random numbers in [0, 1]. In addition, g^* is the best solution in the population, while x_i^* is the best solution in the iteration history of particle i.

The above two iteration equations can be written compactly as a single matrix equation,

$$\begin{pmatrix} x_i \\ v_i \end{pmatrix}^{t+1} = \begin{pmatrix} 1 & 1 \\ -(\alpha\epsilon_1 + \beta\epsilon_2) & 1 \end{pmatrix} \begin{pmatrix} x_i \\ v_i \end{pmatrix}^{t} + \begin{pmatrix} 0 \\ \alpha\epsilon_1 g^* + \beta\epsilon_2 x_i^* \end{pmatrix}. \quad (7.33)$$

The first stability analysis of particle swarm optimization was carried out by Clerc and Kennedy in terms of a simplified dynamical system (Clerc and Kennedy, 2002); however, their analysis was based on a very simplified version, not the above equations. Even so, it gained some greater insight into its stability and parameter ranges. The eigenvalues of their simplified system are

$$\lambda_{1,2} = 1 - \frac{\vartheta}{2} \pm \frac{\sqrt{\vartheta^2 - 4\vartheta}}{2}, \quad (7.34)$$

which leads to a critical behavior at $\vartheta = \alpha + \beta = 4$ (Clerc and Kennedy, 2002).

7.5.3 Stability of the bat algorithm

The full stability analysis of the bat algorithm may be difficult, due to the stochastic switching controlled by varying the pulse emission and loudness of bats (Yang, 2010). However, a slightly simplified version without these variations is tractable for mathematical analysis, and we will briefly outline here the stability analysis of this version (Chen et al., 2018).

To simplify the notations, we do not use bold font for vectors, and the analysis is valid for any individual bat (thus dropping the subscript for bat i). In addition, we also assume the variation of frequency is realized by a simple parameter m_f. With these assumptions and notations, we have

$$v_{k+1} = \theta v_k + (g - x_k)m_f, \quad (7.35)$$

$$x_{k+1} = \delta x_k + \gamma v_{k+1}, \quad (7.36)$$

where θ, δ, and γ are the weight coefficients introduced to generalize the algorithmic equations. In addition, g is the current best solution found by the population at iteration k. Though there are four parameters, only two are key parameters. A detailed analysis shows that $\delta = 1$ and $\gamma = 1$ are the necessary conditions (Chen et al., 2018).

Thus, the reduced dynamical system for the bat algorithm becomes

$$v_{k+1} = -m_f x_k + \theta v_k + m_f g, \quad (7.37)$$

$$x_{k+1} = x_k + \theta v_k + m_f g - m_f x_k. \quad (7.38)$$

This can be rewritten as

$$Y_{k+1} = CY_k + Mg, \quad (7.39)$$

where

$$Y_k = \begin{bmatrix} x_k \\ v_k \end{bmatrix}, \qquad C = \begin{bmatrix} 1 - m_f & \theta \\ -m_f & \theta \end{bmatrix}, \qquad M = \begin{bmatrix} m_f \\ m_f \end{bmatrix}. \qquad (7.40)$$

The behavior of this dynamical system is controlled by matrix C. Clearly, as the iterations proceed and if the algorithm converges, the whole bat population should move towards g (i.e., $x_k \to g$ as $k \to \infty$), leading to zero velocity $v_k \to 0$ as $k \to \infty$. This corresponds to a fixed point

$$Y* = \begin{bmatrix} g \\ 0 \end{bmatrix}. \qquad (7.41)$$

Now the eigenvalues of C can be obtained by

$$\det \begin{vmatrix} 1 - m_f - \lambda & \theta \\ -m_f & \theta - \lambda \end{vmatrix} = 0, \qquad (7.42)$$

which gives

$$(1 - m_f - \lambda)(\theta - \lambda) + m_f \theta = 0, \qquad (7.43)$$

or

$$\lambda^2 + (m_f - \theta - 1)\lambda + \theta = 0. \qquad (7.44)$$

Its solutions are

$$\lambda = \frac{-(m_f - \theta - 1) \pm \sqrt{(m_f - \theta - 1)^2 - 4\theta}}{2}, \qquad (7.45)$$

which can correspond to two eigenvalues λ_1 and λ_2.

From the theory we discussed earlier, the stability of this system requires that the modulus of both eigenvalues must be smaller than unity, which means that $|\lambda| \le 1$. In addition, Vieta's formula for polynomials also requires that $\lambda_1 \cdot \lambda_2 = \theta$.

It is straightforward to derive the final stability conditions where the parameters form a triangular region enclosed by the following inequalities:

$$\begin{cases} -1 \le \theta \le +1, \\ m_f \ge 0, \\ 2\theta - m_f + 2 \ge 0. \end{cases} \qquad (7.46)$$

This analysis clearly shows the algorithm is stable within the parameter ranges of m_f and θ, which has been confirmed by numerical experiments.

In principle, we can follow the similar procedure to analyze the stability of other algorithms, though some linearization and additional assumptions may be needed. It can be expected more theoretical studies will appear on this topic in the near future.

7.6 **Benchmarking and no-free-lunch theorems**

As we have seen so far, theoretical analysis can provide insight into algorithms using rigorous mathematical techniques. However, such analysis may require additional assumptions, and thus their results may not be truly relevant to the actual performance of the algorithm. Alternatively, we can carry out a series of numerical experiments to benchmark the algorithm in terms of its performance, stability and other characteristics.

7.6.1 **Role of benchmarking**

For any new algorithm, an important part of tests and validation is to use benchmark functions to test how the new algorithm may perform in comparison with other algorithms. Such benchmarking is also necessary to gain better understanding and greater insight into the advantages and disadvantages of the algorithm under test. Typically, in the current literature, a set of test functions, ideally with different properties such as mode shapes, are used to test the new algorithm. There are many different test functions (Jamil and Yang, 2013) and other test benchmark suites.

However, there is an important issue concerning such benchmarking processes. Though benchmarking can serve some purpose, it is not actually much used in practice. There are many reasons for this. One reason is that these functions are often well designed and sufficiently smooth, while real-world problems are much more diverse and can be very different from these test functions. Another reason is that these test functions are typically unconstrained or with regular domains, while the problems in real-world applications have many nonlinear complex constraints and the domain can be formed by many isolated regions or islands. Thus, algorithms that work well for test functions may not work well in applications.

From a mathematical perspective, benchmark functions are almost all "well-behaved" functions or toy problems. Imagine a search for maxima in a flat domain with many sharp peaks, where these peaks are Dirac delta functions,

$$\delta(x) = \begin{cases} \infty, & \text{if } x = 0, \\ 0, & \text{if } x \neq 0. \end{cases} \tag{7.47}$$

Even with a unit peak at random location x_0,

$$u_\delta(x) = \begin{cases} 1, & x = x_0, \\ 0, & x \neq x_0, \end{cases} \tag{7.48}$$

it would be equally difficult. It may be argued that Dirac delta-like functions are not smooth, but what happens if we use a smooth approximation

$$f(x) = \frac{1}{\sqrt{\pi a^2}} e^{-x^2/a^2}, \quad 0 < a \ll 1, \tag{7.49}$$

and take the limit $a \to 0$. As a practical example, we can use $a = 0.01$ to $a = 0.0001$ in a domain of $x \in [-1000, 1000]^D$ for $D = 2$. For such needle-like functions, the search domain can be large, but the region with useful "information" such as the non-flat landscape is exceptionally small. In this case, a large fraction of search moves are useless and wasted, and it is understandable that almost all algorithms are ineffective for such function optimization problems.

Alternatively, if we use a vast number of different test functions $f_i(x)$ (where $i = 1, 2, ..., N$), we can validate the algorithm seemingly more extensively. However, when we look at the average performance, it could be quite varied. If we look at the function objective landscape, the average performance seems to work on an averaged landscape, which can be approximated by

$$\Phi(x) = \frac{1}{N} \sum_i^N f_i(x). \tag{7.50}$$

Now if we restrict this set of functions as univariate functions on an interval $[a, b]$, this averaged landscape becomes

$$\bar{\phi} = \frac{1}{b-a} \int_a^b \Phi(x)dx, \tag{7.51}$$

which is a constant, according to the mean-value theorem in calculus. This means the averaged landscape is flat, and consequently no algorithm can perform better than others, due to a lack of information of the objective landscape.

This scenario is very similar to the cases where the no-free-lunch theorem has been suggested. So let us discuss this theorem.

7.6.2 No-free-lunch theorem

There are many different optimization algorithms. A commonly asked question is: Which is the best one to use? Is there a universal tool that can be used to solve all or at least a vast majority of optimization problems? The simple truth is that there are no such algorithms. This is formally proved mathematically by Wolpert and Macready in 1997 in their influential work on the so-called no-free-lunch theorem (Wolpert and Macready, 1997).

Loosely speaking, the main idea of the no-free-lunch theorem can be stated as follows. If an algorithm A can outperform another algorithm B for finding the optima of some objective functions, then there are some other functions on which B will outperform A. If their performance is averaged over *all possible* functions, both A and B can perform equally well over all these functions. If algorithm A is a random search algorithm, this implies that any algorithm can perform equally well as a random search algorithm, if the performance is measured as an average over all possible functions. This seems to be counterintuitive and contradictory to empirical observations. In practice, we know that some algorithms are indeed better than others. For example, the quicksort for sorting numbers is indeed better than a method based on simple pair-wise comparison.

The key to resolve this issue lies in the keywords "all" and "average." In reality, we are always concerned with a particular set of problems, not all problems. We are also concerned with the actual individual performance of solving a particular problem, not the averaged performance over all problems. In fact, for a finite set of algorithms to solve a finite set of problems, this becomes a zero-sum ranking problem (Joyce and Herrmann, 2018; Yang and He, 2019). Thus, some algorithms are better than others when solving a particular set of problems. Thus, the understanding of the no-free-lunch theorem is consistent with empirical observations. Just the perspective and emphasis will be shifted slightly. Now the aim of the research should focus on identifying the most suitable algorithm(s) for a given problem or figuring out the type of problems that a given algorithm can potentially solve.

To gain more insight into the no-free-lunch theorem, let us brief summarize the main steps of Wolpert and Macready's proof here. For technical details, please refer to the original paper by Wolpert and Macready (1997).

Let n_θ be the number of discrete values of parameter θ, which may be due to the finite machine precision or discrete values. Let n_f be the number of discrete values of the objective function f in the optimization problem. Then the number of all possible combinations of objective values is

$$N = n_f^{n_\theta}, \tag{7.52}$$

which is a finite number (though it can be very large). In addition, let $\mu(S_k^y | f, k, A)$ denote the performance measure of algorithm A, iterated k times on an objective function f over the sample set S_k. This performance metric is obtained in a statistical sense, and the superscript y emphasizes the fact that the set is associated with the objective.

The averaged performance over all possible functions for two algorithms A and B is equal; that is,

$$\sum_f \mu(S_k^y | f, k, A) = \sum_f \mu(S_k^y | f, k, B), \tag{7.53}$$

where the time-ordered set $S_k = \{(S_k^x(1), S_k^y(2)), ..., (S_k^x(k), S_k^y(k))\}$ is formed by k distinct visited points with a sample size of k.

The objective or cost function is essentially a mapping $f : \mathcal{X} \mapsto \mathcal{Y}$, which gives all possible problems $\mathcal{F} = \mathcal{Y}^{\mathcal{X}}$ in a finite (but very large) space. There are three assumptions: (1) the search domain is finite, (2) the search sequence generated during iterations is nonrevisiting (no points will be visited again), and (3) the finite sets are closed under permutation, which means any mathematical operations will not lead to a point outside the closed set.

The proof is by induction. First, when $k = 1$, we have $S_1 = \{S_1^x, S_1^y\}$, so the only possible value of S_1^y is $f(S_1^x)$, which means that $\delta(S_1^y, f(S_1^x))$, where δ is the unit

Dirac delta function. Consequently, we have

$$\sum_f \mu(S_1^y | f, k = 1, A) = \sum_f \delta(S_1^y, f(S_1^x)) = |\mathcal{Y}|^{|\mathcal{X}|-1}, \quad (7.54)$$

which is independent of algorithm A. Here, the notation $|\mathcal{Y}|$ denotes the size or cardinality of $\mathcal{Y}$.

Let us assume that the performance is independent of A for $k \geq 1$. That is, the performance metric $\sum_f \mu(d_k^y | f, k, A)$ is independent of A. Then, we try to figure out if it is true for $k + 1$. Since $S_{k+1} = S_k \cup \{x, f(x)\}$ with $S_{k+1}^x(k+1) = x$ and $S_{k+1}^y(k+1) = f(x)$, using the Bayesian theorem, we get

$$\mu(S_{k+1}^y | f, k+1, A) = \mu(S_{k+1}^y(k+1) | S_k, f, k+1, A)\mu(S_k^y | f, k+1, A), \quad (7.55)$$

which leads to

$$\sum_f \mu(S_{k+1}^y | f, k+1, A) = \sum_{f,x} \delta(S_{k+1}^k(k+1), f(x)) \times$$
$$\mu(x | d_k^y, f, k+1, A)\mu(d_k^y | f, k+1, A). \quad (7.56)$$

By using both $\mu(x | d_k, A) = \delta(x, A(S_k))$ and $\mu(S_k | f, k+1, A) = \mu(S_k | f, k, A)$, the above equation becomes

$$\sum_f \mu(S_{k+1}^y | f, k+1, A) = \frac{1}{|\mathcal{Y}|} \sum_f \mu(d_k^y | f, k, A), \quad (7.57)$$

which is also independent of A. This means that the performance averaged over all possible functions is independent of algorithm A itself. In other words, any algorithm (in terms of its performance) is as good as algorithm A (including a pure random search).

As we discussed earlier, even though the no-free-lunch theorem is valid mathematically, its impact on optimization is limited because we are not concerned with all possible functions.

It is worth pointing out that some recent studies suggest that the basic assumptions of the no-free-lunch theorem may not be valid for continuous domains, thus there may be free lunches for continuous optimization (Auger and Teytaud, 2010). In addition, later studies also indicate that free lunch may exist for coevolutionary systems (Wolpert and Macready, 2005) and multiobjective optimization (Corne and Knowles, 2003). For comprehensive reviews on the no-free-lunch theorem and their implications, readers can refer to more advanced literature (Joyce and Herrmann, 2018).

7.6.3 Real-world applications

As we discussed, there are some issues with benchmarking and constraints imposed by the no-free-lunch theorem, and the real acid test of any new algorithm should use real-world optimization problems as benchmarks. Such benchmarks have a diverse range of properties with complex constraints. Therefore, the truly useful benchmarks should be based on realistic design problems with a large number of decision variables, subject to a large number of real constraints.

The good news is that researchers start to use such benchmarks in some extensive studies. For example, in engineering designs, there are the pressure vessel design problem, design of tension and compression springs, speed reducer problems, and many structure design problems (Cagnina et al., 2008; Yang and Gandomi, 2012; Gandomi et al., 2011). There are also some multiobjective design benchmarks such as the biobjective disc brake design (Yang and Deb, 2013). In addition, the well-known traveling salesman problem and scheduling problems all have some realistic benchmarks (Ouaarab et al., 2014; Marichelvam et al., 2014; Osaba et al., 2016, 2017). Furthermore, there are extensive benchmark problems in data mining, image processing, and machine learning (LeCun et al., 2015). These problems need to be solved in a practically acceptable timescale. The results obtained by the algorithm are useful in practice.

Furthermore, even if algorithms may be effective, the handling of complex constraints is also an important part of any implementation. Testing using real-world cases will also test the ability to handle complex constraints at the same time. If an algorithm can be truly effective to solve a wide range of real-world optimization problems, it is no longer needed to do any benchmarking using any toy function problems at all.

7.7 Other approaches

There are other approaches for analyzing algorithms, including Bayesian statistics, variance analysis, and others. We now conclude this chapter by discussing them briefly.

7.7.1 Bayesian approach

Initialization is important for many algorithms, and a good algorithm should "forget" its initial state, which is a long-term property of proper Markov chains. However, when generating new moves, the information gained during the iterations can influence or guide the search moves. This falls into the framework of Bayesian statistics.

To a degree, we can consider the initialization as the uniform prior distribution. During the iteration, the current state and information can be considered as "known" or prior. When new information I_k (data, search move) is available, the posterior probability $P(S_k|I_k)$ should get more focused, given the prior probability $P(S_k)$. From the Bayesian theorem, we have

$$P(S_k|I_k) = \frac{P(I_k|S_k)P(S_k)}{P(I_k)}, \qquad (7.58)$$

where $P(I_k)$ is the marginal likelihood or a scaling constant. As iterations proceed, the posterior distribution should become narrower, based on new data, so that the probability of reaching the global optimality is gradually increased. However, it is not easy to design an algorithm to achieve this, and how to generate moves to maximize this posterior probability is still an open problem.

Another related approach is to use variance analysis. For a diverse population of n solutions, the population variance will change as iterations continue. For differential evolution, Zaharie obtained a formula for population variance var(P_t) at iteration k (Zaharie, 2009), which is given by

$$\text{var}(P_k) = \left[1 - 2F^2 p_m - \frac{p_m(2 - p_m)}{n}\right]^k \text{var}(P_0), \tag{7.59}$$

where P_0 is the initial population. In addition, F is a constant in [0,2] and p_m is the effective mutation probability. This can lead to a critical value for F if the variance has to decrease. However, this method for deriving the variance is only valid for linear systems such as differential evolution. It is not clear if it can be extended to study nonlinear systems such as the firefly algorithm.

7.7.2 Multidisciplinary approaches

We have discussed quite a few different approaches to analyze nature-inspired algorithms in this chapter. One approach usually can only provide insight from one perspective, and different approaches can see different sides of the same algorithm and problem. This indicates that a multidisciplinary approach is necessary to analyze algorithms in a comprehensive way so as to provide a fuller picture. This can involve mathematics, computer science, numerical simulations, and real-world benchmarking. It is hoped that a multidisciplinary framework can be used to study convergence, stability, robustness, parameter tuning, parameter control, sensitivity analysis, initialization, and parallelism in a unified manner. Ultimately, the insight gained from such analyses should be used to guide and potentially design more effective algorithms so as to practically solve a wide range of large-scale problems in optimization, data mining, machine learning, communication, and engineering applications. We can see in this book that nature-inspired computation is a very active area with many applications. There are also some open problems concerning nature-inspired optimization algorithms and their applications. We hope that the work presented in this book can inspire more active research in this area in the near future.

References

Auger, A., Teytaud, O., 2010. Continuous lunches are free plus the design of optimal optimization algorithms. Algorithmica 57 (2), 121–146.

Aytug, H., Bhattacharrya, S., Koehler, G.J., 1996. A Markov chain analysis of genetic algorithms with power of 2 cardinality alphabets. European Journal of Operational Research 96 (1), 195–201.

Cagnina, L.C., Esquivel, S.C., Coello Coello, A.C., 2008. Solving engineering optimization problems with the simple constrained particle swarm optimizer. Informatica 32 (2), 319–326.

Chen, S., Peng, G.H., Xing-Shi Yang, X.S., 2018. Global convergence analysis of the bat algorithm using a Markovian framework and dynamic system theory. Expert Systems with Applications 114 (1), 173–182.

Clerc, M., Kennedy, J., 2002. The particle swarm: explosion, stability, and convergence in a multidimensional complex space. IEEE Transactions on Evolutionary Computation 6 (1), 58–73.

Corne, D., Knowles, J., 2003. Some multiobjective optimizers are better than others. Evolutionary Computation 4 (2), 2506–2512.

Gandomi, A.H., Yang, X.S., Alavi, A.H., 2011. Mixed variable structural optimization using firefly algorithm. Computers & Structures 89 (23–24), 2325–2336.

Ghate, A., Smith, R., 2008. Adaptive search with stochastic acceptance probability for global optimization. Operations Research Letters 36 (3), 285–290.

Granas, A., Dugundji, J., 2003. Fixed Point Theory. Springer-Verlag, New York.

Greenhalgh, D., Marshal, S., 2000. Convergence criteria for genetic algorithm. SIAM Journal on Computing 30 (1), 269–282.

Grindstead, C.M., Snell, J.L., 1997. Introduction to Probability, second ed. American Mathematical Society, Providence, Rhode Island.

Gutjahr, W.J., 2010. Convergence analysis of metaheuristics. In: Matheuristics. In: Annals of Information Systems, vol. 10 (1), pp. 159–187.

Hasselblatt, B., Katok, A., 2003. A First Course in Dynamics. Cambridge University Press, Cambridge, UK.

He, X.S., Wang, F., Wang, Y., Yang, X.S., 2018. Global convergence analysis of cuckoo search using Markov theory. In: Yang, X.S. (Ed.), Nature-Inspired Algorithms and Applied Optimization, vol. 744. Springer Nature, Cham, Switzerland, pp. 53–67.

Jamil, M., Yang, X.S., 2013. A literature survey of benchmark functions for global optimisation problems. International Journal of Mathematical Modelling and Numerical Optimisation 4 (2), 150–194.

Joyce, T., Herrmann, J.M., 2018. A review of no free lunch theorems, and their implications for metaheuristic optimisation. In: Yang, X.S. (Ed.), Nature-Inspired Algorithms and Applied Optimization. Springer, Cham, Switzerland, pp. 27–52.

Kennedy, J., Eberhart, R., 1995. Particle swarm optimization. In: Proceedings of the IEEE International Conference on Neural Networks. IEEE, Piscataway, NJ, USA, pp. 1942–1948.

Khalil, H., 1996. Nonlinear Systems, third ed. Prentice Hall, New Jersey.

Khamsi, M.A., Kirk, W.A., 2001. An Introduction to Metric Space and Fixed Point Theory. John Wiley & Sons, New York.

LeCun, Y., Bengio, Y., Hinton, G.E., 2015. Deep learning. Nature 521 (7553), 436–444.

Marichelvam, M.K., Prabaharan, T., Yang, X.S., 2014. A discrete firefly algorithm for the multi-objective hybrid flowshop scheduling problems. IEEE Transactions on Evolutionary Computation 18 (2), 301–305.

Osaba, E., Yang, X.S., Diaz, F., Lopez-Garcia, P., Carballedo, R., 2016. An improved discrete bat algorithm for symmetric and asymmetric travelling salesman problems. Engineering Applications of Artificial Intelligence 48 (1), 59–71.

Osaba, E., Yang, X.S., Diaz, F., Onieva, E., Masegosa, A., Perallos, A., 2017. A discrete firefly algorithm to solve a rich vehicle routing problem modelling a newspaper distribution system with recycling policy. Soft Computing 21 (18), 5295–5308.

Ouaarab, A., Ahiod, B., Yang, X.S., 2014. Discrete cuckoo search algorithm for the travelling salesman problem. Neural Computing and Applications 24 (7–8), 1659–1669.

Solis, F., Wets, R., 1981. Minimization by random search techniques. Mathematics of Operations Research 6 (1), 19–30.

Suzuki, J., 1995. A Markov chain analysis on simple genetic algorithms. IEEE Transactions on Systems, Man and Cybernetics 25 (4), 655–659.

Wolpert, D.H., Macready, W.G., 1997. No free lunch theorems for optimization. IEEE Transactions on Evolutionary Computation 1 (1), 67–82.

Wolpert, D.H., Macready, W.G., 2005. Coevolutionary free lunches. IEEE Transactions on Evolutionary Computation 9 (6), 721–735.

Yang, X.S., 2010. A new metaheuristic bat-inspired algorithm. In: Cruz, C., González, J.R., Pelta, D.A., Terrazas, G. (Eds.), Nature Inspired Cooperative Strategies for Optimization (NISCO 2010). In: Studies in Computational Intelligence, vol. 284. Springer, Berlin, Germany, pp. 65–74.

Yang, X.S., Deb, S., 2013. Multiobjective cuckoo search for design optimization. Computers & Operations Research 40 (6), 1616–1624.

Yang, X.S., Gandomi, A.H., 2012. Bat algorithm: a novel approach for global engineering optimization. Engineering Computations 29 (5), 464–483.

Yang, X.S., He, X.S., 2019. Mathematical Foundations of Nature-Inspired Algorithms. Springer Briefs in Optimization. Springer, Cham, Switzerland.

Zaharie, D., 2009. Influence of crossover on the behavior of the differential evolution algorithm. Applied Soft Computing 9 (3), 1126–1138.

Applications

Fine-tuning restricted Boltzmann machines using quaternion-based flower pollination algorithm

8

Leandro Aparecido Passos, Gustavo Henrique de Rosa, Douglas Rodrigues, João Paulo Papa

São Paulo State University, Department of Computing, Bauru, Brazil

CONTENTS

8.1 Introduction

Machine learning is a reality present in diverse organizations and people's quotidian lives. Amongst the machine learning subtopics on the rise, deep learning has obtained much recognition due to its capacity in solving several problems. Areas such as computer vision, automatic speech recognition, and natural language processing have significantly benefited from deep learning techniques. One example can be seen as an unsupervised learning process, where the idea is to learn decent features that best represent a given problem and then classify it into different groups. Some of the most well-known techniques include convolutional neural networks (CNNs)

Nature-Inspired Computation and Swarm Intelligence. https://doi.org/10.1016/B978-0-12-819714-1.00019-1

(LeCun et al., 1998), restricted Boltzmann machines (RBMs) (Ackley et al., 1988; Hinton, 2012), deep belief networks (DBNs) (Hinton et al., 2006), and deep Boltzmann machines (DBMs) (Salakhutdinov and Hinton, 2009), to name a few.

Despite the success obtained by the models mentioned above, they still suffer drawbacks related to proper selection of their hyperparameters. Such selection carries with it a considerable burden, demanding the user a prior knowledge of the nature of the technique as well as the problem to be solved. Thus, an autonomous method capable of finding the hyperparameters that maximize the learning performance is extremely desirable.

Metaheuristic algorithms have become a viable alternative to solve optimization problems due to their simple implementation. Thus, algorithms based on natural or physical phenomena have been highlighted in problems of choosing suitable hyperparameters in deep learning techniques, since they can be modeled as an optimization task. Papa et al. (2015a) employed harmony search in the context of metaparameter fine-tuning concerning RBMs, discriminative RBMs (Papa et al., 2015b), and DBNs (Papa et al., 2015c). Rosa et al. (2016) addressed the firefly algorithm to fine-tune DBN metaparameters and the harmony search to fine-tune CNNs (Rosa et al., 2015). Finally, Passos et al. (2018) proposed a similar approach, comparing several metaheuristic techniques to the task of metaparameter fine-tuning concerning DBMs, infinity RBMs (Passos and Papa, 2017; Passos et al., 2019a), and RBM-based models in general (Passos and Papa, 2018).

Although such approaches performed very well in the task of RBM-based model hyperparameter fine-tuning, some of them still face issues related to the fitness function landscape, i.e., the convergence gets stuck at a local optimum. Hence, the idea of finding a method to drive such function landscapes more smooth sounds seductive. Supposedly, quaternion properties are capable of performing such a task.

Recently, metaheuristic algorithms combined with quaternion algebra emerged in the literature. The quaternionic algebra extends the complex numbers by representing a number using four components instead of two. Most nature-inspired algorithms are Euclidean-based, having their fitness landscape more complicated as the dimensional space increases. However, the main advantages of quaternion algebra concern improving the algorithm performance by smoothing the fitness landscape, which supposedly would help to avoid getting trapped in local optima. Another motivation behind these algebra concerns performing rotations with minimal computation. Fister et al. (2013) presented a modified version of the firefly algorithm based on quaternions, and also proposed a similar approach to the bat algorithm (Fister et al., 2015). Papa et al. (2016) introduced a harmony search approach based on quaternion algebra and later on applied it to fine-tune DBN hyperparameters (Papa et al., 2017). In the same context, Rosa et al. (2017) employed the quaternion algebra to the FPA. Fernandes and Papa (2017) proposed a quaternion-based ensemble pruning strategy using metaheuristic algorithms to minimize the optimum-path forest classifier error.

In this chapter we evaluate the QFPA (Rosa et al., 2017), a quaternion-based version of FPA (Yang, 2012; Rodrigues et al., 2018) in the task of RBM hyperparameter optimization in the context of binary image reconstruction. We present a discussion about the viability in using such an approach against seven naïve metaheuristic techniques, i.e., the backtracking search optimization algorithm (BSA) (Civicioglu, 2013), the bat algorithm (BA) (Yang and Gandomi, 2012), cuckoo search (CS) (Yang and Deb, 2009), the firefly algorithm (FA) (Yang, 2010), FPA (Yang, 2012), adaptive differential evolution (JADE) (Zhang and Sanderson, 2009), and particle swarm optimization (PSO) (Kennedy and Eberhart, 1995), as well as two quaternion-based techniques, i.e., QBA (Fister et al., 2015) and QBSA (Passos et al., 2019b), and a random search. The experimental section comprised three public datasets, as well as a statistical evaluation through the Wilcoxon signed-rank test.

The remainder of this chapter is organized as follows. Section 8.2 introduces the theoretical background concerning RBMs, quaternionic representation, FPA, and QFPA. Sections 8.3 and 8.4 present the methodology and the experimental results, respectively. Finally, Section 8.5 states conclusions and future works.

8.2 Theoretical background

This section presents a theoretical background concerning RBMs, the quaternionic hypercomplex representation, FPA, and the quaternion-based FPA.

8.2.1 Restricted Boltzmann machines

One can understand RBMs as energy-based stochastic neural networks, mainly constituted of two layers of neurons (visible and hidden), where an unsupervised algorithm conducts its learning procedure.

A naïve architecture of an RBM comprises a visible layer $\mathbf{v}$ with m units and a hidden layer $\mathbf{h}$ with n units. Moreover, a real-valued matrix $\mathbf{W}_{m \times n}$ models the weights between the visible and hidden neurons, where w_{ij} stands for the weight between the visible unit v_i and the hidden unit h_j. Fig. 8.1 depicts the vanilla RBM architecture.

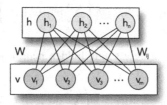

FIGURE 8.1

Vanilla RBM architecture.

Firstly, let us assume both $\mathbf{v}$ and $\mathbf{h}$ are binary-valued units. That is, $\mathbf{v} \in \{0, 1\}^m$ and $\mathbf{h} \in \{0, 1\}^n$. The energy function of an RBM is given by the following equation:

$$E(\mathbf{v}, \mathbf{h}) = -\sum_{i=1}^{m} a_i v_i - \sum_{j=1}^{n} b_j h_j - \sum_{i=1}^{m}\sum_{j=1}^{n} v_i h_j w_{ij}, \tag{8.1}$$

where $\mathbf{a}$ and $\mathbf{b}$ stand for the biases of visible and hidden units, respectively.

The probability of a joint configuration $(\mathbf{v}, \mathbf{h})$ is computed as follows:

$$P(\mathbf{v}, \mathbf{h}) = \frac{1}{Z} e^{-E(\mathbf{v}, \mathbf{h})}, \tag{8.2}$$

where Z stands for the so-called partition function, which is a normalization factor estimated over all possible configurations involving visible and hidden units. Furthermore, the marginal probability of a visible (input) vector is given by

$$P(\mathbf{v}) = \frac{1}{Z} \sum_{\mathbf{h}} e^{-E(\mathbf{v}, \mathbf{h})}. \tag{8.3}$$

One can see the RBM architecture as a bipartite graph, leading to mutually independent activations of both visible and hidden units. This assumption points to the following conditional probabilities:

$$P(\mathbf{v}|\mathbf{h}) = \prod_{i=1}^{m} P(v_i|\mathbf{h}), \tag{8.4}$$

where $P(\mathbf{v}|\mathbf{h})$ is the probability of the visible layer given the hidden states,

$$P(\mathbf{h}|\mathbf{v}) = \prod_{j=1}^{n} P(h_j|\mathbf{v}), \tag{8.5}$$

and $P(\mathbf{h}|\mathbf{v})$ is the probability of the hidden layer given the visible states.

Moreover, we can define the probability of activating a single visible neuron given the hidden states as follows:

$$P(v_i = 1|\mathbf{h}) = \phi\left(\sum_{j=1}^{n} w_{ij} h_j + a_i\right), \tag{8.6}$$

and the probability of activating a single hidden neuron given the visible states as follows:

$$P(h_j = 1|\mathbf{v}) = \phi\left(\sum_{i=1}^{m} w_{ij} v_i + b_j\right), \tag{8.7}$$

where $\phi(\cdot)$ stands for the logistic-sigmoid function.

Let $\theta = (W, a, b)$ be the set of parameters of an RBM, which is learned through a training algorithm that aims at maximizing the product of probabilities given all the available training data $\mathcal{V}$. The following equation describes this procedure:

$$\arg\max_{\Theta} \prod_{\mathbf{v} \in \mathcal{V}} P(\mathbf{v}). \tag{8.8}$$

Finally, it is possible to solve the aforementioned equation using the following derivatives over the matrix of weights W and biases a and b at iteration t:

$$\mathbf{W}^{t+1} = \mathbf{W}^t + \underbrace{\eta(P(\mathbf{h}|\mathbf{v})\mathbf{v}^T - P(\tilde{\mathbf{h}}|\tilde{\mathbf{v}})\tilde{\mathbf{v}}^T) + \Phi}_{=\Delta\mathbf{W}^t}, \tag{8.9}$$

$$\mathbf{a}^{t+1} = \mathbf{a}^t + \underbrace{\eta(\mathbf{v} - \tilde{\mathbf{v}}) + \alpha\Delta\mathbf{a}^{t-1}}_{=\Delta\mathbf{a}^t}, \tag{8.10}$$

$$\mathbf{b}^{t+1} = \mathbf{b}^t + \underbrace{\eta(P(\mathbf{h}|\mathbf{v}) - P(\tilde{\mathbf{h}}|\tilde{\mathbf{v}})) + \alpha\Delta\mathbf{b}^{t-1}}_{=\Delta\mathbf{b}^t}, \tag{8.11}$$

where η stands for the learning rate, α denotes the momentum, $\tilde{\mathbf{v}}$ stands for the reconstruction of the visible layer given $\mathbf{h}$, and $\tilde{\mathbf{h}}$ denotes an estimation of the hidden vector $\mathbf{h}$ given $\tilde{\mathbf{v}}$. Note that $P(\tilde{\mathbf{h}}|\tilde{\mathbf{v}})$ and $\tilde{\mathbf{v}}$ can be obtained by the contrastive divergence (CD) (Hinton, 2002) or the persistent CD (PCD) (Tieleman, 2008) algorithms,[1] which basically perform Gibbs sampling using the training data as the visible units. In a nutshell, Eqs. (8.9), (8.10), and (8.11) employ the well-known gradient descent as the optimization algorithm. The additional term Φ in Eq. (8.9) is used to control the values of matrix W during the convergence process, and it is described as follows:

$$\Phi = -\lambda\mathbf{W}^t + \alpha\Delta\mathbf{W}^{t-1}, \tag{8.12}$$

where λ stands for the weight decay.

8.2.2 Quaternionic hypercomplex representation

A quaternion q is composed of real and complex numbers, i.e., $q = x_0 + x_1 i + x_2 j + x_3 k$, where $x_0, x_1, x_2, x_3 \in \Re$ and i, j, k are imaginary numbers (also known as "fundamental quaternions units"). This assumption holds the following set of equations:

$$ij = k, \tag{8.13}$$

$$jk = i, \tag{8.14}$$

$$ki = j, \tag{8.15}$$

$$ji = -k, \tag{8.16}$$

---

[1] Note that $\tilde{\mathbf{v}}$ and $\tilde{\mathbf{h}}$ are obtained by a sampling step over $\mathbf{h}$ and $\tilde{\mathbf{v}}$, respectively.

$$kj = -i, \tag{8.17}$$

$$ik = -j, \tag{8.18}$$

and

$$i^2 = j^2 = k^2 = -1. \tag{8.19}$$

Essentially, a quaternion q is a four-dimensional space representation over the real numbers, i.e., $\Re^4$.

Given two arbitrary quaternions $q_1 = x_0 + x_1 i + x_2 j + x_3 k$ and $q_2 = y_0 + y_1 i + y_2 j + y_3 k$, the quaternion algebra defines a set of main operations (Eberly, 2002). The addition operation, for instance, can be defined by

$$\begin{aligned} q_1 + q_2 &= (x_0 + x_1 i + x_2 j + x_3 k) + (y_0 + y_1 i + y_2 j + y_3 k) \\ &= (x_0 + y_0) + (x_1 + y_1)i + (x_2 + y_2)j + (x_3 + y_3)k, \end{aligned} \tag{8.20}$$

while the subtraction is defined as follows:

$$\begin{aligned} q_1 - q_2 &= (x_0 + x_1 i + x_2 j + x_3 k) - (y_0 + y_1 i + y_2 j + y_3 k) \\ &= (x_0 - y_0) + (x_1 - y_1)i + (x_2 - y_2)j + (x_3 - y_3)k. \end{aligned} \tag{8.21}$$

Another important operation is the norm, which maps a given quaternion to a real-valued number as follows:

$$N(q_1) = N(x_0 + x_1 i + x_2 j + x_3 k) = \sqrt{x_0^2 + x_1^2 + x_2^2 + x_3^2}. \tag{8.22}$$

Finally, Fister et al. (2013, 2015) introduced two other operations, q_{rand} and q_{zero}. The former initializes a given quaternion with values drawn from a Gaussian distribution, and is defined as follows:

$$q_{rand}() = \{x_i = \mathcal{N}(0, 1) | i \in \{0, 1, 2, 3\}\}. \tag{8.23}$$

The latter function initializes a quaternion with zero values as follows:

$$q_{zero}() = \{x_i = 0 | i \in \{0, 1, 2, 3\}\}. \tag{8.24}$$

8.2.3 Flower pollination algorithm

FPA is a metaheuristic optimization technique, developed by Yang (2012), which tries to mimic the biological pollination process of flowering plants. Nevertheless, there are some ground rules that need to be established in order to oversee the algorithm.

1. Firstly, we consider biotic pollination (cross-pollination) as a process of global pollination, which can be carried out by pollinator agents, such as birds, insects, and others. Their movement obeys the well-known Lévy flights.

2. Secondly, we apply the abiotic pollination (self-pollination) as a local pollination method. One can see that this technique is used by blooming plants, where pollen is carried by diffusion and wind, and not by a pollinator agent.
3. Some pollinators, such as insects, can develop a flower constancy, substantially equivalent to a reproduction probability proportional to the similarity of the two flowers involved.
4. Finally, there is a switch probability $p \in [0, 1]$ which controls the interaction between local and global pollination. Moreover, this switch is slightly biased towards local pollination.

Considering the rules mentioned above, we can compose a set of equations to conduct the algorithm. For example, regarding the global pollination step, insects can often fly and move over a much longer range (Yang, 2012), thus carrying flower pollen gametes over a longer distance. Therefore, we can mathematically describe rules 1 and 3 as follows:

$$x_i^{(t+1)} = x_i^{(t)} + L(\lambda, s, \alpha)(\hat{g} - x_i^{(t)}) \qquad (8.25)$$

and

$$L(\lambda, s, \alpha) \sim \frac{\lambda \cdot \Gamma(\lambda) \cdot \sin(\lambda)}{\pi} \cdot \frac{\alpha}{s^{1+\lambda}}, \quad |s| \to \infty, \qquad (8.26)$$

where $x_i^t \in \Re^D$ stands for pollen i (solution vector) at iteration t, $\hat{g}$ denotes the current best solution among all solutions, $L(\cdot) \in \Re^D$ is the Lévy flight step size (pollination's strength), $\Gamma(\lambda)$ stands for the gamma function with index λ, α is a control parameter for the tail distribution ($\alpha = 1$), and s is the step size. According to Mantegna (1994), for large steps $s \gg s_0 > 0$, a linear transformation can be used to solve it, i.e.,

$$s = \frac{U}{|V|^{\lambda-1}}, \qquad (8.27)$$

where U and V are drawn from a Gaussian distribution with zero mean and a standard deviation σ_u and σ_v:

$$\sigma_u = \left[\frac{\Gamma(1+\lambda)}{\lambda\Gamma((1+\lambda)/2)} \frac{\sin(\frac{\pi\lambda}{2})}{2^{(\lambda-1)/2}} \right], \sigma_v = 1. \qquad (8.28)$$

As mentioned above, insects may fly over longer distances with various distance steps, making Lévy flights an exciting technique to mimic this characteristic. Therefore, we can represent rules 2 and 3 for local pollination as follows:

$$x_i^{(t+1)} = x_i^{(t)} + \epsilon(x_j^{(t)} - x_k^{(t)}), \qquad (8.29)$$

where x_j^t and x_k^t stand for the pollen from flowers j and k,[2] respectively, such that $j \neq k$. This behavior is responsible for mimicking the flower constancy in a limited

---

[2] Note that flowers j and k are from the same species.

Algorithm 1 Flower pollination algorithm.

---

input   : p, η, number of flowers m, dimensions D, and iterations T.
output   : Global best position $\widehat{g}$.
auxiliaries: Fitness functions $\mathbf{f} \in \Re^m$ and variables acc, $maxfit$, $globalfit$, and $maxindex$.

1 **for** *each flower i ($\forall i = 1, \ldots, m$)* **do**
2   **for** *each dimension j ($\forall j = 1, \ldots, D$)* **do**
3          $x_i^j \leftarrow U\{0, 1\}$;
4      $f_i \leftarrow -\infty$;
5 *$globalfit \leftarrow -\infty$;*
6 **for** *each iteration t ($t = 1, \ldots, T$)* **do**
7   **for** *each flower i ($\forall i = 1, \ldots, m$)* **do**
8   **for** *each dimension j ($\forall j = 1, \ldots, D$)* **do**
9              $rand \leftarrow U\{0, 1\}$;
            if *rand $< p$* **then**
10                 $x_i^j \leftarrow x_i^j + \eta \oplus \text{Lévy}(\lambda)$
11   **else**
12                 $x_i^j \leftarrow x_i^j + \epsilon(\hat{x}_i^j - x_i^k)$

13   Evaluate all solutions;
    for *each flower i ($\forall i = 1, \ldots, m$)* **do**
14   **if** *($acc > f_i$)* **then**
15             $f_i \leftarrow acc$;
            for *each dimension j ($\forall j = 1, \ldots, D$)* **do**
16                 $\hat{x}_i^j \leftarrow x_i^j(t)$;

17     $\left[maxfit, maxindex \right] \leftarrow max(\mathbf{f})$;
    if *($maxfit > globalfit$)* **then**
18         $globalfit \leftarrow maxfit$;
        for *each dimension j ($\forall j = 1, \ldots, d$)* **do**
19             $\widehat{g}^j \leftarrow x_{maxindex}^j(t)$;

---

neighborhood. Mathematically speaking, if $\mathbf{x}_j^t$ and $\mathbf{x}_k^t$ are species equivalent to or drawn from the same population and ϵ is drawn from a uniform distribution,[3] its formulation becomes a local random walk. A switch probability (rule 4) or proximity probability p is used to simulate the local and global flower pollination. For better

---

[3] A uniform distribution is bounded by the [0, 1] interval.

comprehension, one can refer to Algorithm 1, as it explains the whole procedure in more precise steps.

8.2.4 Quaternion-based flower pollination algorithm

QFPA maps the usual real-valued optimization problem into a hypercomplex space. Such conversion is presumed to fulfill a smoother fitness landscape, mitigating local optimal points and helping to accomplish better results.

Acknowledging the standard FPA, where each feasible solution is represented in an $\Re^n$ search space, the quaternion-based algorithm encodes each solution as a tensor in an $\Re^{4 \times n}$ space. In other words, each real-valued decision variable in FPA is encoded by a quaternion in QFPA.

Essentially, we can define the QFPA procedure as follows. First, solutions are initialized using a Gaussian distribution.[4] Although quaternions use their own space in the search process, we need to map them back to a real-valued number as the fitness function was designed to work with real numbers only. Such mapping is easily computed by using Eq. (8.22). Finally, the standard FPA is employed, but now in a hypercomplex space, meaning that quaternionic algebra takes over the standard one. In other words, QFPA searches for the quaternions whose norms minimize the fitness function, while standard FPA searches for real-valued numbers whose values minimize the fitness function. For the sake of explanation, Fig. 8.2 illustrates the encoding process of QFPA.

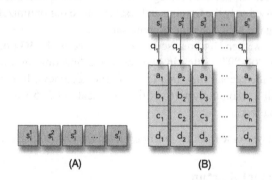

(A) (B)

FIGURE 8.2

Modeling a solution $s_i \in \mathcal{P}$: (A) vector (FPA), and (B) tensor (QFPA).

8.3 Methodology

In this section, we introduce the intended procedure for RBM parameter calibration, and we describe the employed datasets and the experimental setup.

---

[4] Each real part of the quaternions is set to a Gaussian-distributed value.

8.3.1 Modeling RBM parameter optimization

We propose to model the problem of choosing suitable parameters acknowledging RBMs in the task of binary image reconstruction. As mentioned in Section 8.2.1, the learning step has four parameters: the learning rate η, the weight decay λ, the momentum α, and the number of hidden units n. Hence, we have a four-dimensional search space with three real-valued variables, as well as the integer-valued number of hidden units. In a nutshell, the stated approach aims at electing the assortment of RBM parameters that minimizes the mean square error (MSE) of the reconstructed images over the training set. Subsequently, the selected set of parameters is applied to reconstruct the unseen images of the test set.

8.3.2 Datasets

We employed three datasets, as described below:

- MNIST dataset[5]: This is composed of images of handwritten digits. The original version contains a training set with $60,000$ images from digits "0"–"9," as well as a test set with $10,000$ images.[6] Due to the high computational burden for RBM parameter selection, we decided to use the original test set together with a reduced version of the training set.[7]
- CalTech 101 Silhouettes dataset[8]: This is based on the former Caltech 101 dataset, and it comprises silhouettes of images from 101 classes with a resolution of 28×28. We have used only the training and test sets since our optimization model aims at minimizing the MSE over the training set.
- Semeion Handwritten Digit dataset[9]: This is formed by $1,593$ images from handwritten digits "0"–"9" written in two ways: the first time in a normal way (accurately) and the second time in a fast way (no accuracy). In the end, they were stretched with a resolution of 16×16 in grayscale of 256 values and then each pixel was binarized.

Fig. 8.3 displays some training examples from the above datasets.

8.3.3 Experimental setup

In order to provide a statistical analysis through Wilcoxon signed-rank test (Wilcoxon, 1945), we conducted a two-fold cross-validation with 20 runs. Additionally, for every metaheuristic, five agents (particles) were used over 50 convergence iterations.

---

[5] http://yann.lecun.com/exdb/mnist/.

[6] The images are originally available in grayscale with a resolution of 28×28.

[7] The original training set was reduced to 2% of its former size, which corresponds to $1,200$ images.

[8] https://people.cs.umass.edu/~marlin/data.shtml.

[9] https://archive.ics.uci.edu/ml/datasets/Semeion+Handwritten+Digit.

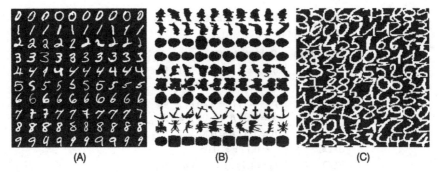

FIGURE 8.3

Some training examples from (A) MNIST, (B) CalTech 101 Silhouettes, and (C) Semeion datasets.

To provide a thorough comparison between metaheuristics, we have chosen different techniques, ranging from swarm-based to evolutionary-inspired ones, in the context of RBM fine-tuning:

- adaptive differential evolution with optional external archive (JADE) (Zhang and Sanderson, 2009);
- BSA (Civicioglu, 2013);
- BA (Yang and Gandomi, 2012);
- CS (Yang and Deb, 2010; Rodrigues et al., 2013);
- FA (Yang, 2010);
- FPA (Yang et al., 2014; Rodrigues et al., 2016);
- PSO (Kennedy and Eberhart, 2001).

Note that for the standard FPA, we will also present its quaternion-based version, being preceded by a Q prefix (QFPA). The same approach was conducted considering BA and BSA, thus called QBA (Fister et al., 2015) and QBSA (Passos et al., 2019b), respectively. Table 8.1 exhibits the parameter configuration for every metaheuristic technique.[10] We omitted the quaternion-based algorithms from Table 8.1, as their parameters are the same as for the original versions.

Ultimately, we have set each RBM parameter according to the following ranges: $n \in [5, 100]$, $\eta \in [0.1, 0.9]$, $\lambda \in [0.1, 0.9]$, and $\alpha \in [0.00001, 0.01]$. Hence, this indicates we have used such ranges to initialize the optimization techniques. We also have employed $T = 10$ as the number of epochs for RBM learning weights procedure with minibatches of size 20. In order to present a more finicky experimental validation, all RBMs were trained with CD (Hinton, 2002) and persistent CD (PCD) (Tieleman, 2008).

---

[10] Note that these values were empirically chosen according to their authors' definitions.

Table 8.1 Metaheuristic algorithms' parameter configuration.

Algorithm	Parameters
BA	$f_{min} = 0 \mid f_{max} = 100 \mid A = 1.5 \mid r = 0.5$
BSA	$mix\_rate = 1.0,\ F = 3$
CS	$\beta = 1.5 \mid p = 0.25 \mid \alpha = 0.8$
FA	$\alpha = 0.2 \mid \beta = 1.0 \mid \gamma = 1.0$
FPA	$\beta = 1.5 \mid p = 0.8$
JADE	$c = 0.1 \mid g = 0.05$
PSO	$c_1 = 1.7 \mid c_2 = 1.7 \mid w = 0.7$

8.4 Experimental results

This section introduces the experimental results regarding QFPA applied to the task of RBM hyperparameter fine-tuning. Tables 8.2, 8.3, and 8.4 present the average MSE and the standard deviation regarding MNIST, CalTech 101 Silhouettes, and Semeion datasets, respectively. Bold values stand for the best results according to the Wilcoxon signed-rank test.

With respect to the MNIST dataset, one can observe from Table 8.2 that the best results concerning the CD algorithm were obtained exclusively with FPA, while BA, BSA, JADE, and the three quaternion-based algorithms, i.e., QBA, QBSA, and QFPA, achieved the best results when trained with the PCD algorithm. Figs. 8.4 and 8.5 depict the optimization convergence over the MNIST dataset considering CD and PCD algorithms, respectively. Even though QFPA obtained similar results to the other quaternion-based techniques, Fig. 8.5A depicts that the QFPA convergence was faster during the training process. Additionally, QFPA training process is faster than QBSA. Moreover, both demanded, on average, half of QBA's time, as shown in Table 8.5.

Table 8.3 highlights that BSA obtained the best overall results using the CD algorithm on the CalTech 101 Silhouettes dataset. Considering the quaternion-based approaches, QFPA and QBSA obtained the best results, with QFPA being approximately 5% faster than QBSA, according to Table 8.5.

Figs. 8.6A and 8.7A depict an interesting observation, where a random search optimization performs a faster convergence than every other technique. A feasible explanation may lie in the fitness function landscape, which may require a broader range for the learning rate parameter. In a nutshell, smaller values of the learning rate would possibly allow the algorithms to explore regions arbitrarily found by a random search.

Results for the Semeion Handwritten Digit dataset are presented in Table 8.4, where BSA achieved the most accurate results. Concerning the quaternion-based techniques, QBA obtained the best results when trained with the CD algorithm, at a time consuming cost of 35% more than QFPA, as shown in Table 8.5. Regarding the convergence steps, Figs. 8.8 and 8.9 depict the optimization convergence over the Semeion Handwritten Digit dataset considering CD and PCD algorithms, respec-

Table 8.2 Average MSE values considering the MNIST dataset.

Algorithm	Statistics	BA	BSA	CS	FA	FPA	JADE	PSO	QBA	QBSA	QFPA	RANDOM
CD	Mean	0.08762	0.08763	0.08763	0.08764	**0.08759**	0.08761	0.08763	0.08763	0.08763	0.08763	0.08763
	SD	0.00007	0.00005	0.00005	0.00007	0.00009	0.00007	0.00006	0.00006	0.00006	0.00007	0.00006
PCD	Mean	**0.08763**	**0.08762**	0.08764	0.08765	0.08761	**0.08763**	0.08763	**0.08763**	**0.08764**	**0.08764**	0.08764
	SD	0.00006	0.00007	0.00007	0.00006	0.00007	0.00006	0.00007	0.00006	0.00006	0.00006	0.00005

Table 8.3 Average MSE values considering the CalTech 101 Silhouettes dataset.

Algorithm	Statistics	BA	BSA	CS	FA	FPA	JADE	PSO	QBA	QBSA	QFPA	RANDOM
CD	Mean	0.15923	**0.15600**	0.15923	0.16003	0.15607	0.15608	0.15612	0.16017	0.15769	0.15843	0.15676
	SD	0.00167	0.00154	0.00171	0.00156	0.00230	0.00184	0.00189	0.00136	0.00237	0.00249	0.00162
PCD	Mean	0.16009	0.15775	0.15993	0.15957	0.15810	0.15790	0.15862	0.16039	0.15931	0.15994	0.15846
	SD	0.00076	0.00151	0.00103	0.00118	0.00192	0.00135	0.00170	0.00065	0.00215	0.00133	0.00122

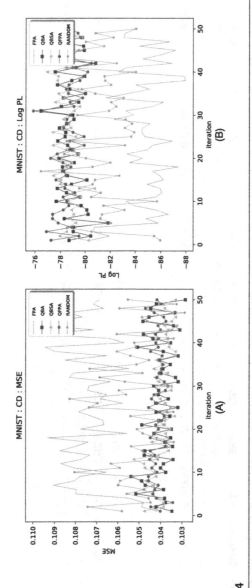

FIGURE 8.4

(A) Mean square error and (B) log pseudolikelihood concerning the MNIST dataset training step using CD.

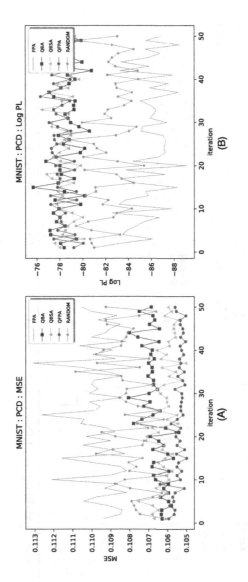

FIGURE 8.5

(A) Mean square error and (B) log pseudolikelihood concerning the MNIST dataset training step using PCD.

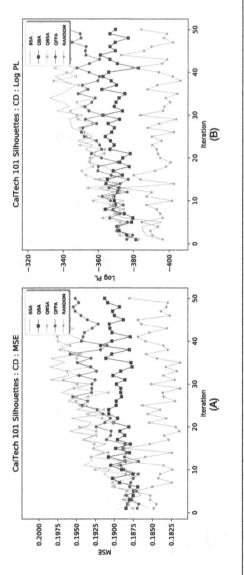

FIGURE 8.6

(A) Mean square error and (B) log pseudolikelihood concerning the CalTech 101 Silhouettes dataset training step using CD.

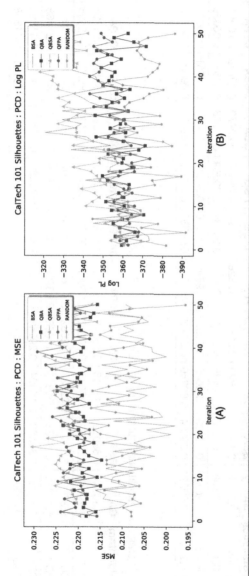

FIGURE 8.7

(A) Mean square error and (B) log pseudolikelihood concerning the CalTech 101 Silhouettes dataset training step using PCD.

Table 8.4 Average MSE values considering the Semeion Handwritten Digit dataset.

Algorithm	Statistics	BA	BSA	CS	FA	FPA	JADE	PSO	QBA	QBSA	QFPA	RANDOM
CD	Mean	0.20743	0.19571	0.20529	0.20638	**0.19526**	0.19893	0.19850	0.20772	0.20716	0.20951	0.19711
	SD	0.00449	0.00365	0.00495	0.00492	0.00456	0.00789	0.00669	0.00503	0.00583	0.00059	0.00313
PCD	Mean	0.20763	0.20003	0.20648	0.20895	0.19895	0.20166	0.20310	0.20691	0.20961	0.20912	0.20362
	SD	0.00274	0.00254	0.00356	0.00209	0.00223	0.00532	0.00393	0.00476	0.00036	0.00232	0.00184

Table 8.5 Average computational burden (in hours).

Dataset	Algorithm	BA	BSA	CS	FA	FPA	JADE	PSO	QBA	QBSA	QFPA	RANDOM
MNIST	CD	1.30341	0.58982	0.18108	0.45145	0.58773	0.59969	1.28356	1.79883	0.83189	0.81733	0.14433
	PCD	1.29018	0.59709	0.27049	0.89217	0.66907	0.97534	1.36263	1.60102	0.85778	0.82695	0.16496
Caltech	CD	5.37060	3.75112	0.92937	2.02582	3.47451	4.94216	5.26207	5.69820	3.74185	3.60963	0.63309
	PCD	5.46213	3.13693	0.60878	3.15928	3.63909	2.58896	5.06949	5.68607	3.84492	3.61468	0.69592
Semeion	CD	0.59225	0.31366	0.15404	0.29369	0.40037	0.32241	0.75088	0.63015	0.41325	0.41810	0.08471
	PCD	0.59998	0.32504	0.10765	0.44271	0.39609	0.13222	0.69011	0.67257	0.40143	0.44398	0.08063

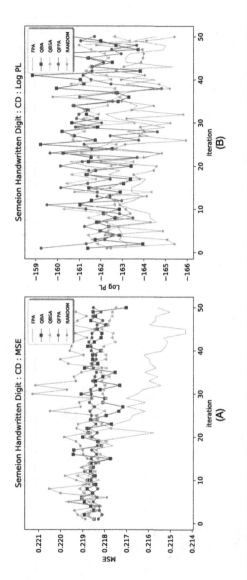

FIGURE 8.8

(A) Mean square error and (B) log pseudolikelihood concerning the Semeion Handwritten Digit dataset training step using CD.

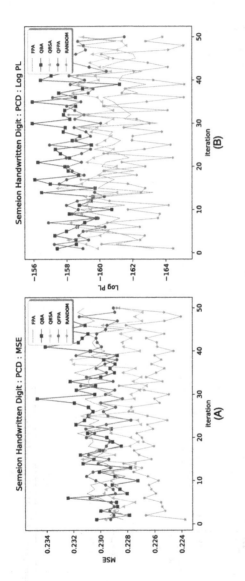

FIGURE 8.9

(A) Mean square error and (B) log pseudolikelihood concerning the Semeion Handwritten Digit dataset training step using PCD.

tively. As expected, the FPA training process converged faster than other techniques, as depicted in Fig. 8.8A.

The computational burden consumed by each technique is presented in Table 8.5. Although CS was the fastest technique, it did not achieve the best results in any of the datasets. Regarding QFPA, one can observe it was the fastest quaternion-based technique in most of the cases. Moreover, QFPA execution is even faster than the naïve version of BA and PSO considering all datasets, as well as BSA and JADE, trained with the CD algorithm, concerning Caltech 101 Silhouettes dataset. Note that the time is presented in hours.

Results obtained in this section lead to three main assumptions: (i) FPA is assumed to be the most suitable technique to the task of RBM hyperparameter fine-tuning, achieving the best results in two out of three datasets with satisfactory time burden; (ii) although QFPA did not achieve the best results, it is still a valid option for the aforementioned task, since the results are compatible with the other quaternion-based techniques, even better and faster in some cases; and (iii) a more in-depth analysis, evaluating wider ranges in the search space of the metaparameters would probably sway the results.

8.5 **Conclusion**

This chapter introduced QFPA for the task of RBM hyperparameter fine-tuning. Experimental results conducted over three well-known public datasets, i.e., MNIST, CalTech 101 Silhouettes, and Semeion Handwritten Digit, demonstrated that despite QFPA did not achieve the best results, it is still competitive with the other quaternion-based techniques.

Regarding future works, we intend to perform similar experiments, evaluating both a wider range concerning the RBM metaparameters search space, i.e., learning rate, weight decay, momentum, and the number of hidden units, and the performance of the model for RBM classification tasks.

References

Ackley, D., Hinton, G., Sejnowski, T.J., 1988. A learning algorithm for Boltzmann machines. In: Waltz, D., Feldman, J. (Eds.), Connectionist Models and Their Implications: Readings from Cognitive Science. Ablex Publishing Corp., Norwood, NJ, USA, pp. 285–307.

Civicioglu, P., 2013. Backtracking search optimization algorithm for numerical optimization problems. Applied Mathematics and Computation 219 (15), 8121–8144.

Eberly, D., 2002. Quaternion Algebra and Calculus, vol. 26. Magic Software, Inc..

Fernandes, S.E.N., Papa, J.P., 2017. Pruning optimum-path forest ensembles using quaternion-based optimization.

Fister, I., Yang, X.S., Brest, J., Fister, I., 2013. Modified firefly algorithm using quaternion representation. Expert Systems with Applications 40 (18), 7220–7230.

Fister, I., Brest, J., Fister, I., Yang, X.S., 2015. Modified bat algorithm with quaternion representation. In: IEEE Congress on Evolutionary Computation, pp. 491–498.

Hinton, G.E., 2002. Training products of experts by minimizing contrastive divergence. Neural Computation (ISSN 0899-7667) 14 (8), 1771–1800. Cambridge, MA, USA.

Hinton, G.E., 2012. A practical guide to training restricted Boltzmann machines. In: Montavon, G., Orr, G., Müller, K.R. (Eds.), Neural Networks: Tricks of the Trade. In: Lecture Notes in Computer Science, vol. 7700. Springer, Berlin, Heidelberg, pp. 599–619.

Hinton, G.E., Osindero, S., Teh, Y.W., 2006. A fast learning algorithm for deep belief nets. Neural Computation (ISSN 0899-7667) 18 (7), 1527–1554. Cambridge, MA, USA.

Kennedy, J., Eberhart, R., 1995. Particle Swarm Optimization. IEEE, Washington, DC, pp. 1942–1948.

Kennedy, J., Eberhart, R.C., 2001. Swarm Intelligence. Morgan Kaufmann Publishers Inc., San Francisco, USA.

LeCun, Y., Bottou, L., Bengio, Y., Haffner, P., 1998. Gradient-based learning applied to document recognition. Proceedings of the IEEE 86 (11), 2278–2324.

Mantegna, R.N., 1994. Fast, accurate algorithm for numerical simulation of Lévy stable stochastic processes. Physical Review E 49, 4677–4683.

Papa, J.P., Rosa, G.H., Costa, K.A.P., Marana, A.N., Scheirer, W., Cox, D.D., 2015a. On the model selection of Bernoulli restricted Boltzmann machines through harmony search. In: Proceedings of the Genetic and Evolutionary Computation Conference. ACM, New York, NY, USA, pp. 1449–1450.

Papa, J.P., Rosa, G.H., Marana, A.N., Scheirer, W., Cox, D.D., 2015b. Model selection for discriminative restricted Boltzmann machines through meta-heuristic techniques. Journal of Computational Science (ISSN 1877-7503) 9, 14–18.

Papa, J.P., Scheirer, W., Cox, D.D., 2015c. Fine-tuning deep belief networks using harmony search. Applied Soft Computing (ISSN 1568-4946).

Papa, J.P., Pereira, D., Baldassin, A., Yang, X., 2016. On the harmony search using quaternions. In: Artificial Neural Networks in Pattern Recognition. Springer International Publishing, pp. 126–137.

Papa, J.P., Rosa, G.H., Pereira, D.R., Yang, X., 2017. Quaternion-based deep belief networks fine-tuning. Applied Soft Computing 60, 328–335.

Passos, L.A., Papa, J.P., 2017. Fine-tuning infinity restricted Boltzmann machines. In: 2017 30th SIBGRAPI Conference on Graphics, Patterns and Images (SIBGRAPI), pp. 63–70.

Passos, L.A., Papa, J.P., 2018. On the training algorithms for restricted Boltzmann machine-based models.

Passos, L.A., Rodrigues, D.R., Papa, J.P., 2018. Fine tuning deep Boltzmann machines through meta-heuristic approaches. In: 2018 IEEE 12th International Symposium on Applied Computational Intelligence and Informatics (SACI), 000419.

Passos, L.A., de Souza Jr, L.A., Mendel, R., Ebigbo, A., Probst, A., Messmann, H., Palm, C., Papa, J.P., 2019a. Barrett's esophagus analysis using infinity restricted Boltzmann machines. Journal of Visual Communication and Image Representation.

Passos, L.A., Rodrigues, D., Papa, J.a.P., 2019b. Quaternion-based backtracking search optimization algorithm. In: 2019 IEEE Congress on Evolutionary Computation (CEC), pp. 1–8.

Rodrigues, D., Pereira, L.A.M., Almeida, T.N.S., Papa, J.P., Souza, A.N., Ramos, C.O., Yang, X.S., 2013. BCS: a binary cuckoo search algorithm for feature selection. In: IEEE International Symposium on Circuits and Systems, pp. 465–468.

Rodrigues, D., Silva, G.F.A., Papa, J.P., Marana, A.N., Yang, X.S., 2016. EEG-based person identification through binary flower pollination algorithm. Expert Systems with Applications 62, 81–90.

Rodrigues, D., Rosa, Gustavo, Passos, L.A., Papa, J.P., 2018. Adaptive improved flower pollination algorithm for global optimization. In: Nature-Inspired Computation in Data Mining and Machine Learning. Elsevier.

Rosa, G.H., Papa, J.P., Marana, A.N., Scheirer, W., Cox, D.D., 2015. Fine-tuning convolutional neural networks using harmony search. In: Pardo, A., Kittler, J. (Eds.), Progress in Pattern Recognition, Image Analysis, Computer Vision, and Applications. In: Lecture Notes in Computer Science, vol. 9423. Springer International Publishing, pp. 683–690.

Rosa, G., Papa, J.P., Costa, K., Passos, L.A., Pereira, C., Yang, X.S., 2016. Learning parameters in deep belief networks through firefly algorithm. In: IAPR Workshop on Artificial Neural Networks in Pattern Recognition, pp. 138–149.

Rosa, G.H., Afonso, L.C.S., Baldassin, A., Papa, J.P., Yang, X., 2017. Quaternionic flower pollination algorithm. In: Computer Analysis of Images and Patterns. Springer International Publishing, pp. 47–58.

Salakhutdinov, R., Hinton, G., 2009. Deep Boltzmann machines. In: Artificial Intelligence and Statistics, pp. 448–455.

Tieleman, T., 2008. Training restricted Boltzmann machines using approximations to the likelihood gradient. In: Proceedings of the 25th International Conference on Machine Learning. ICML '08. ACM, New York, NY, USA, pp. 1064–1071.

Wilcoxon, F., 1945. Individual comparisons by ranking methods. Biometrics Bulletin 1 (6), 80–83.

Yang, X.S., 2010. Firefly algorithm, stochastic test functions and design optimisation. International Journal Bio-Inspired Computing 2 (2), 78–84.

Yang, X.S., 2012. Flower pollination algorithm for global optimization. In: Proceedings of the 11th International Conference on Unconventional Computation and Natural Computation. UCNC'12. Springer-Verlag, Berlin, Heidelberg, pp. 240–249.

Yang, X.S., Deb, S., 2009. Cuckoo search via Lévy flights. In: Proc. of World Congress on Nature & Biologically Inspired Computing. NaBIC 2009, 2009, pp. 210–214.

Yang, X.S., Deb, S., 2010. Engineering optimisation by cuckoo search. International Journal of Mathematical Modelling and Numerical Optimisation 1, 330–343.

Yang, X.S., Gandomi, A.H., 2012. Bat algorithm: a novel approach for global engineering optimization. Engineering Computations 29 (5), 464–483.

Yang, S.S., Karamanoglu, M., He, X., 2014. Flower pollination algorithm: a novel approach for multiobjective optimization. Engineering Optimization 46 (9), 1222–1237.

Zhang, J., Sanderson, A.C., 2009. Jade: adaptive differential evolution with optional external archive. IEEE Transactions on Evolutionary Computation 13 (5), 945–958.

Traveling salesman problem: a perspective review of recent research and new results with bio-inspired metaheuristics

9

Eneko Osaba[a], **Xin-She Yang**[b], **Javier Del Ser**[a,c]

[a]*TECNALIA, Basque Research and Technology Alliance (BRTA), Derio, Spain*
[b]*Middlesex University London, School of Science and Technology, London, United Kingdom*
[c]*University of the Basque Country (UPV/EHU), Bilbao, Spain*

CONTENTS

Nature-Inspired Computation and Swarm Intelligence. https://doi.org/10.1016/B978-0-12-819714-1.00020-8

9.1 Introduction

In the current operations research and optimization communities, routing problems are one of the most studied paradigms. Two principal reasons that make this topic a paramount one in the field are (i) their inherent practical nature and social interest, which allow routing problems to be applicable not only in leisure or tourism scenarios, but also in situations related to logistics and business, and (ii) their complexity, making such problems very difficult to optimally solve even for medium-sized datasets. Arguably, the modeling and formulation of this kind of problems draws inspiration from real-world logistic and transportation situations, directly implying a social and/or business benefit in the case of its proper solution. Furthermore, the efficient addressing of these problems usually supposes a tough challenge for the scientific community because of their NP-hard nature. This fact leads related researchers to the adoption of diverse artificial intelligence solvers, aiming at solving them in a computationally affordable fashion. The problem gets even more involved when bearing in mind the rich literature with regard to different formulations of variants. Among this wide variety of problems, the traveling salesman problem (TSP) (Lawler et al., 1985) and the vehicle routing problem (VRP) (Christofides, 1976) are widely recognized as the most studied ones. This study is focused in the first of these problems, the TSP.

In line with this, many optimization approaches have been proposed along the years for dealing with the VRP. The three most studied and well-established schemes are exact methods (Laporte, 1992a,b), heuristics (Vaghela et al., 2018; Pozna et al., 2010), and metaheuristics. The present research is focused on the latter ones, which have demonstrated a remarkable efficiency for properly solving routing problems, especially in the last decade. The most recognized methods in this category could be simulated annealing (SA) (Kirkpatrick et al., 1983) and tabu search (TS) (Glover, 1989) as local search-based solvers, and ant colony optimization (ACO) (Bell and McMullen, 2004; Yu et al., 2009), particle swarm optimization (PSO) (Kennedy et al., 1995; Tang et al., 2015), and the genetic algorithm (GA) (Goldberg, 1989; De Jong, 1975) as population-based methods. In addition to these classical and recognized approaches, the design and implementation of new metaheuristics is a hot topic in the related operations research and optimization communities. As a result of this scientific trend, lots of successful solvers have been proposed in recent years, such as the bat algorithm (BA) (Yang, 2010), the firefly algorithm (FA) (Yang, 2009), the gravitational search algorithm (Rashedi et al., 2009; David et al., 2013), and fireworks algorithm optimization (FAO) (Tan and Zhu, 2010), among many others.

The main contribution of this work can be divided into three different points. First, we devote a comprehensive section for outlining the research carried out in recent years around the TSP problem, focusing our effort on its solving through the use of metaheuristic algorithms. Secondly, we take a step further over the state of the art in the elaboration of a new research direction: the hybridization of novelty search (NS) mechanisms and bio-inspired computation algorithms for solving the TSP. NS (Lehman and Stanley, 2008) was proposed in 2008 as a way to enhance the exploratory ability of population-based algorithmic solvers. After showing its great performance applied to several optimization problems, we hypothesize its promising performance also for the TSP. To this end, we have developed different versions of well-known swarm intelligence methods, namely, PSO, FA, and BA, and we evaluate in this chapter the performance of these metaheuristics embedding the NS mechanism on their basic scheme. We introduce through the chapter the descriptions on how these methods have been modeled for tackling the problem at hand and how the NS has been adapted for this discrete scenario. In order to assess the performance of each implemented solver, outcomes obtained in 15 instances are compared and discussed. Finally, an important additional contribution is our personal envisioned status of this field, which we present in the form of challenges and open opportunities that should be addressed in the near future.

The rest of this chapter is structured as follows. In Section 9.2 the TSP and some of its most important variants are described and mathematically formulated. Section 9.3 elaborates on the first contribution of the chapter by analyzing the recent research done around the TSP. In Section 9.4, the concepts behind NS are introduced, placing emphasis on how we have hybridized metaheuristic solvers with this mechanism. Considered heuristic solvers and their implementation details are described in Section 9.5. The experimental setup is detailed in Section 9.6, along with a discussion on the obtained results. Research opportunities for the area are highlighted in Section 9.7. Finally, Section 9.8 concludes the chapter with a general outlook for the wide audience.

9.2 **Problem statement**

In this chapter, our experimentation with NS and the chosen bio-inspired optimization methods is done using the basic version of the TSP. As can be read in many scientific works, the canonical TSP can be represented as a complete graph $\mathcal{G} \doteq (\mathcal{V}, \mathcal{A})$, where $\mathcal{V} \doteq \{v_1, v_2, \ldots, v_N\}$ illustrates the vertex group that represents the nodes of the graph, and $\mathcal{A} \doteq \{(v_i, v_j) : v_i, v_j \in \mathcal{V} \times \mathcal{V}, i, j \in \{1, \ldots, N\} \times \{1, \ldots, N\}, i \neq j\}$ is the group of edges linking every pair of nodes in $\mathcal{V}$. Moreover, each edge (v_i, v_j) has an associated cost $c_{ij} \in \mathbb{R}^+$, denoting the traveling weight of this arc. Because of the symmetric nature of the basic TSP, it is ensured that $c_{ij} = c_{ji}$, meaning that the cost of going from one v_i to another v_j is equal to the reverse trip (v_j, v_i).

Thus, the principal optimization objective of the TSP pivots on the discovery of a route that visits each node once and only once (i.e., a Hamiltonian cycle in graph G), minimizing the total cost of the whole route. This genetic problem can be mathematically formulated as

$$\underset{\mathbf{X}}{\text{minimize}} \quad f(\mathbf{X}) = \sum_{\substack{i=1 \\ i \neq j}}^{N} \sum_{j=1}^{N} c_{ij} x_{ij} \qquad (9.1a)$$

$$\text{subject to} \quad \sum_{\substack{j=1 \\ i \neq j}}^{N} x_{ij} = 1, \quad \forall j \in \{1, \dots, N\}, \tag{9.1b}$$

$$\sum_{\substack{i=1 \\ i \neq j}}^{N} x_{ij} = 1, \quad \forall i \in \{1, \dots, N\}, \tag{9.1c}$$

$$\sum_{\substack{i \in S \\ j \in S \\ i \neq j}} x_{ij} \geq 1, \quad \forall S \subset \mathcal{V}, \tag{9.1d}$$

where $\mathbf{X} \doteq [x_{ij}]$ is an $N \times N$ binary matrix whose entry $x_{ij} \in \{0, 1\}$ takes value 1 if edge (i, j) is used in the solution. Furthermore, the objective function is represented in Eq. (9.1a) as the sum of costs associated to all the edges in the solution. Moreover, Eqs. (9.1b) and (9.1c) depict that each vertex must be visited once and only once. Lastly, (9.1d) guarantees the absence of subtours and forces that any subset of nodes S has to be abandoned at least one time. This restriction is needed to prevent the existence of subtours on the whole route.

Apart from this canonical formulation of the TSP, many different variants have been modeled over the years, aiming at adapting and addressing different characteristics present in the logistics and transportation world. We list here some of the most famous advanced variants of the TSP:

- *The asymmetric TSP* (ATSP) (Asadpour et al., 2017; Svensson, 2018): The central characteristic of the TSP is that, although there may be arcs where $c_{ij} = c_{ji}$, in general $c_{ij} \neq c_{ji}$.
- *The multiple TSP* (M-TSP) (Kitjacharoenchai et al., 2019; Rostami et al., 2015): In the M-TSP, a set of m exact salesmen are available, which should visit a set of n cities starting and ending at the same city.
- *The TSP with time windows* (TSPTW) (Roberti and Wen, 2016; Fachini and Armentano, 2018): In this variant, the traveling salesman should visit each vertex respecting a time window fixed by each separated node.
- *The time-dependent TSP* (TDTSP) (Arigliano et al., 2019; Furini et al., 2016): The basic idea behind the TDTSP is that the cost of traveling between two different nodes is time-dependent. In other words, travel times could significantly change over the day, for example, during *peak* and *off-peak* hours.
- *The generalized TSP* (GTSP) (Smith and Imeson, 2017; Helsgaun, 2015): In the GTSP, the set of nodes is partitioned into different clusters. The main goal of this formulation is to find a minimum cost tour passing through exactly one node from each cluster.

Other interesting research activity can be detected around the rich TSP (R-TSP), also known as multiattribute TSP (Caceres-Cruz et al., 2015). This type of problems are specific cases of the TSP with complex formulations and multiple restrictions. The principal characteristic of an R-TSP is its complex formulation, which is composed by multiple constraints. This feature directly leads to an increased complexity of resolution, which entails to a major scientific challenge at the same time. These problems are especially important in the current community because they model many real-world problems. Accordingly, the efficient solution of the R-TSP can be useful in many valuable real-world applications. Some remarkable examples can be found in Lahyani et al. (2017), Osaba et al. (2015), or Maity et al. (2019).

As can be seen, the number of TSP variants proposed in the literature is overwhelming, making the listing of all interesting and valuable formulation in this section infeasible. For this reason, we have outlined some of the most commonly used ones, with the intention of settling the idea that there is a vibrant scientific activity behind this problem.

9.3 **Recent advances in the traveling salesman problem**

Since its formulation, the TSP has become one of the most employed benchmarking problems in performance analysis of discrete optimization algorithms. A plethora of methods have been applied to the TSP and its variants in recent decades. We can highlight classical methods such as GA (Grefenstette et al., 1985; Larrañaga et al., 1999), TS (Fiechter, 1994; Knox, 1994; Gendreau et al., 1998), or SA (Malek et al., 1989; Aarts et al., 1988). Besides these classical algorithms, more recent and effective approaches have been extensively used for solving the TSP, such as ACO (Dorigo and Gambardella, 1997; Jun-man and Yi, 2012), PSO (Clerc, 2004; Shi et al., 2007), or the variable neighborhood search (Carrabs et al., 2007; Burke et al., 2001). In addition to these well-known methods, the TSP and its multiple derivations have been the focus of many benchmarking studies for measuring the quality of many recently proposed nature-inspired methods. Some examples are FA (Kumbharana and Pandey, 2013), CS (Ouaarab et al., 2014), ICA (Yousefikhoshbakht and Sedighpour, 2013), the well-reputed artificial bee colony (ABC) (Karaboga and Gorkemli, 2011), and honey bee mating optimization (Marinakis et al., 2011).

As can be seen, the TSP has been extensively used by operation research and computational intelligence researchers since its formulation for different purposes. The state of the art around this problem is such wide that, in this section, we will focus our attention on highlighting the research and advances conducted in the last few years. Being aware that the related literature is bigger than what is represented in this systematic review, we refer interested readers to surveys such as (Potvin, 1993; Laporte, 1992a; Bellmore and Nemhauser, 1968; Lust and Teghem, 2010; Matai et al., 2010).

9.3.1 **TSP and genetic algorithms**

GA has been adopted by many authors of the current scientific community for the TSP and its variants. In Dong and Cai (2019), for example, we can find an interesting GA for solving the challenging large-scale colored balanced TSP. In Lo et al. (2018), Lo et al. explored the adaptation of GA to the real-life oriented multiple TSP. Additional common practice related to GA applied to TSP is the formulation of novel operators, as can be seen in Hussain et al. (2019), in which a new crossover operator is proposed. In Roy et al. (2019) a so-called multiparent crossover is formulated, which shares some notions with that proposed in Sakai et al. (2018), called edge assembly crossover. An additional valuable example of this practice can be found in Wang et al. (2016), in which a multioffspring GA is proposed. Authors of that research claim that, in the basic versions of GA, the number of generated offsprings is the same as the number of parents. Thus, they explore the concept that, for the survival and diversity of the species, it should be desirable to generate a greater number of offsprings. In Hussain et al. (2017), Hussain et al. built an effective combination function called modified cycle crossover operator. In Liu and Li (2018), authors present a new method to initialize the population of GA for the TSP, called greedy permuting method. Another initialization strategy is developed in Deng et al. (2015), which rests its inspiration in the well-known k-mean algorithm. As can be easily checked, the literature around TSP and this successful method is abundant nowadays, being the subject of a myriad of works year by year. Interested readers are referred to additional outstanding works such as Bolaños et al. (2015), Groba et al. (2015), and Contreras-Bolton and Parada (2015).

9.3.2 TSP and simulated annealing

Despite being a classic method, SA is still the focus of much research around the TSP and its variants. In Ezugwu et al. (2017), for example, Ezugwu et al. proposed a hybrid metaheuristic for solving the TSP based on SA and the recently proposed symbiotic organisms search method. Zhan et al. (2016) presented the coined list-based SA, the main basis of which rests on a new mechanism for controlling the temperature parameter. The implemented method counts with a list of temperatures, the maximum of which is used by a Metropolis acceptance criterion to decide whether to accept a candidate solution. Furthermore, the temperature list is dynamically adapted according to the solution space of the problem. In the short paper published by Osaba et al. (2016a), an evolutionary SA is developed for the TSP and compared with additional metaheuristics such as the TS. An additional interesting study can be found in Wu and Gao (2017), in which the performance of the basic SA is improved through the use of a greedy search mechanism for properly dealing with the large-scale TSP. A very recent study can be found in Zhou et al. (2019), in which SA is employed for increasing the population diversity of a gene expression programming method, aiming to improve the ability of the search. Additional valuable related research can be found in papers such as Liu and Zhang (2018), Xu et al. (2017), and Makuchowski (2018).

9.3.3 TSP and tabu search

Among the classic methods, TS is probably the one that has suffered most during the course of time. In the past decades, TS was a successful method considered a cornerstone in the combinatorial optimization and TSP scientific community. Over the years, sophisticated methods laid aside the TS, and today it is difficult to find remarkable studies around the figure of the TS. Among these few studies, we can find as one of the most representative ones the research conducted by Lin et al. (2016), who proposed a TS-SA hybrid solver for the symmetric TSP. One of the characteristics of this hybrid scheme is the development of a dynamic neighborhood structure, the principal goal of which is the enhancement of the search efficiency of the method, by means of the randomness reduction of the conventional 2-opt neighborhood. In Osaba et al. (2018d), TS is employed as part of a pool of metaheuristics for solving the open-path asymmetric green TSP. The main objective of this multiattribute TSP variant is to find a route between a fixed origin and destination, visiting a group of intermediate points exactly once, minimizing the CO_2 emitted by the car and the total distance traveled. An additional analysis can be found in Xu et al. (2015). Among other aspects, authors of that study not only explore the efficiency of four different version of TS using different tabu mechanisms, but also the synergy between TS and ACO in a hybrid method.

9.3.4 TSP and ant colony optimization

Conversely to TS, one of the most used methods in recent years in the TSP community is ACO, as can be seen in works such as Ariyasingha and Fernando (2015) and Mahi et al. (2015). In the first of these works, the recently proposed multiobjective ACO is used for solving the multiobjective TSP under different configurations, using two, three, and four objectives, and different numbers of ants and iterations. In the second research, a hybrid approach is presented which employs PSO for optimizing the parameters that affect performance of the ACO algorithm. Additionally, a 3-opt heuristic is endowed to the proposed method for improving local solutions. A similar solver is implemented in Gülcü et al. (2018), called PACO-3Opt. This hybrid parallel and cooperative method, which counts with multiple colonies and a master–slave paradigm, employs also the 3-opt function for avoiding local minima. An additional well-reputed study can be found in Mavrovouniotis et al. (2016). The principal value of this work is its application to the dynamic TSP. For properly dealing with the unstable nature of this instance of the problem, authors endowed the ACO with a local search operator (called unstring and string) which iteratively takes the best solution found by the algorithm and removes/inserts cities in such a way that the solution quality is improved. The same dynamic formulation of the problem is also used in Chowdhury et al. (2018), implementing a variant of ACO in combination with an adaptive large neighborhood search. Moreover, a note-worthy multiobjective version of ACO is presented in Zhang et al. (2016) for solving the biobjective TSP. One of the essential characteristics of the proposed algorithm is the initialization of a pheromone matrix with the prior knowledge of a *Physarum*-inspired mathematical model. Additional interesting works focusing on ACO can be found in Pang et al. (2015), Eskandari et al. (2019), and Zaidi and Gupta (2018). For readers interested in memetic or hybrid approaches, studies such as Sahana et al. (2018), Liao and Liu (2018), and Dahan et al. (2019) are highly recommended.

9.3.5 **TSP and particle swarm optimization**

Since its introduction in 1995 by Eberhart and Kennedy, PSO has become the most used technique in the swarm intelligence field, and one of its main influential representatives. PSO was developed under the inspiration of the behavior of bird flocks, fish schools, and human communities, and although it was not initially designed to be applied to discrete problems, several modifications have made it possible. Regarding the TSP, lots of papers have been devoted to its application in the last decade; we highlight contributions such as Wang et al. (2003), Clerc (2004), and Pang et al. (2004). Focusing our attention on the research conducted in the most recent years, we can highlight the work introduced in Zhong et al. (2018), in which PSO in combination with a Metropolis acceptance criterion is implemented. The fundamental reason of this merge is to enhance PSO to escape from premature convergence, endowing the method with a sophisticated mechanism to decide whether or not to accept newly produced solutions. A highly interesting research was published by Marinakis et al. (2015), in which a probabilistic TSP was tackled through an adaptive multiswarm PSO. In that adaptive PSO, random values are assigned in the initial phase of the search. After that, these parameters are dynamically optimized simultaneously with the optimization of the objective function of the problem. An additional improved PSO is proposed in Khan et al. (2018) for solving the imprecise cost matrix TSP. Main modifications of the modeled PSO consist on adopting the swap sequence, swap operation, and different velocity update rules. Another interesting example of enhanced PSO can be found in Yu et al. (2015), in which the multiobjective TSP is solved using the so-called set-based comprehensive learning PSO. Further recently published works include Wang and Xu (2017), Chang (2016), and Akhand et al. (2016).

9.3.6 **TSP and bat algorithm**

Focusing our attention on recently proposed nature-inspired methods, and starting with the well-known BA, we can find an interesting recent study by Al-Sorori et al. (2016), in which a hybrid approach is proposed in combination with the genetic operators crossover and mutation, and using the 2-opt and 3-opt operators as local search mechanisms for improving searching performance and speeding up the convergence. Another interesting approach is proposed in Saji and Riffi (2016), whose principal contribution is the velocity scheme used, represented as the number of permutations needed for a bat to reach the best candidate of the swarm. A similar alternative was presented in Jiang (2016), employing nearest neighbor tour construction heuristics for initializing the population and the 2-opt edge-exchange algorithm for the local search step of the method. Among all these papers, we should highlight the research proposed in Osaba et al. (2016b), which is not only considered as the first adaptation of BA to the TSP and asymmetric TSP problems, but also the most cited one. Several aspects make this study interesting, such as the use of the well-known Hamming distance as distance function or its *inclination* mechanism, allowing the method to modify the solution space scheme along the running.

9.3.7 **TSP and firefly algorithm**

If we turn our attention to FA, we can highlight the research recently proposed in Mohsen and Al-Sorori (2017). Authors of this study clearly based their work on the previously published study by Al-Sorori et al. (2016), endowing to the adapted FA both crossover and mutation mechanisms, and employing both 3-opt and 2-opt functions for enhancing the convergence and search performance of the method. An additional hybrid scheme was also proposed in Teng and Li (2018), in which FA is combined with GA. Authors of that work redefine the distance of FA by introducing a swap operator and swap sequence to avoid algorithm easily falling into local optima. A more elaborated study is presented in Li et al. (2015), which is focused on the solution of the multiple TSP, this being a generalization of the TSP in which more than one salesman is allowed to be used in the solution. In Chuah et al. (2017), a swap-based FA is developed, which bases its movement strategy on the widely employed swap function. Furthermore, authors of this paper integrate their FA with nearest-neighborhood initialization, a reset strategy, and a fixed-radius near-neighbor 2-opt operator. Two further interesting and valuable studies are presented in Zhou et al. (2015) and Jie et al. (2017), The former one has the particularity of adopting the dynamic mechanism based on a neighborhood search algorithm, while the second one is combined with a *k-opt* algorithm. Additional recent papers focusing on applications of FA can be found in Saraei and Mansouri (2019), Wang et al. (2018b), and Jati et al. (2013).

9.3.8 **TSP and cuckoo search**

Regarding CS, arguably the most valuable research published recently is the one published by Ouaarab et al. (2014), which is considered by the community as the first application of CS to the TSP. This paper has served as inspiration for subsequent research, such as that described in Ouaarab et al. (2015). In that paper, a random-key CS is proposed, which develops a simplified random-key encoding scheme to pass from a continuous space to a combinatorial space. Especially interesting is the work proposed in Tzy-Luen et al. (2016), in which a subpopulation-based parallel CS on Open Multiprocessing (OpenMP) is implemented for solving the TSP. Lin et al. developed the so-called genotype–phenotype CS in Lin et al. (2017), the essential contribution of which is the representation scheme used for building the solutions. Furthermore, the CS has been present in combination with other methods, such as the studies shown in Hasan (2018) and Kumar et al. (2015), in which the CS is implemented combined with ACO. Another example of this trend can be found in Min et al. (2017), who hybridized CS with the Metropolis acceptance criterion of an SA algorithm, in order to allow accepting inferior solutions with certain probability.

9.3.9 **TSP and artificial bee colony**

Since its inception in 2007 by Karaboga and Basturk, the ABC (Karaboga and Basturk, 2007) has also been adapted for solving combinatorial optimization problems such as the TSP. In recent years, specifically, it has been the focus point of some valuable studies within the TSP community. Very recent is the research proposed by one of the designers of the technique, Karaboga, along with Gorkemli in Karaboga and Gorkemli (2019), in which new improved versions of the discrete ABC were introduced for solving the symmetric TSP. Very recent is also the work that can be found in Choong et al. (2019), in which a hyperheuristic method called modified choice function is implemented for properly regulating the choice of the neighborhood search operators used by the onlooker and employed bees. An additional valuable study was presented in Zhong et al. (2017), introducing a hybrid ABC algorithm which adopts the threshold acceptance criterion method as accepting mechanism. Especially valuable is the work introduced by Venkatesh and Singh (2019), in which the challenging generalized covering TSP is tackled. To do that, authors developed an ABC with dynamic degrees of perturbations, where the degree to which a solution is modified for generating new bees is reduced along the execution. Singh also participated in the conduction of the work presented in Pandiri and Singh (2018), in which a hyperheuristic-based ABC was designed for facing a k-interconnected multidepot TSP. Further remarkable studies can be found in Khan and Maiti (2019), Hu et al. (2016), and Meng et al. (2016).

9.3.10 TSP and imperialist competitive algorithm

The imperialist competitive algorithm (ICA) is a multipopulation metaheuristic introduced in 2007 which finds its inspiration in the concept of imperialism, dividing the whole population in independent empires which fight with each other aiming at conquering the weakest colonies of the rest of the empire (Atashpaz-Gargari and Lucas, 2007). The solver has also been prolific in the TSP community, being used in many reputed studies published recently. We can find in Yousefikhoshbakht and Sedighpour (2013) the first adaptation of this sophisticated method, which has served as a main inspiration for many authors and works, such as Xu et al. (2014). In Ardalan et al. (2015), an improved version of ICA was presented for dealing with the generalized TSP. Authors of that work improved the basic version of ICA with some mechanisms such as a novel encoding scheme, an assimilation policy procedure, destruction/construction operators, and imperialist development plans. Furthermore, the Taguchi method is employed for properly configuring some of the most crucial parameters of the algorithm. Chen et al. (2017) proposed a hybrid method combining ICA with a policy learning function. The central idea behind this hybridization is to permit weak colonies to generate increasingly promising offspring by learning the policies of strong individuals. A brief adaptation of ICA can also be found in Osaba et al. (2018e), as part of a pool of metaheuristics for solving the TSP and the ATSP. Interested readers on this specific metaheuristic are referred to Firoozkooh (2011), Haleh and Esmaeili Aliabadi (2015), and Yousefikhoshbakht and Dolatnejad (2016).

9.3.11 **TSP and other nature-inspired metaheuristics**

Regarding the nature-inspired community, the proposal of PSO and ACO two decades ago decisively influenced the creation of a surfeit of methods, which clearly inherit their essential philosophy. For the design and proposal of these novel approaches, many different inspirational sources have been considered, such as (1) the behavioral patterns of animals such as buffaloes or whales, (2) social and political behaviors as hierarchical societies, and (3) physical processes such as optics systems, electromagnetic theory, or gravitational dynamics.

For this reason, in the current community a countless number of methods of this kind can be found. Along this section, some of the most successful metaheuristic solvers in the TSP community have been reviewed. In any case, we are perfectly aware that the whole community is composed of a plethora of additional methods, usually less often used than the ones outlined here. Furthermore, despite the comprehensive nature of this section, we are also conscious about the difficulty of congregating all the related works published. For this reason, we have only considered these ones that are strictly related with the TSP community, and which have been published in recognized scientific databases.

In any case, a reader may think of certain methods that deserve mention, or even a whole section. Seeking the completeness of this study, in the last part of this section we show a table summarizing additional methods that have been used in recent years for solving the TSP (Table 9.1). In this table we depict the name of the method, its main inspiration, and some related works.

9.4 **Novelty search**

The main objective of NS is to enhance the diversity capacity of a populations-based metaheuristic. To do that, this mechanism finds novel solutions in the behavioral space instead of the search space. Usually, candidates that comprise a population tend to congregate in the same region of the solution space. Conversely, this tendency does not happen in the behavioral space, which is structured employing the Euclidean distance. In this way, we can measure numerically the novelty of a candidate $\mathbf{x}$ using the following formula:

$$\rho(\mathbf{c}) = \frac{1}{k} \sum_{i=1}^{k} d(\mathbf{c}, \boldsymbol{\mu}_i), \tag{9.2}$$

where $d(\cdot, \cdot)$ represents the Euclidean distance. Additionally, k is the number of neighbor solutions chosen from the subset of neighbor candidates selected from the subset of neighbors $\mathcal{N} = \{\mu_1, \mu_2, \ldots, \mu_k\} \subseteq \mathcal{P}$ (i.e., the neighborhood size). This last parameter is problem-dependent and should be established empirically. Additionally, the selection of individuals is conducted using the distance metric, which also depends on the problem.

Table 9.1 Summary of additional nature-inspired methods and their application to the TSP.

Algorithms	Main inspiration	Refs.
Flower pollination algorithm (Yang, 2012)	Pollination process of flowers	Zhou et al. (2017); Strange (2017)
Harmony search (Geem et al., 2001)	Mimicking the improvisation of music players	Boryczka and Szwarc (2019b,a)
Fireworks algorithm (Tan and Zhu, 2010)	Fireworks explosions and location of sparks	Luo et al. (2018); Taidi et al. (2017)
African buffalo optimization (Odili et al., 2015)	The organizational ability of African buffaloes	Odili and Mohmad Kahar (2016); Odili et al. (2017)
Brain storm optimization (Shi, 2011)	Human brainstorming process	Xu et al. (2018); Hua et al. (2016)
Golden ball metaheuristic (Osaba et al., 2014b)	Teams and players organization in the soccer world	Osaba et al. (2014a); Sayoti and Riffi (2015)
Penguin search optimization (Gheraibia and Moussaoui, 2013)	Collaborative hunting strategy of penguins	Mzili et al. (2015, 2017)
Honey bee mating optimization (Haddad et al., 2006)	Honey bee mating process	Odili et al. (2016); Marinakis et al. (2011)
Whale optimization algorithm (Mirjalili and Lewis, 2016)	Social behavior of humpback whales	Gupta et al. (2018)
Water cycle algorithm (Eskandar et al., 2012)	Natural surface runoff of water	Osaba et al. (2018e)
Swallow swarm optimization (Neshat et al., 2013)	Reproduce the behavior of swallow swarms	Bouzidi and Riffi (2017)
Black hole algorithm (Hatamlou, 2013)	Black hole phenomenon in the open space	Hatamlou (2018)
Hydrological cycle algorithm (Wedyan et al., 2017)	Movement of water drops in natural cycle	Wedyan et al. (2018)
Dragonfly algorithm (Mirjalili, 2016)	Swarming behavior of dragonflies	Hammouri et al. (2018)
Pigeon-inspired optimization (Duan and Qiao, 2014)	Homing characteristics of pigeons	Zhong et al. (2019)

It is important to highlight that although NS has demonstrated a great efficiency in many works published up to now (Liapis et al., 2015; Gomes et al., 2015; Fister et al., 2019; López-López et al., 2018), the strategy for properly adapting this mechanism to a problem is still weakly defined, and it is subject to the problem at hand (Fister et al., 2018).

In the research that we are presenting in this chapter, NS has been applied in the same way for the three implemented bio-inspired metaheuristics. Something crucial when implementing NS is the modeling of a proper distance metric. In this study, the function selected in the Hamming distance $D_H(\cdot, \cdot)$, which is detailed in the following section. Additionally, a subset $\mathcal{B}$ is considered, in which all the discarded and replaced candidates are inserted every generation. This way, the size of $\mathcal{B}$ is the same as the main population of the solver.

Conceptually, the subset $\mathcal{B}$ is comprised of the solutions which are potentially *novel*, and prone to be reintroduced in the main population. Thus, when an evolved candidate c_i is better than the individual which it is going to replace, it is directly inserted into the principal population, while the replaced solution is introduced in $\mathcal{B}$. On the other hand, if the trial candidate is not better than its preceding version, the former is inserted into $\mathcal{B}$. Additionally, once the tth generation comes to its end, if r_{NS} (a value drawn from a normal probability distribution) is lower than the parameter $NS_P \in [0.0, 1.0]$, the NS mechanism is conducted. In this research, we have set $NS_P = 0.25$ after a comprehensive empirical analysis.

It is also noteworthy that there is not a specific scientific consensus about the proper number of solutions that should be reintroduced in the main population throughout NS, and how they should replace the existing individuals. In this regard, researchers advocate to adapt these criteria depending on the problem at hand. In this specific work, we have set the number of reinserted candidates to eight. These solutions replace the worst individuals in terms of fitness of the main population. Moreover, these candidates are selected from $\mathcal{B}$ based on their distance regarding the whole swarm. Thus, the eight solutions having a greater diversity with respect to the population are those chosen for reinsertion.

Lastly, the main contribution that we propose in our implemented NS procedure consist on a novel neighborhood changing procedure. Specifically, every time a candidate c is inserted in $\mathcal{B}$, its movement function $\Psi(\cdot, \cdot)$ is modified. Hence, when a candidate is reintroduced in the principal population, it can explore the solution space using different strategies. This simple mechanism enhances both the diversity of the swarm and the exploratory capacity of the algorithm.

9.5 Proposed bio-inspired methods

We propose in this work the combination of three different bio-inspired metaheuristic methods and the NS mechanism. Before specifying the details of each solver, we introduce here some crucial aspects for properly understanding the research conducted.

These aspects are related to solution representation and the metrics used for measuring the difference between different candidates.

When solving the TSP, the way in which the routes are encoded can follow diverse strategies. In this work, the frequently referenced path encoding has been used. Thus, each individual is represented as a permutation of numbers depicting the sequential order in which the nodes are visited. For instance, in a given 10-node dataset, a possible solution could be encoded as $\mathbf{x} = [8, 9, 1, 4, 3, 5, 2, 6, 7, 0]$, meaning that node 8 is visited first, followed by nodes 9, 1, and so forth. Each candidate adopts this approach. Additionally, the objective function employed is the total cost of a complete path given in Eq. (9.1a).

Probably, the most crucial issue when adapting PSO, FA, and BA to a discrete problem such as the TSP is to design the functions resembling how candidates move around the solution space, while guaranteeing their efficient contribution to the search problem under study. For conducting these movements, three well-known movement operators have been used depending on the distance between individuals:

- *Insertion:* This is one of the most frequently used functions for solving combinatorial optimization problems of different nature. Specifically, it selects and extracts one randomly chosen node from the route. Afterwards, this node is reinserted in the route in a randomly selected position.
- *Swapping function*: This well-known function is also widely employed in lots of research studies (Tarantilis, 2005). In this case, two nodes of a solution are selected randomly, and they swap their position.
- *2-opt:* This operator, first proposed in Lin (1965), has been extensively applied in different kinds of routing problems such as the TSP (Tarantilis and Kiranoudis, 2007; Bianchessi and Righini, 2007). The main design principle behind this operator is to randomly eliminate two arcs within the existing route, in order to create two new arcs, avoiding the generation of subtours.

At this point, it is interesting to clarify that *insertion* has been considered as the main operator for all metaheuristic methods; *swapping* and 2-opt, however, compose the pool of functions that NS considers for the reinsertion of candidates.

Finally, for assessing the distance between two different individuals (routes), the well-known Hamming distance $D_H(\cdot, \cdot)$ has been adopted. This function is calculated as the number of noncorresponding elements in the sequence of both individuals; e.g., if the following vectors represent two feasible routes:

$$\mathbf{x}^p = [8, 9, 1, 4, 3, 5, 2, 6, 7, 0],$$
$$\mathbf{x}^{p'} = [8, 7, 1, 4, 3, 5, 0, 6, 2, 9],$$

their Hamming distance $D_H(\mathbf{x}^p, \mathbf{x}^{p'})$ would be equal to 4. Once the distance between the two individuals has been computed, the movement is performed. We now introduce the metaheuristic algorithms under consideration.

9.5.1 Bat algorithm

BA was proposed by Yang (2010), and it is based on the echolocation behavior of microbats, which can find their prey and discriminate different kinds of insects even in complete darkness. As can be read in several surveys (Yang and He, 2013a; Chawla and Duhan, 2015), this method has been extensively adapted for dealing with very diverse optimization fields and problems. The fact that many studies can be found in the literature purely focused on BA confirms that it attracts a lot of interest from the community (Saad et al., 2019; Lu and Jiang, 2019; Osaba et al., 2019; Chen et al., 2018).

BA was first proposed for solving continuous optimization problems (Yang, 2010). Thus, a discrete adaptation must be conducted for properly accommodating its scheme to the combinatorial nature of the problem tackled in this study. In the literature, several adaptations of this kind can be found (Osaba et al., 2018b; Cai et al., 2019). First, each bat in the population represents a feasible solution of the TSP. Moreover, both loudness A_i and pulse emissions r_i concepts have been modeled analogously to the naïve BA. In order to simplify the approach, no frequency parameter has been considered. Besides that, velocity v_i has been used adopting the Hamming distance as its similarity function as $v_p^t = \text{rand}[1, D_H(\mathbf{c}_p, \mathbf{c}^{best})]$. In other words, the velocity v_i of the pth bat in the population at generation t is a random number, which follows a discrete uniform distribution between 1 and the Hamming distance between $\mathbf{c}_p$ and the best bat of the swarm $\mathbf{c}^{best}$.

With all this, $\mathbf{c}_p$ moves towards $\mathbf{c}^{best}$ at generation t as

$$\mathbf{c}_p(t+1) = \Psi\left(\mathbf{c}_p(t), \min\left\{V, v_p^t\right\}\right), \tag{9.3}$$

where $\Psi(\mathbf{c}, Z) \in \{insertion, swapping, 2\text{-}opt\}$ is the movement operator, parametrized by the Z times this function is applied to $\mathbf{c}$. After Z trials, the best considered movement is chosen as output.

9.5.2 Firefly algorithm

The first version of FA was developed by Yang (2008, 2009), and it was based on the idealized behavior of the flashing characteristics of fireflies. As we have pointed out for BA, FA has been the focus of many recent comprehensive surveys (Tilahun and Ngnotchouye, 2017; Fister et al., 2013, 2014; Yang and He, 2013b, 2018). Furthermore, it has been recently applied in many different problems and knowledge fields (Osaba et al., 2017; Danraka et al., 2019; Matthopoulos and Sofianopoulou, 2018; Osaba et al., 2018c).

Because the canonical FA was initially designed for dealing with continuous optimization problems, some modifications have also been done for its proper adaptation. Thus, as in the previously described BA, each firefly of the swarm represents a possible solution for the TSP. Moreover, light absorption has been considered, which is an essential concept for adjusting fireflies' attractiveness. For the movement of the fireflies around the solution space the same logic shown in Eq. (9.3) has been followed.

Finally, for measuring the similarity between two individuals, the $D_H(\cdot, \cdot)$ has also been used.

9.5.3 Particle swarm optimization

PSO is one of the most used swarm intelligence metaheuristics, and it has been adapted to both continuous (Precup and David, 2019) and discrete problems (Wu et al., 2019; Qiu and Xiang, 2019) in very recent years. Works such as Zhong et al. (2007) have inspired us for the discrete PSO developed in this research. Also in this adaptation each individual of the population (particles) represents a feasible solution for the faced problem, while the calculation of the velocity $v_i^{(t)}$ and movement functions have been considered as for the previously described solvers. Furthermore, the movement criterion represented in Eq. (9.3) has also been used for driving the movement of particles. Lastly, $D_H(\cdot, \cdot)$ has also been taken as distance function.

9.6 Experimentation and results

The performance of the three developed solvers has been gauged through 15 contrasted TSP datasets, all of them drawn from the famous TSPLIB repository (Reinelt, 1991). The size of the considered datasets is between 30 and 124 nodes. Taking as inspiration the good practices proposed in Osaba et al. (2018a), similar functions and parameters have been considered in all solvers, aiming at obtaining fair and rigorous insights. Additionally, 20 independent runs have been executed for each (*dataset, technique*) combination. Thus, we provide statistically reliable findings on the performance of each method. The population size has been established as 50 individuals for each method. In FA, the value of the light absorption coefficient is configured as $\gamma = 0.95$, whereas for BA $\alpha = \beta = 0.98$, $A_i^0 = 1.0$ (*loudness*), and $r_i^0 = 0.1$ (*rate*).

In Table 9.2 the outcomes obtained by each method are shown. For properly understanding the influence of NS in each metaheuristic scheme, we show not only the outcomes of each method using this mechanism (represented with the subscript NS), but also the results of the basic versions. Average (Avg) and standard deviation (SD) are provided for each (*problem, technique*) combination. Moreover, we also included in the table the mean generation number t_{conv} for which the best solution was met for every technique and problem instance. We represent this value in hundreds. Furthermore, we depict in bold the best outcome obtained for each metaheuristic, in order to facilitate the visual analysis of the influence of the NS. All the tests conducted in this work have been performed on an Intel Core i7-7600U, and Java has been used as the programming language.

Additionally, being aware that the comparison between the selected schemes PSO, FA, and BA is not the focus of this study, a statistical test has been carried out with the obtained results for the sake of completeness. To do that, and following the guidelines in Derrac et al. (2011), the Friedman nonparametric test for multiple comparison has been conducted, which allows to check if there are significant differences in

Table 9.2 Obtained optimization results using BA, FA, and PSO for the TSP in combination with the NS mechanism.

Instance		PSO			PSO$_{NS}$			FA			FA$_{NS}$			BA			BA$_{NS}$		
Name	Optima	Avg	SD	t_{conv}	Avg	SD	t_{conv}	Avg	SD	t_{conv}	Avg	SD	t_{conv}	Avg	SD	t_{conv}	Avg	SD	t_{conv}
Oliver30	420	420.3	0.47	0.21	420.4	0.49	0.10	421.0	0.80	0.04	420.4	0.58	0.03	421.2	1.69	0.18	420.2	0.43	0.11
Eilon50	425	435.4	4.49	0.80	432.2	3.89	0.40	439.3	2.91	0.14	439.4	2.33	0.11	436.0	5.33	0.71	432.0	3.88	0.38
Eil51	426	437.1	4.10	0.78	434.5	5.51	0.39	442.5	3.08	0.15	440.0	2.22	0.14	437.1	4.85	0.76	433.5	2.61	0.35
Berlin52	7542	7667.3	89.00	1.07	7699.8	148.69	0.49	7678.1	51.64	0.29	7593.2	25.98	0.16	7711.4	118.35	1.05	7620.5	100.98	0.49
St70	675	693.6	7.55	1.86	689.8	10.09	1.30	702.7	3.92	0.32	697.3	3.40	0.26	696.7	8.53	1.97	688.4	4.53	1.05
Eilon75	535	565.1	5.04	1.96	549.4	7.37	1.25	572.6	2.53	0.38	569.2	2.99	0.27	564.3	6.39	2.08	555.2	6.85	1.27
Eil76	538	566.2	7.34	2.47	557.4	7.87	1.24	572.4	3.27	0.42	568.7	2.79	0.23	565.4	7.68	2.28	557.8	7.87	1.05
KroA100	21282	22335.0	372.26	8.00	21907.6	495.90	5.03	22586.1	77.63	1.14	22429.9	77.35	0.47	22528.1	524.92	7.62	21740.0	246.23	5.27
KroB100	22140	23457.0	412.72	6.34	22743.8	259.96	5.70	23663.9	172.21	1.05	23346.2	147.47	0.58	23393.3	374.39	7.64	22795.2	343.61	5.04
KroC100	20749	22064.4	430.82	7.50	21388.8	347.78	5.74	22197.0	117.20	0.97	21900.3	117.55	0.46	22135.8	286.75	7.56	21347.7	430.22	5.10
KroD100	21294	22684.9	292.39	6.86	21866.2	338.32	5.45	22634.3	104.09	1.19	22312.3	85.21	0.48	22561.2	373.45	7.84	22040.2	514.55	4.81
KroE100	22008	23362.7	537.66	7.03	22586.2	286.88	5.35	23453.4	126.53	1.18	23248.9	112.23	0.45	23560.5	384.86	6.50	22649.8	393.95	5.76
Eil101	629	673.8	7.47	5.11	635.0	5.84	3.97	670.2	4.21	1.00	662.3	3.10	0.58	670.5	11.41	5.84	654.6	4.99	3.85
Pr107	44303	46592.5	563.92	9.25	45746.3	1056.43	5.86	46336.4	224.35	1.35	45941.3	90.77	0.42	46727.4	897.94	10.80	45709.8	981.18	6.47
Pr124	59030	64150.5	1635.70	14.85	60387.7	898.76	11.12	64505.9	332.04	1.43	62552.8	204.86	0.84	64436.7	1985.18	14.84	60554.1	1179.60	11.39
								FRIEDMAN'S NONPARAMETRIC TEST											
Rank		4.0333			2.1			4.5333			4			4.7			1.8333		

the results obtained by all reported methods. Thus, in the last row of Table 9.2, we have displayed the mean ranking returned by this nonparametric test for each of the compared algorithms and scenarios (the lower the rank, the better the performance). Additionally, the Friedman statistic obtained is 35.914. The confidence interval has been set at 99%, with 9.236 being the critical point in a χ^2 distribution with five degrees of freedom. Since $35.914 > 5.991$, it can be concluded that there are significant differences among the results.

Several conclusions can be drawn from the results obtained through this preliminary experimentation. First of all, it can be seen how the employment of the NS mechanism strongly improves the results in all the three used metaheuristic schemes. In the case of PSO, NS improves the average quality of the outcomes in 13 out of 15 instances. For FA, this improvement is observed for all 15 datasets. Finally, BA reaches better solutions in 14 of the 15 cases. These findings support the hypothesis that the NS procedure enhances the exploratory capacity of the three considered bio-inspired metaheuristics thanks to the diversification it injects in the population.

Equally interesting is the phenomenon that can be observed regarding the convergence behavior. In such a case, it can be seen how the introduction of the NS mechanism in the metaheuristic scheme supposes also an improvement on the convergence, reaching the final solution in a lower number of generations, and lower computation effort. Along with the improvement of the results, this feature supposes a huge advantage for the NS procedure.

As final reflection, and being something easily observable through the results obtained by the Friedman nonparametric test, we can highlight that the methods using NS are the ones that reached better outcomes. In this sense, BA_{NS} has emerged as the best alternative, followed by PSO_{NS} and FA_{NS}. Likewise, BA_{NS} is the solver that presents the best convergence behavior among the six implemented techniques. In any case, as mentioned before, the comparison between the different metaheuristics falls outside the scope of this experimentation.

9.7 Research opportunities and open challenges

In light of the literature overview made in Section 9.3 and the novel experimentation conducted exploring the synergies between bio-inspired computation and NS, it is unquestionable that the TSP is a topic that still attracts remarkable attention from the related community, being the scope of abundant research material. The current state of the computation and the multiple resources in the hands of practitioners open the opportunity of facing new challenges in the field. In this context, we foresee promising research directions along diverse axes, among which we pause at the following ones.

- As outlined in Section 9.3, an ample collection of classical and sophisticated solvers have been proposed in both past and recent literature for efficiently solving the TSP and its variants. One of the main challenges that the community should face urgently is to slow down the elaboration of additional novel methods. Despite the existence of a wide variety of well-reputed methods, part of the community continues scrutinizing the natural world seeking to formulate new metaheuristics mimicking some new biological phenomena. Some recent examples con be found in recent studies such as Kaveh and Zolghadr (2016), Arora and Singh (2019), and Wang et al. (2018a). These novel methods do not only offer a step forward for the community, but also augment the skepticism of critical researchers. These practitioners are continuously questioning the need of new methods, which apparently are very similar to previously published ones. In contrast to this trend, the whole community should pull in the same direction, trying to adapt the existing methods to more complex formulations of the TSP, and explore the different synergies that can arise between different approaches or mechanisms.

- Related to the previous challenge, currently, the TSP is still conceived by the community as a benchmarking or an academic problem with a very limited applicability to real-world problems. Trying to deal with this stigma, practitioners in the field should work on the formulation of richer and more complex formulations of the TSP, aiming to adapt the problem to real logistic and transportation problems. This research trend, which is currently receiving some attention from researchers, has led to the coining of the term *rich* or *multiattribute* TSP. As pointed out in Section 9.2, these problems are attracting the interest of the scientific community for their closer match to realistic situations. Despite this growing activity, the research behind these specific formulations is still not remarkable. This is, in part, because part of the community is working in the branch mentioned in the previous challenge. Thus, through this chapter, we call for a profound reflection around not only the formulation of new complex formulations of the TSP, but also the exploration of new ways for their solution, such as hybridized and memetic metaheuristics.
- Finally, we highly encourage involved researchers to consider the tackling of TSP variants of large size. Many of the studies that can be found in the current literature deal with controlled problem datasets of small or medium size (in terms of number of nodes). The experimental part of this study is also an example of this tendency. Notwithstanding, real-world problems are prone to have a higher magnitude, supposing a challenge for both researchers and their proposed solvers. In fact, large-scale variants not only hinder the efficiency of many of the often used methods, but they also suppose a compromise for the convergence of the solvers. In this context, the consideration of new optimization approaches, such as the ones referred to as *large-scale global optimization* techniques, can unchain unprecedented benefits for this field. Some methods that can be considered for being applied are SHADE-ILS (Molina et al., 2018) or multiple offspring sampling (La-Torre et al., 2012). Additional interesting research trends related to computational efficiency can be found in the area of cooperative co-evolutionary algorithms (Ma et al., 2018). Lastly, an additional promising alternative could be the design and implementation of self-adaptive solvers (Kramer, 2008).

9.8 Conclusions

This chapter has focused on the review of the well-known TSP. In the first part of this work, we have briefly introduced this famous problem, along with some of its most valuable variants. After that, we have made a systematic overview of the recent history of this problem, describing some of the most remarkable studies published in recent years. To do that, our attention has gravitated around both classical (SA, TS, GA, etc.) and sophisticated (BA, ICA, FA) metaheuristic solvers. After this literature review, we have presented an experimental study focused on the hybridization of the NS mechanism and three different bio-inspired computation schemes: PSO, FA, and BA. The performance of the implemented solvers has been tested over a benchmark comprised of 15 well-known datasets. The main conclusions drawn from this first study support the hypothesis that NS is a promising mechanism for being considered for the solution of the TSP, proving that this procedure helps metaheuristics to improve the quality of their reached results.

After these preliminary tests, we have concluded our research by sharing our envisioned future of the related community. To do that, we have pinpointed several inspiring opportunities and their related challenges, which should gather most of the research efforts made in the coming years. Among the future lines of research we foresee, we advocate the facing of bigger and more applicable datasets, using alternative methods not yet deeply explored, or synergistic hybridization of solvers proposed by the related experts along the years. Arguably, we foresee an exciting and still prolific future for the TSP community, adding new alluring nodes to visit in this endless path that TSP research is.

Acknowledgment

Eneko Osaba and Javier Del Ser would like to thank the Basque Government for its funding support through the EMAITEK program.

References

Aarts, E.H., Korst, J.H., van Laarhoven, P.J., 1988. A quantitative analysis of the simulated annealing algorithm: a case study for the traveling salesman problem. Journal of Statistical Physics 50 (1–2), 187–206.

Akhand, M., Hossain, S., Akter, S., 2016. A comparative study of prominent particle swarm optimization based methods to solve traveling salesman problem. International Journal of Swarm Intelligence and Evolutionary Computation 5 (139), 2.

Al-Sorori, W., Mohsen, A., et al., 2016. An improved hybrid bat algorithm for traveling salesman problem. In: International Conference on Bio-Inspired Computing: Theories and Applications, pp. 504–511.

Ardalan, Z., Karimi, S., Poursabzi, O., Naderi, B., 2015. A novel imperialist competitive algorithm for generalized traveling salesman problems. Applied Soft Computing 26, 546–555.

Arigliano, A., Ghiani, G., Grieco, A., Guerriero, E., Plana, I., 2019. Time-dependent asymmetric traveling salesman problem with time windows: properties and an exact algorithm. Discrete Applied Mathematics 261, 28–39.

Ariyasingha, I., Fernando, T., 2015. Performance analysis of the multi-objective ant colony optimization algorithms for the traveling salesman problem. Swarm and Evolutionary Computation 23, 11–26.

Arora, S., Singh, S., 2019. Butterfly optimization algorithm: a novel approach for global optimization. Soft Computing 23 (3), 715–734.

Asadpour, A., Goemans, M.X., Madry, A., Gharan, S.O., Saberi, A., 2017. An o (log n/log log n)-approximation algorithm for the asymmetric traveling salesman problem. Operations Research 65 (4), 1043–1061.

Atashpaz-Gargari, E., Lucas, C., 2007. Imperialist competitive algorithm: an algorithm for optimization inspired by imperialistic competition. In: 2007 IEEE Congress on Evolutionary Computation, pp. 4661–4667.

Bell, J.E., McMullen, P.R., 2004. Ant colony optimization techniques for the vehicle routing problem. Advanced Engineering Informatics 18 (1), 41–48.

Bellmore, M., Nemhauser, G.L., 1968. The traveling salesman problem: a survey. Operations Research 16 (3), 538–558.

Bianchessi, N., Righini, G., 2007. Heuristic algorithms for the vehicle routing problem with simultaneous pick-up and delivery. Computers & Operations Research 34 (2), 578–594.

Bolaños, R., Echeverry, M., Escobar, J., 2015. A multiobjective non-dominated sorting genetic algorithm (NSGA-II) for the multiple traveling salesman problem. Decision Science Letters 4 (4), 559–568.

Boryczka, U., Szwarc, K., 2019a. An effective hybrid harmony search for the asymmetric travelling salesman problem. Engineering Optimization, 1–17.

Boryczka, U., Szwarc, K., 2019b. The harmony search algorithm with additional improvement of harmony memory for asymmetric traveling salesman problem. Expert Systems with Applications 122, 43–53.

Bouzidi, S., Riffi, M.E., 2017. Discrete swallow swarm optimization algorithm for travelling salesman problem. In: Proceedings of the 2017 International Conference on Smart Digital Environment, pp. 80–84.

Burke, E.K., Cowling, P.I., Keuthen, R., 2001. Effective local and guided variable neighbourhood search methods for the asymmetric travelling salesman problem. In: Applications of Evolutionary Computing. Springer, pp. 203–212.

Caceres-Cruz, J., Arias, P., Guimarans, D., Riera, D., Juan, A.A., 2015. Rich vehicle routing problem: survey. ACM Computing Surveys (CSUR) 47 (2), 32.

Cai, Y., Qi, Y., Cai, H., Huang, H., Chen, H., 2019. Chaotic discrete bat algorithm for capacitated vehicle routing problem. International Journal of Autonomous and Adaptive Communications Systems (IJAACS) 12 (2), 91–108.

Carrabs, F., Cordeau, J.F., Laporte, G., 2007. Variable neighborhood search for the pickup and delivery traveling salesman problem with LIFO loading. INFORMS Journal on Computing 19 (4), 618–632.

Chang, J.C., 2016. Modified particle swarm optimization for solving traveling salesman problem based on a Hadoop MapReduce framework. In: 2016 International Conference on Applied System Innovation (ICASI), pp. 1–4.

Chawla, M., Duhan, M., 2015. Bat algorithm: a survey of the state-of-the-art. Applied Artificial Intelligence 29 (6), 617–634.

Chen, M.H., Chen, S.H., Chang, P.C., 2017. Imperial competitive algorithm with policy learning for the traveling salesman problem. Soft Computing 21 (7), 1863–1875.

Chen, S., Peng, G.H., He, X.S., Yang, X.S., 2018. Global convergence analysis of the bat algorithm using a Markovian framework and dynamical system theory. Expert Systems with Applications 114, 173–182.

Choong, S.S., Wong, L.P., Lim, C.P., 2019. An artificial bee colony algorithm with a modified choice function for the traveling salesman problem. Swarm and Evolutionary Computation 44, 622–635.

Chowdhury, S., Marufuzzaman, M., Tunc, H., Bian, L., Bullington, W., 2018. A modified ant colony optimization algorithm to solve a dynamic traveling salesman problem: a case study with drones for wildlife surveillance. Journal of Computational Design and Engineering.

Christofides, N., 1976. The vehicle routing problem. RAIRO. Recherche Opérationnelle 10 (V1), 55–70.

Chuah, H.S., Wong, L.P., Hassan, F.H., 2017. Swap-based discrete firefly algorithm for traveling salesman problem. In: International Workshop on Multi-Disciplinary Trends in Artificial Intelligence, pp. 409–425.

Clerc, M., 2004. Discrete particle swarm optimization, illustrated by the traveling salesman problem. In: New Optimization Techniques in Engineering. Springer, pp. 219–239.

Contreras-Bolton, C., Parada, V., 2015. Automatic combination of operators in a genetic algorithm to solve the traveling salesman problem. PLoS ONE 10 (9), e0137724.

Dahan, F., El Hindi, K., Mathkour, H., AlSalman, H., 2019. Dynamic flying ant colony optimization (DFACO) for solving the traveling salesman problem. Sensors 19 (8), 1837.

Danraka, S.S., Yahaya, S.M., Usman, A.D., Umar, A., Abubakar, A.M., 2019. Discrete firefly algorithm based feature selection scheme for improved face recognition. Computing & Information Systems 23 (2).

David, R.C., Precup, R.E., Petriu, E.M., Rădac, M.B., Preitl, S., 2013. Gravitational search algorithm-based design of fuzzy control systems with a reduced parametric sensitivity. Information Sciences 247, 154–173.

De Jong, K., 1975. Analysis of the behavior of a class of genetic adaptive systems. Ph.D. thesis. University of Michigan, Michigan, USA.

Deng, Y., Liu, Y., Zhou, D., 2015. An improved genetic algorithm with initial population strategy for symmetric TSP. Mathematical Problems in Engineering 2015.

Derrac, J., García, S., Molina, D., Herrera, F., 2011. A practical tutorial on the use of nonparametric statistical tests as a methodology for comparing evolutionary and swarm intelligence algorithms. Swarm and Evolutionary Computation 1 (1), 3–18.

Dong, X., Cai, Y., 2019. A novel genetic algorithm for large scale colored balanced traveling salesman problem. Future Generations Computer Systems 95, 727–742.

Dorigo, M., Gambardella, L.M., 1997. Ant colony system: a cooperative learning approach to the traveling salesman problem. IEEE Transactions on Evolutionary Computation 1 (1), 53–66.

Duan, H., Qiao, P., 2014. Pigeon-inspired optimization: a new swarm intelligence optimizer for air robot path planning. International Journal of Intelligent Computing and Cybernetics 7 (1), 24–37.

Eskandar, H., Sadollah, A., Bahreininejad, A., Hamdi, M., 2012. Water cycle algorithm – a novel meta-heuristic optimization method for solving constrained engineering optimization problems. Applied Soft Computing 110 (111), 151–166.

Eskandari, L., Jafarian, A., Rahimloo, P., Baleanu, D., 2019. A modified and enhanced ant colony optimization algorithm for traveling salesman problem. In: Mathematical Methods in Engineering. Springer, pp. 257–265.

Ezugwu, A.E.S., Adewumi, A.O., Frîncu, M.E., 2017. Simulated annealing based symbiotic organisms search optimization algorithm for traveling salesman problem. Expert Systems with Applications 77, 189–210.

Fachini, R.F., Armentano, V.A., 2018. Exact and heuristic dynamic programming algorithms for the traveling salesman problem with flexible time windows. Optimization Letters, 1–31.

Fiechter, C.N., 1994. A parallel tabu search algorithm for large traveling salesman problems. Discrete Applied Mathematics 51 (3), 243–267.

Firoozkooh, I., 2011. Using imperial competitive algorithm for solving traveling salesman problem and comparing the efficiency of the proposed algorithm with methods in use. Australian Journal of Basic and Applied Sciences 5, 540–543.

Fister, I., Fister Jr, I., Yang, X.S., Brest, J., 2013. A comprehensive review of firefly algorithms. Swarm and Evolutionary Computation 13, 34–46.

Fister, I., Yang, X.S., Fister, D., 2014. Firefly algorithm: a brief review of the expanding literature. In: Cuckoo Search and Firefly Algorithm. Springer, pp. 347–360.

Fister, I., Iglesias, A., Galvez, A., Del Ser, J., Osaba, E., 2018. Using novelty search in differential evolution. In: International Conference on Practical Applications of Agents and Multi-Agent Systems, pp. 534–542.

Fister, I., Iglesias, A., Galvez, A., Del Ser, J., Osaba, E., Fister Jr, I., Perc, M., Slavinec, M., 2019. Novelty search for global optimization. Applied Mathematics and Computation 347, 865–881.

Furini, F., Persiani, C.A., Toth, P., 2016. The time dependent traveling salesman planning problem in controlled airspace. Transportation Research. Part B: Methodological 90, 38–55.

Geem, Z.W., Kim, J.H., Loganathan, G.V., 2001. A new heuristic optimization algorithm: harmony search. Simulation 76 (2), 60–68.

Gendreau, M., Laporte, G., Semet, F., 1998. A tabu search heuristic for the undirected selective travelling salesman problem. European Journal of Operational Research 106 (2), 539–545.

Gheraibia, Y., Moussaoui, A., 2013. Penguins search optimization algorithm (PeSOA). In: International Conference on Industrial, Engineering and Other Applications of Applied Intelligent Systems, pp. 222–231.

Glover, F., 1989. Tabu search, part I. ORSA Journal on Computing 1 (3), 190–206.

Goldberg, D., 1989. Genetic Algorithms in Search, Optimization, and Machine Learning. Addison-Wesley Professional.

Gomes, J., Mariano, P., Christensen, A.L., 2015. Devising effective novelty search algorithms: a comprehensive empirical study. In: Proceedings of the 2015 Annual Conference on Genetic and Evolutionary Computation, pp. 943–950.

Grefenstette, J., Gopal, R., Rosmaita, B., Van Gucht, D., 1985. Genetic algorithms for the traveling salesman problem. In: Proceedings of the First International Conference on Genetic Algorithms and Their Applications, pp. 160–168.

Groba, C., Sartal, A., Vázquez, X.H., 2015. Solving the dynamic traveling salesman problem using a genetic algorithm with trajectory prediction: an application to fish aggregating devices. Computers & Operations Research 56, 22–32.

Gülcü, Ş., Mahi, M., Baykan, Ö.K., Kodaz, H., 2018. A parallel cooperative hybrid method based on ant colony optimization and 3-opt algorithm for solving traveling salesman problem. Soft Computing 22 (5), 1669–1685.

Gupta, R., Shrivastava, N., Jain, M., Singh, V., Rani, A., 2018. Greedy WOA for travelling salesman problem. In: International Conference on Advances in Computing and Data Sciences, pp. 321–330.

Haddad, O.B., Afshar, A., Mariño, M.A., 2006. Honey-bees mating optimization (HBMO) algorithm: a new heuristic approach for water resources optimization. Water Resources Management 20 (5), 661–680.

Haleh, H., Esmaeili Aliabadi, D., 2015. Improvement of imperialist colony algorithm by employment of imperialist learning operator and implementing in travel salesman problem. Journal of Development & Evolution Management 1394 (22), 55–61.

Hammouri, A.I., Samra, E.T.A., Al-Betar, M.A., Khalil, R.M., Alasmer, Z., Kanan, M., 2018. A dragonfly algorithm for solving traveling salesman problem. In: 2018 8th IEEE International Conference on Control System, Computing and Engineering, pp. 136–141.

Hasan, L.S., 2018. Solving traveling salesman problem using cuckoo search and ant colony algorithms. Journal of Al-Qadisiyah for Computer Science and Mathematics 10 (2), 59.

Hatamlou, A., 2013. Black hole: a new heuristic optimization approach for data clustering. Information Sciences 222, 175–184.

Hatamlou, A., 2018. Solving travelling salesman problem using black hole algorithm. Soft Computing 22 (24), 8167–8175.

Helsgaun, K., 2015. Solving the equality generalized traveling salesman problem using the Lin–Kernighan–Helsgaun algorithm. Mathematical Programming Computation 7 (3), 269–287.

Hu, G., Chu, X., Niu, B., Li, L., Lin, D., Liu, Y., 2016. An augmented artificial bee colony with hybrid learning for traveling salesman problem. In: International Conference on Intelligent Computing, pp. 636–643.

Hua, Z., Chen, J., Xie, Y., 2016. Brain storm optimization with discrete particle swarm optimization for TSP. In: 2016 12th International Conference on Computational Intelligence and Security (CIS), pp. 190–193.

Hussain, A., Muhammad, Y.S., Nauman Sajid, M., Hussain, I., Mohamd Shoukry, A., Gani, S., 2017. Genetic algorithm for traveling salesman problem with modified cycle crossover operator. Computational Intelligence and Neuroscience 2017.

Hussain, A., Muhammad, Y.S., Sajid, M.N., 2019. A simulated study of genetic algorithm with a new crossover operator using traveling salesman problem. Journal of Mathematics (ISSN 1016-2526) 51 (5), 61–77.

Jati, G.K., Manurung, R., Suyanto, 2013. Discrete firefly algorithm for traveling salesman problem: a new movement scheme. Swarm Intelligence and Bio-Inspired Computation: Theory and Applications, 295–312.

Jiang, Z., 2016. Discrete bat algorithm for traveling salesman problem. In: 2016 3rd International Conference on Information Science and Control Engineering (ICISCE), pp. 343–347.

Jie, L., Teng, L., Yin, S., 2017. An improved discrete firefly algorithm used for traveling salesman problem. In: International Conference on Swarm Intelligence, pp. 593–600.

Jun-man, K., Yi, Z., 2012. Application of an improved ant colony optimization on generalized traveling salesman problem. Energy Procedia 17, 319–325.

Karaboga, D., Basturk, B., 2007. A powerful and efficient algorithm for numerical function optimization: artificial bee colony (ABC) algorithm. Journal of Global Optimization 39 (3), 459–471.

Karaboga, D., Gorkemli, B., 2011. A combinatorial artificial bee colony algorithm for traveling salesman problem. In: International Symposium on Innovations in Intelligent Systems and Applications, pp. 50–53.

Karaboga, D., Gorkemli, B., 2019. Solving traveling salesman problem by using combinatorial artificial bee colony algorithms. International Journal on Artificial Intelligence Tools 28 (01), 1950004.

Kaveh, A., Zolghadr, A., 2016. A novel meta-heuristic algorithm: tug of war optimization. International Journal of Optimization in Civil Engineering 6 (4), 469–492.

Kennedy, J., Eberhart, R., et al., 1995. Particle swarm optimization. In: Proceedings of IEEE International Conference on Neural Networks, vol. 4, pp. 1942–1948.

Khan, I., Maiti, M.K., 2019. A swap sequence based artificial bee colony algorithm for traveling salesman problem. Swarm and Evolutionary Computation 44, 428–438.

Khan, I., Pal, S., Maiti, M.K., 2018. A modified particle swarm optimization algorithm for solving traveling salesman problem with imprecise cost matrix. In: 2018 4th International Conference on Recent Advances in Information Technology (RAIT), pp. 1–8.

Kirkpatrick, S., Gellat, C., Vecchi, M., 1983. Optimization by simulated annealing. Science 220 (4598), 671–680.

Kitjacharoenchai, P., Ventresca, M., Moshref-Javadi, M., Lee, S., Tanchoco, J.M., Brunese, P.A., 2019. Multiple traveling salesman problem with drones: mathematical model and heuristic approach. Computers & Industrial Engineering 129, 14–30.

Knox, J., 1994. Tabu search performance on the symmetric traveling salesman problem. Computers & Operations Research 21 (8), 867–876.

Kramer, O., 2008. Self-Adaptive Heuristics for Evolutionary Computation, vol. 147. Springer.

Kumar, S., Kurmi, J., Tiwari, S.P., 2015. Hybrid ant colony optimization and cuckoo search algorithm for travelling salesman problem. International Journal of Scientific and Research Publications 5 (6), 1–5.

Kumbharana, S.N., Pandey, G.M., 2013. Solving travelling salesman problem using firefly algorithm. International Journal for Research in Science & Advanced Technologies 2 (2), 53–57.

Lahyani, R., Khemakhem, M., Semet, F., 2017. A unified matheuristic for solving multi-constrained traveling salesman problems with profits. EURO Journal on Computational Optimization 5 (3), 393–422.

Laporte, G., 1992a. The traveling salesman problem: an overview of exact and approximate algorithms. European Journal of Operational Research 59 (2), 231–247.

Laporte, G., 1992b. The vehicle routing problem: an overview of exact and approximate algorithms. European Journal of Operational Research 59 (3), 345–358.

Larrañaga, P., Kuijpers, C.M.H., Murga, R.H., Inza, I., Dizdarevic, S., 1999. Genetic algorithms for the travelling salesman problem: a review of representations and operators. Artificial Intelligence Review 13 (2), 129–170.

LaTorre, A., Muelas, S., Peña, J.M., 2012. Multiple offspring sampling in large scale global optimization. In: 2012 IEEE Congress on Evolutionary Computation, pp. 1–8.

Lawler, E.L., Lenstra, J.K., Kan, A.R., Shmoys, D.B., 1985. The Traveling Salesman Problem: a Guided Tour of Combinatorial Optimization, Vol. 3. Wiley, New York.

Lehman, J., Stanley, K.O., 2008. Exploiting open-endedness to solve problems through the search for novelty. In: ALIFE, pp. 329–336.

Li, M., Ma, J., Zhang, Y., Zhou, H., Liu, J., 2015. Firefly algorithm solving multiple traveling salesman problem. Journal of Computational and Theoretical Nanoscience 12 (7), 1277–1281.

Liao, E., Liu, C., 2018. A hierarchical algorithm based on density peaks clustering and ant colony optimization for traveling salesman problem. IEEE Access 6, 38921–38933.

Liapis, A., Yannakakis, G.N., Togelius, J., 2015. Constrained novelty search: a study on game content generation. Evolutionary Computation 23 (1), 101–129.

Lin, S., 1965. Computer solutions of the traveling salesman problem. Bell System Technical Journal 44 (10), 2245–2269.

Lin, Y., Bian, Z., Liu, X., 2016. Developing a dynamic neighborhood structure for an adaptive hybrid simulated annealing–tabu search algorithm to solve the symmetrical traveling salesman problem. Applied Soft Computing 49, 937–952.

Lin, M., Zhong, Y., Liu, B., Lin, X., 2017. Genotype-phenotype cuckoo search algorithm for traveling salesman problem. Computer Engineering and Applications 2017 (24), 28.

Liu, J., Li, W., 2018. Greedy permuting method for genetic algorithm on traveling salesman problem. In: 8th International Conference on Electronics Information and Emergency Communication, pp. 47–51.

Liu, C., Zhang, Y., 2018. Research on MTSP problem based on simulated annealing. In: Proceedings of the 2018 International Conference on Information Science and System, pp. 283–285.

Lo, K.M., Yi, W.Y., Wong, P.K., Leung, K.S., Leung, Y., Mak, S.T., 2018. A genetic algorithm with new local operators for multiple traveling salesman problems. International Journal of Computational Intelligence Systems 11 (1), 692–705.

López-López, V.R., Trujillo, L., Legrand, P., 2018. Novelty search for software improvement of a slam system. In: Proceedings of the Genetic and Evolutionary Computation Conference Companion, pp. 1598–1605.

Lu, Y., Jiang, T., 2019. Bi-population based discrete bat algorithm for the low-carbon job shop scheduling problem. IEEE Access 7, 14513–14522.

Luo, H., Xu, W., Tan, Y., 2018. A discrete fireworks algorithm for solving large-scale travel salesman problem. In: 2018 IEEE Congress on Evolutionary Computation (CEC), pp. 1–8.

Lust, T., Teghem, J., 2010. The multiobjective traveling salesman problem: a survey and a new approach. In: Advances in Multi-Objective Nature Inspired Computing. Springer, pp. 119–141.

Ma, X., Li, X., Zhang, Q., Tang, K., Liang, Z., Xie, W., Zhu, Z., 2018. A survey on cooperative coevolutionary algorithms. IEEE Transactions on Evolutionary Computation 23 (3), 421–441.

Mahi, M., Baykan, Ö.K., Kodaz, H., 2015. A new hybrid method based on particle swarm optimization, ant colony optimization and 3-opt algorithms for traveling salesman problem. Applied Soft Computing 30, 484–490.

Maity, S., Roy, A., Maiti, M., 2019. A rough multi-objective genetic algorithm for uncertain constrained multi-objective solid travelling salesman problem. Granular Computing 4 (1), 125–142.

Makuchowski, M., 2018. Effective algorithm of simulated annealing for the symmetric traveling salesman problem. In: International Conference on Dependability and Complex Systems, pp. 348–359.

Malek, M., Guruswamy, M., Pandya, M., Owens, H., 1989. Serial and parallel simulated annealing and tabu search algorithms for the traveling salesman problem. Annals of Operations Research 21 (1), 59–84.

Marinakis, Y., Marinaki, M., Dounias, G., 2011. Honey bees mating optimization algorithm for the Euclidean traveling salesman problem. Information Sciences 181 (20), 4684–4698.

Marinakis, Y., Marinaki, M., Migdalas, A., 2015. Adaptive tunning of all parameters in a multi-swarm particle swarm optimization algorithm: an application to the probabilistic traveling salesman problem. In: Optimization, Control, and Applications in the Information Age. Springer, pp. 187–207.

Matai, R., Singh, S., Mittal, M.L., 2010. Traveling salesman problem: an overview of applications, formulations, and solution approaches. In: Traveling Salesman Problem, Theory and Applications. IntechOpen.

Matthopoulos, P.P., Sofianopoulou, S., 2018. A firefly algorithm for the heterogeneous fixed fleet VRP. International Journal of Industrial and Systems Engineering.

Mavrovouniotis, M., Müller, F.M., Yang, S., 2016. Ant colony optimization with local search for dynamic traveling salesman problems. IEEE Transactions on Cybernetics 47 (7), 1743–1756.

Meng, L., Yin, S., Hu, X., 2016. A new method used for traveling salesman problem based on discrete artificial bee colony algorithm. Telkomnika 14 (1), 342.

Min, L., Bixiong, L., Xiaoyu, L., 2017. Hybrid discrete cuckoo search algorithm with metropolis criterion for traveling salesman problem. Journal of Nanjing University (Natural Science) 5, 17.

Mirjalili, S., 2016. Dragonfly algorithm: a new meta-heuristic optimization technique for solving single-objective, discrete, and multi-objective problems. Neural Computing & Applications 27 (4), 1053–1073.

Mirjalili, S., Lewis, A., 2016. The whale optimization algorithm. Advances in Engineering Software 95, 51–67.

Mohsen, A.M., Al-Sorori, W., 2017. A new hybrid discrete firefly algorithm for solving the traveling salesman problem. In: Applied Computing and Information Technology. Springer, pp. 169–180.

Molina, D., LaTorre, A., Herrera, F., 2018. Shade with iterative local search for large-scale global optimization. In: 2018 IEEE Congress on Evolutionary Computation (CEC), pp. 1–8.

Mzili, I., Bouzidi, M., Riffi, M.E., 2015. A novel hybrid penguins search optimization algorithm to solve travelling salesman problem. In: 2015 Third World Conference on Complex Systems (WCCS), pp. 1–5.

Mzili, I., Riffi, M.E., Benzekri, F., 2017. Hybrid penguins search optimization algorithm and genetic algorithm solving traveling salesman problem. In: International Conference on Advanced Information Technology, Services and Systems, pp. 461–473.

Neshat, M., Sepidnam, G., Sargolzaei, M., 2013. Swallow swarm optimization algorithm: a new method to optimization. Neural Computing & Applications 23 (2), 429–454.

Odili, J.B., Mohmad Kahar, M.N., 2016. Solving the traveling salesman's problem using the African buffalo optimization. Computational Intelligence and Neuroscience 2016, 3.

Odili, J.B., Kahar, M.N.M., Anwar, S., 2015. African buffalo optimization: a swarm-intelligence technique. Procedia Computer Science 76, 443–448.

Odili, J.B., Kahar, M.N., Noraziah, A., 2016. Solving traveling salesman's problem using African buffalo optimization, honey bee mating optimization & Lin–Kerninghan algorithms. World Applied Sciences Journal 34 (7), 911–916.

Odili, J., Kahar, M.N.M., Anwar, S., Ali, M., 2017. Tutorials on African buffalo optimization for solving the travelling salesman problem. International Journal of Software Engineering and Computer Systems 3 (3), 120–128.

Osaba, E., Díaz, F., Carballedo, R., Onieva, E., Perallos, A., 2014a. Focusing on the golden ball metaheuristic: an extended study on a wider set of problems. The Scientific World Journal 2014.

Osaba, E., Diaz, F., Onieva, E., 2014b. Golden ball: a novel meta-heuristic to solve combinatorial optimization problems based on soccer concepts. Applied Intelligence 41 (1), 145–166.

Osaba, E., Onieva, E., Diaz, F., Carballedo, R., Lopez, P., Perallos, A., 2015. An asymmetric multiple traveling salesman problem with backhauls to solve a dial-a-ride problem. In: 2015 IEEE 13th International Symposium on Applied Machine Intelligence and Informatics (SAMI), pp. 151–156.

Osaba, E., Carballedo, R., López-García, P., Diaz, F., 2016a. Comparison between golden ball metaheuristic, evolutionary simulated annealing and tabu search for the traveling salesman problem. In: Proceedings of the 2016 on Genetic and Evolutionary Computation Conference Companion, pp. 1469–1470.

Osaba, E., Yang, X.S., Diaz, F., Lopez-Garcia, P., Carballedo, R., 2016b. An improved discrete bat algorithm for symmetric and asymmetric traveling salesman problems. Engineering Applications of Artificial Intelligence 48, 59–71.

Osaba, E., Yang, X.S., Diaz, F., Onieva, E., Masegosa, A.D., Perallos, A., 2017. A discrete firefly algorithm to solve a rich vehicle routing problem modelling a newspaper distribution system with recycling policy. Soft Computing 21 (18), 5295–5308.

Osaba, E., Carballedo, R., Diaz, F., Onieva, E., Masegosa, A., Perallos, A., 2018a. Good practice proposal for the implementation, presentation, and comparison of metaheuristics for solving routing problems. Neurocomputing 271, 2–8.

Osaba, E., Carballedo, R., Yang, X.S., Fister Jr, I., Lopez-Garcia, P., Del Ser, J., 2018b. On efficiently solving the vehicle routing problem with time windows using the bat algorithm with random reinsertion operators. In: Nature-Inspired Algorithms and Applied Optimization. Springer, pp. 69–89.

Osaba, E., Del Ser, J., Camacho, D., Galvez, A., Iglesias, A., Fister, I., 2018c. Community detection in weighted directed networks using nature-inspired heuristics. In: International Conference on Intelligent Data Engineering and Automated Learning, pp. 325–335.

Osaba, E., Del Ser, J., Iglesias, A., Bilbao, M.N., Fister, I., Galvez, A., 2018d. Solving the open-path asymmetric green traveling salesman problem in a realistic urban environment. In: International Symposium on Intelligent and Distributed Computing, pp. 181–191.

Osaba, E., Del Ser, J., Sadollah, A., Bilbao, M.N., Camacho, D., 2018e. A discrete water cycle algorithm for solving the symmetric and asymmetric traveling salesman problem. Applied Soft Computing 71, 277–290.

Osaba, E., Yang, X.S., Fister Jr, I., Del Ser, J., Lopez-Garcia, P., Vazquez-Pardavila, A.J., 2019. A discrete and improved bat algorithm for solving a medical goods distribution problem with pharmacological waste collection. Swarm and Evolutionary Computation 44, 273–286.

Ouaarab, A., Ahiod, B., Yang, X.S., 2014. Discrete cuckoo search algorithm for the travelling salesman problem. Neural Computing & Applications 24 (7–8), 1659–1669.

Ouaarab, A., Ahiod, B., Yang, X.S., 2015. Random-key cuckoo search for the travelling salesman problem. Soft Computing 19 (4), 1099–1106.

Pandiri, V., Singh, A., 2018. A hyper-heuristic based artificial bee colony algorithm for k-interconnected multi-depot multi-traveling salesman problem. Information Sciences 463, 261–281.

Pang, W., Wang, K.p., Zhou, C.g., Dong, L.j., 2004. Fuzzy discrete particle swarm optimization for solving traveling salesman problem. In: The Fourth International Conference on Computer and Information Technology. CIT'04, 2004, pp. 796–800.

Pang, S., Ma, T., Liu, T., 2015. An improved ant colony optimization with optimal search library for solving the traveling salesman problem. Journal of Computational and Theoretical Nanoscience 12 (7), 1440–1444.

Potvin, J.Y., 1993. State-of-the-art survey—the traveling salesman problem: a neural network perspective. ORSA Journal on Computing 5 (4), 328–348.

Pozna, C., Precup, R.E., Tar, J.K., Škrjanc, I., Preitl, S., 2010. New results in modelling derived from Bayesian filtering. Knowledge-Based Systems 23 (2), 182–194.

Precup, R.E., David, R.C., 2019. Nature-Inspired Optimization Algorithms for Fuzzy Controlled Servo Systems. Butterworth-Heinemann.

Qiu, C., Xiang, F., 2019. Feature selection using a set based discrete particle swarm optimization and a novel feature subset evaluation criterion. Intelligent Data Analysis 23 (1), 5–21.

Rashedi, E., Nezamabadi-Pour, H., Saryazdi, S., 2009. GSA: a gravitational search algorithm. Information Sciences 179 (13), 2232–2248.

Reinelt, G., 1991. TSPLIB: a traveling salesman problem library. ORSA Journal on Computing 3 (4), 376–384.

Roberti, R., Wen, M., 2016. The electric traveling salesman problem with time windows. Transportation Research. Part E, Logistics and Transportation Review 89, 32–52.

Rostami, A.S., Mohanna, F., Keshavarz, H., Hosseinabadi, A.A.R., 2015. Solving multiple traveling salesman problem using the gravitational emulation local search algorithm. Applied Mathematics & Information Sciences 9 (2), 1–11.

Roy, A., Manna, A., Maity, S., 2019. A novel memetic genetic algorithm for solving traveling salesman problem based on multi-parent crossover technique. Decision Making: Applications in Management and Engineering.

Saad, A., Dong, Z., Buckham, B., Crawford, C., Younis, A., Karimi, M., 2019. A new Kriging–Bat algorithm for solving computationally expensive black-box global optimization problems. Engineering Optimization 51 (2), 265–285.

Sahana, S.K., et al., 2018. An improved modular hybrid ant colony approach for solving traveling salesman problem. GSTF Journal on Computing (JoC) 1 (2).

Saji, Y., Riffi, M.E., 2016. A novel discrete bat algorithm for solving the travelling salesman problem. Neural Computing & Applications 27 (7), 1853–1866.

Sakai, M., Hanada, Y., Orito, Y., 2018. Edge assembly crossover using multiple parents for traveling salesman problem. In: 2018 Joint 10th International Conference on Soft Computing and Intelligent Systems, pp. 474–477.

Saraei, M., Mansouri, P., 2019. HMFA: a hybrid mutation-base firefly algorithm for travelling salesman problem. In: Fundamental Research in Electrical Engineering. Springer, pp. 413–427.

Sayoti, F., Riffi, M., 2015. Random-keys golden ball algoritm for solving traveling salesman problem. International Review on Modelling and Simulations (IREMOS) 8 (1), 84–89.

Shi, Y., 2011. Brain storm optimization algorithm. In: International Conference in Swarm Intelligence, pp. 303–309.

Shi, X.H., Liang, Y.C., Lee, H.P., Lu, C., Wang, Q., 2007. Particle swarm optimization-based algorithms for TSP and generalized TSP. Information Processing Letters 103 (5), 169–176.

Smith, S.L., Imeson, F., 2017. GLNS: an effective large neighborhood search heuristic for the generalized traveling salesman problem. Computers & Operations Research 87, 1–19.

Strange, R., 2017. Discrete Flower Pollination Algorithm for Solving the Symmetric Traveling Salesman Problem. Ph.D. thesis.

Svensson, O., 2018. Algorithms for the asymmetric traveling salesman problem. In: 38th IARCS Annual Conference on Foundations of Software Technology and Theoretical Computer Science.

Taidi, Z., Benameur, L., Chentoufi, J.A., 2017. A fireworks algorithm for solving travelling salesman problem. International Journal of Computational Systems Engineering 3 (3), 157–162.

Tan, Y., Zhu, Y., 2010. Fireworks algorithm for optimization. In: International conference in swarm intelligence, pp. 355–364.

Tang, K., Li, Z., Luo, L., Liu, B., 2015. Multi-strategy adaptive particle swarm optimization for numerical optimization. Engineering Applications of Artificial Intelligence 37, 9–19.

Tarantilis, C.D., 2005. Solving the vehicle routing problem with adaptive memory programming methodology. Computers & Operations Research 32 (9), 2309–2327.

Tarantilis, C., Kiranoudis, C., 2007. A flexible adaptive memory-based algorithm for real-life transportation operations: two case studies from dairy and construction sector. European Journal of Operational Research 179 (3), 806–822.

Teng, L., Li, H., 2018. Modified discrete firefly algorithm combining genetic algorithm for traveling salesman problem. Telkomnika 16 (1), 424–431.

Tilahun, S.L., Ngnotchouye, J.M.T., 2017. Firefly algorithm for discrete optimization problems: a survey. KSCE Journal of Civil Engineering 21 (2), 535–545.

Tzy-Luen, N., Keat, Y.T., Abdullah, R., 2016. Parallel cuckoo search algorithm on OpenMP for traveling salesman problem. In: 2016 3rd International Conference on Computer and Information Sciences (ICCOINS), pp. 380–385.

Vaghela, K.N., Tanna, P.J., Lathigara, A.M., 2018. Job scheduling heuristics and simulation tools in cloud computing environment: a survey. International Journal of Advanced Networking and Applications 10 (2), 3782–3787.

Venkatesh, P., Singh, A., 2019. An artificial bee colony algorithm with variable degree of perturbation for the generalized covering traveling salesman problem. Applied Soft Computing.

Wang, Y., Xu, N., 2017. A hybrid particle swarm optimization method for traveling salesman problem. International Journal of Applied Metaheuristic Computing (IJAMC) 8 (3), 53–65.

Wang, K.P., Huang, L., Zhou, C.G., Pang, W., 2003. Particle swarm optimization for traveling salesman problem. In: Proceedings of the 2003 International Conference on Machine Learning and Cybernetics. IEEE Cat. No. 03EX693, vol. 3, pp. 1583–1585.

Wang, J., Ersoy, O.K., He, M., Wang, F., 2016. Multi-offspring genetic algorithm and its application to the traveling salesman problem. Applied Soft Computing 43, 415–423.

Wang, G.G., Gao, X.Z., Zenger, K., Coelho, L.d.S., 2018a. A novel metaheuristic algorithm inspired by rhino herd behavior. In: Proceedings of The 9th EUROSIM Congress on Modelling and Simulation, vol. 142, pp. 1026–1033.

Wang, Y., Wang, Q.P., Wang, X.F., 2018b. Solving traveling salesman problem based on improved firefly algorithm. Computer Systems & Applications 8, 37.

Wedyan, A., Whalley, J., Narayanan, A., 2017. Hydrological cycle algorithm for continuous optimization problems. Journal of Optimization 2017.

Wedyan, A., Whalley, J., Narayanan, A., 2018. Solving the traveling salesman problem using hydrological cycle algorithm. American Journal of Operations Research 8 (03), 133.

Wu, X., Gao, D., 2017. A study on greedy search to improve simulated annealing for large-scale traveling salesman problem. In: International Conference on Swarm Intelligence, pp. 250–257.

Wu, X., Shen, X., Zhang, L., 2019. Solving the planning and scheduling problem simultaneously in a hospital with a bi-layer discrete particle swarm optimization. Mathematical Biosciences and Engineering 16 (2), 831–861.

Xu, S., Wang, Y., Huang, A., 2014. Application of imperialist competitive algorithm on solving the traveling salesman problem. Algorithms 7 (2), 229–242.

Xu, D., Weise, T., Wu, Y., Lässig, J., Chiong, R., 2015. An investigation of hybrid tabu search for the traveling salesman problem. Bio-Inspired Computing-Theories and Applications, 523–537.

Xu, M., Li, S., Guo, J., 2017. Optimization of multiple traveling salesman problem based on simulated annealing genetic algorithm. In: MATEC Web of Conferences, vol. 100, 02025.

Xu, Y., Wu, Y., Fu, Y., Wang, X., Lu, A., 2018. Discrete brain storm optimization algorithm based on prior knowledge for traveling salesman problems. In: 2018 13th IEEE Conference on Industrial Electronics and Applications (ICIEA), pp. 2740–2745.

Yang, X.S., 2008. Nature-Inspired Metaheuristic Algorithms. Luniver Press, UK.

Yang, X.S., 2009. Firefly algorithms for multimodal optimization. In: Stochastic Algorithms: Foundations and Applications. Springer, pp. 169–178.

Yang, X.S., 2010. A new metaheuristic bat-inspired algorithm. In: Nature Inspired Cooperative Strategies for Optimization (NICSO 2010). Springer, pp. 65–74.

Yang, X.S., 2012. Flower pollination algorithm for global optimization. In: International Conference on Unconventional Computing and Natural Computation, pp. 240–249.

Yang, X.S., He, X., 2013a. Bat algorithm: literature review and applications. International Journal of Bio-Inspired Computation 5 (3), 141–149.

Yang, X.S., He, X., 2013b. Firefly algorithm: recent advances and applications. International Journal of Swarm Intelligence (IJSI) 1, 36–50.

Yang, X.S., He, X.S., 2018. Why the firefly algorithm works? In: Nature-Inspired Algorithms and Applied Optimization. Springer, pp. 245–259.

Yousefikhoshbakht, M., Dolatnejad, A., 2016. An efficient combined meta-heuristic algorithm for solving the traveling salesman problem. BRAIN. Broad Research in Artificial Intelligence and Neuroscience 7 (3), 125–138.

Yousefikhoshbakht, M., Sedighpour, M., 2013. New imperialist competitive algorithm to solve the travelling salesman problem. International Journal of Computer Mathematics 90 (7), 1495–1505.

Yu, B., Yang, Z.Z., Yao, B., 2009. An improved ant colony optimization for vehicle routing problem. European Journal of Operational Research 196 (1), 171–176.

Yu, X., Chen, W.N., Hu, X.m., Zhang, J., 2015. A set-based comprehensive learning particle swarm optimization with decomposition for multiobjective traveling salesman problem. In: Proceedings of the 2015 Annual Conference on Genetic and Evolutionary Computation, pp. 89–96.

Zaidi, T., Gupta, P., 2018. Traveling salesman problem with ant colony optimization algorithm for cloud computing environment. International Journal of Grid and Distributed Computing 11 (8), 13–22.

Zhan, S.h., Lin, J., Zhang, Z.j., Zhong, Y.w., 2016. List-based simulated annealing algorithm for traveling salesman problem. Computational Intelligence and Neuroscience 2016, 8.

Zhang, Z., Gao, C., Lu, Y., Liu, Y., Liang, M., 2016. Multi-objective ant colony optimization based on the Physarum-inspired mathematical model for bi-objective traveling salesman problems. PLoS ONE 11 (1), e0146709.

Zhong, W.h., Zhang, J., Chen, W.n., 2007. A novel discrete particle swarm optimization to solve traveling salesman problem. In: IEEE Congress on Evolutionary Computation, pp. 3283–3287.

Zhong, Y., Lin, J., Wang, L., Zhang, H., 2017. Hybrid discrete artificial bee colony algorithm with threshold acceptance criterion for traveling salesman problem. Information Sciences 421, 70–84.

Zhong, Y., Lin, J., Wang, L., Zhang, H., 2018. Discrete comprehensive learning particle swarm optimization algorithm with metropolis acceptance criterion for traveling salesman problem. Swarm and Evolutionary Computation 42, 77–88.

Zhong, Y., Wang, L., Lin, M., Zhang, H., 2019. Discrete pigeon-inspired optimization algorithm with metropolis acceptance criterion for large-scale traveling salesman problem. Swarm and Evolutionary Computation 48, 134–144.

Zhou, L., Ding, L., Qiang, X., Luo, Y., 2015. An improved discrete firefly algorithm for the traveling salesman problem. Journal of Computational and Theoretical Nanoscience 12 (7), 1184–1189.

Zhou, Y., Wang, R., Zhao, C., Luo, Q., Metwally, M.A., 2017. Discrete greedy flower pollination algorithm for spherical traveling salesman problem. Neural Computing & Applications, 1–16.

Zhou, A.H., Zhu, L.P., Hu, B., Deng, S., Song, Y., Qiu, H., Pan, S., 2019. Traveling-salesman-problem algorithm based on simulated annealing and gene-expression programming. Information 10 (1), 7.

Clustering with nature-inspired metaheuristics

10

Szymon Łukasik[a,b], **Piotr A. Kowalski**[a,b]

[a]*AGH University of Science and Technology,
Faculty of Physics and Applied Computer Science, Kraków, Poland*
[b]*Systems Research Institute, Polish Academy of Sciences, Warsaw, Poland*

CONTENTS

10.1 Introduction

Unsupervised learning is aimed at revealing unknown patterns in datasets without preexisting labels. Such patterns may correspond to outliers or groups of data elements sharing similar properties. The latter corresponds to the well-known task of cluster analysis/clustering, and it manifests itself in a variety of domains, ranging from marketing (Dietrich et al., 2016) to the analysis of power systems (Radionov et al., 2015). Practically finding clusters can for example uncover groups of customers sharing same behavioral patterns or pixels of an aerial image representing similar objects.

Fig. 10.1 demonstrates the result of clustering for a two-dimensional representation of the well-known benchmark *seeds* dataset. It demonstrates both algorithmic difficulty of dividing the dataset into disjoint groups and also its subjectivity with regard to the number of clusters suitable for the considered task.

Existing classical approaches to cluster analysis belong to one of the following classes:

- hierarchical clustering,
- partitional clustering,

Nature-Inspired Computation and Swarm Intelligence. https://doi.org/10.1016/B978-0-12-819714-1.00021-X

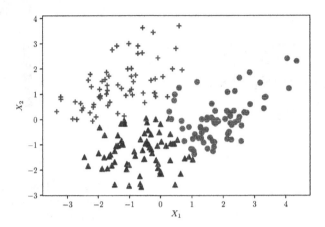

FIGURE 10.1

Results of clustering of the *seeds* dataset projected into two-dimensional space.

- density-based clustering,
- grid-based clustering,
- model-based clustering.

In hierarchical clustering a series of partitions takes place – either starting from one cluster encompassing the whole dataset or m singleton clusters (with m indicating the dataset's size). Clusters are merged or divided according to the distance between them. Several methods of hierarchical clustering differ in the way this distance is defined (Azzalini and Scarpa, 2012).

Partitional clustering relies on decomposing the dataset directly into disjoint groups using some optimization criteria, typically associated with local properties of the dataset. The most representative technique of this type is the k-means algorithm (MacQueen, 1967). It allocates cluster centers in a way that the sum of squared distances from all data points to the closest cluster center is minimized (locally).

Density-based algorithms, such as DBSCAN (Ester et al., 1996), associate clusters with dense areas in the feature space of the considered dataset. Besides grouping, it may lead to a possibility of additional data cleanup where low-density areas are treated as noise.

Grid-based clustering tackles the problem through data quantization – into separate cells and performing clustering on a lower level. A typical representative of this strategy is the WaveCluster approach, which in addition uses wavelet transformation to convert spatial data into frequency domain data (Sheikholeslami et al., 2000).

Model-based methods in turn optimize the fit between the given data and some predefined mathematical model. It can be either a mixture of probability distributions or its nonparametric formulation (Kulczycki and Charytanowicz, 2010). For a more detailed overview of classical clustering methods, readers can refer to Fahad et al. (2014).

Clustering is also tackled with metaheuristics, which lies in the scope of this contribution. Despite the high complexity of some problems, these algorithms have the ability to include various aspects of clustering tasks (such as constraints, different optimization criteria, etc.). This makes such algorithms very flexible clustering problem solvers.

This chapter will discuss the applications of nature-inspired metaheuristics. In the next section we will present a formal definition of clustering and review solution representation and possible optimization criteria. This section will also provide examples of existing nature-inspired clustering algorithms. In the subsequent section we will demonstrate the results of our investigations on the clustering algorithms based on techniques mimicking natural phenomena. We will conclude this chapter by the analysis of future research prospects in the area of nature-inspired clustering.

10.2 Clustering with metaheuristics

Let as assume that we are dealing with the dataset $Y = \{y_1, y_2, ...y_M\}$, consisting of M elements with N features. Clustering can be formally defined as a task of assigning to each element a membership to one of the clusters $CL = \{CL_1, CL_2, ..., CL_C\}$, that is, finding an assignment

$$f : Y \rightarrow CL. \tag{10.1}$$

The following part of this chapter is devoted to the methods of representing such problem in the optimization domain. It involves first selecting a representation of clustering solution and then formulating clustering as an optimization task, i.e., the task of finding a value of x_n within the feasible n-dimensional search space S, denoted as x^*, such that

$$x^* = \underset{x \in S}{\arg\min} \, h(x), \tag{10.2}$$

assuming that the goal is to minimize the cost function h.

For the purpose of optimization clustering solutions can be represented directly – as the vector of object–cluster associations. It means that the solution is then encoded using m integers ranging from 1 to C. It makes a clustering problem a combinatorial optimization task. An example of such approach involving the use of genetic algorithms as an optimization tool can be found in Murthy and Chowdhury (1996). The second method to represent clustering solutions is through the boundaries of clusters. Examples of studies using this paradigm can be found in Das et al. (2009). The third method involves encoding a solution through cluster centroids defined, for a nonempty cluster CL_i, as follows:

$$u_i = \frac{1}{M_i} \sum_{y_j \in CL_i} y_j, \quad i = 1, ..., C, \tag{10.3}$$

where M_i constitutes cardinality of cluster i. Similarly U represents the center of gravity of the whole dataset, i.e.,

$$U = \frac{1}{M} \sum_{j=1}^{M} y_j.$$ (10.4)

It can be supplemented with an additional shape parameter.

Typically clustering involves either a priori, directly defining the number of clusters or defining some parameter associated with the number of clusters. In the case of metaheuristic clustering, one could define a representation with a variable number of groups. An example of such approach can be found in Bandyopadhyay and Maulik (2002).

As an optimization criterion one could choose to minimize the within-cluster sum of squares (WCSS):

$$V_W = \sum_{c=1}^{C} \sum_{y_i \in CL_c} \| y_i - u_c \|^2,$$ (10.5)

with $\| \cdot \|$ being the L^2-norm (Euclidean distance) between the two vectors. It is the optimization criterion of the classic k-means algorithm. Most metaheuristic approaches however use one of the internal validity indices, commonly used to evaluate clustering quality, such as the Dunn index (Dunn, 1974), the Davies–Bouldin index (Davies and Bouldin, 1979), or the Calinski–Harabasz index (Caliński and Harabasz, 1974). A more extensive experimental study on the properties of clustering indices was presented in Arbelaitz et al. (2013).

The objective of this section is to present the most common strategies for solving clustering problems with heuristic algorithms. For a recent study proving wide-ranging references to over 60 clustering approaches based on metaheuristics readers can refer to José-García and Gómez-Flores (2016). An interesting outlook taking into account also multiobjective formulations of the clustering task can be found in Nanda and Panda (2014). In the following section we will provide illustrative examples of employing three selected nature-inspired techniques for cluster analysis.

10.3 Use cases

10.3.1 Problem definition

Let us assume that the clustering solution is represented by a vector of cluster centers $x_p = [u_1, u_2, ..., u_C]$. Consequently the dimensionality D of solution vector is equal to $C \times N$ as each cluster center u_i is a vector of N elements.

As a function evaluating solution quality we are using the following formula:

$$f(x_p) = \frac{1}{I_{CH,p}} + \#_{CL_{i,p}=\emptyset, \, i=1,...,C},$$ (10.6)

where $I_{CH,p}$ corresponds to the value of the Calinski–Harabasz index calculated for solution p. It is obtained as

$$I_{CH} = \frac{V_B}{V_W} \frac{M - C}{C - 1}, \tag{10.7}$$

where V_B and V_W denote overall between-cluster and within-cluster variance (defined in Eq. (10.5)), respectively. These are calculated according to the following formulas:

$$V_B = \sum_{c=1}^{C} \#CL_c \|u_c - U\|^2 \tag{10.8}$$

and

$$V_W = \sum_{c=1}^{C} \sum_{y_i \in CL_c} \|y_i - u_c\|^2. \tag{10.9}$$

The value denoted as $\#_{CL_{i,p}=\emptyset, i=1,...,C}$ in Eq. (10.6) corresponds to the number of empty clusters identified in the x_p clustering solution. It effectively penalizes solutions which do not include a desirable number of clusters.

10.3.2 Optimization algorithms

For the purpose of this chapter we will present the results of employing the krill herd algorithm (KHA), the flower pollination algorithm (FPA), and the cuttlefish optimization algorithm (COA) for the clustering task. Their basic schemes will be presented in this subsection.

KHA represents a global optimization procedure mimicking social interactions identified in the Antarctic krill swarms. This procedure was originally introduced by Gandomi and Alavi (2012). KHA is based on modeling the behavior of individual krill within a large herd, which can span hundreds of meters in a natural habitat (Kowalski and Łukasik, 2015).

Algorithm 1 starts (iteration $k = 0$) via the initialization of data structures, i.e., vectors describing individuals $x_i(k = 0)$, as well as the whole population consisting of P krills. Initializing the data structure representing a single krill is equivalent to placing it in a certain place of the solution space. For this purpose, it is usually recommended to employ random number generation following a uniform distribution. Similarly to other algorithms inspired by nature, each individual represents one possible solution of the problem under consideration. After the initialization, the algorithm continues through a series of iterations. Each of them starts with calculating the fitness function value $K(x_i(k))$ for each individual of the population. This is equivalent to the calculation of the optimized function value for the coordinates of the krill's position. Then, for each individual of the population, the vector which indicates its displacement in the solution space is calculated. The movement of the individual krill

---

Algorithm 1 Krill herd algorithm pseudocode.

InitializeParameters (D^{max}, N^{max}, etc.);
$k \leftarrow 1$;
for $i = 1$ *to* P **do**
    GenerateSolution ($x_i(k)$);
    evaluate and update best solutions;
    $K(x_i(0)) \leftarrow$ EvaluateQuality ($x_i(0)$);
$x^* \leftarrow$ SaveBest ($x^*(0)$);
main loop;
repeat
    for $i = 1$ *to* P **do**
        Perform motion calculation and genetic operators;
        $N_i \leftarrow$ MotionInducedOthers ();
        $F_i \leftarrow$ Foraging ();
        $D_i \leftarrow$ RandomDiffusion ();
        Crossover ();
        Mutation ();
        UpdateSolution ($x_m(k)$);
        Evaluate and update best solutions;
        $f(x_i(k)) \leftarrow$ EvaluateQuality ($x_i(k)$);
    $x^* \leftarrow$ SaveBest $x^*(k)$
    stop condition $\leftarrow$ CheckStopCondition ();
    $k \leftarrow k + 1$;
until *stop condition = false*;
return $f(x^*(k))$, $x^*(k)$, k;

---

is described by an equation relying on three factors:

$$\frac{dX_i}{dt} = N_i + F_i + D_i, \tag{10.10}$$

where N_i refers to the movement induced by the influence of other krills, F_i is a movement associated with the search for food, and D_i constitutes a random diffusion factor. Finally it is worth to observe that KHA utilizes standard evolutionary operators such as mutation and crossover. More information on the movements strategies in KHA, parameter values, etc., can be found in Kowalski and Łukasik (2015).

FPA tries to mimic in the optimization domain a set of complex mechanisms crucial to the success of plants' reproductive strategies. A single flower or pollen gamete constitutes a solution of the optimization problem, with the whole flower population being actually used. Their constancy will be understood as solution fitness. Pollen is transferred in the course of two operations used interchangeably, that is, global and local pollination. The first one employs pollinators to carry pollen to long distances towards individuals characterized by higher fitness. Local pollination on the other hand occurs within a limited range of the individual flower thanks to pollination mediators like wind or water (Łukasik and Kowalski, 2015).

FPA's formal description using the aforementioned notation and ideas will be presented below (Algorithm 2). It can be seen that global pollination occurs with probability $prob$, defined by the so-called switch probability. If this phase is omitted local pollination takes place instead. The first type of pollination refers to the pollinator's movement towards the best solution $x^*(k)$ found by the algorithm, with s representing a D-dimensional step vector following a Lévy distribution, i.e.,

$$L(s) \sim \frac{\lambda \Gamma(\lambda) \sin(\pi \lambda/2)}{\pi s^{1+\lambda}}, \ (s \gg s_0 > 0), \tag{10.11}$$

and Γ being the standard gamma function. Yang et al. (2013) suggested the following parameter values: $\lambda = 1.5$ and $s_0 = 0.1$. Practical methods for obtaining step sizes s following this distribution, by means of the Mantegna algorithm, are given in Yang (2014). Local pollination includes two randomly selected members of the population and is performed via movement towards them, with randomly selected step size ϵ. Finally, the algorithm is terminated when the number of iteration k reaches a predetermined limit defined by K (Łukasik and Kowalski, 2015).

COA is a global optimization procedure inspired by the natural behavioral activity of cuttlefish swarms. This procedure was introduced by Adel Sabry Eesa, Adnan Mohsin Abdulazeez Brifcani, and Zeynep Orman, in their paper (Eesa et al., 2013).

The algorithm is inspired by the change of color by the cuttlefish – a predatory cephalopod with eight arms and two tentacles. Cuttlefish can change color, both to hide from the threat and to lure a partner during the mating season. The patterns and colors of cuttlefish are produced by light reflected from different layers of cells (chromatophores, leucophores, and iridophores) put together and this combination of cells allows the cuttlefish to take on a large number of patterns and colors. The algorithm was designed based on two processes, i.e., reflection and visibility, which serve as components allowing to generate and analyze new solutions.

The way of finding a new solution in the optimization task by means of reflection and visibility is described as follows:

$$x_{new} = reflection + visibility. \tag{10.12}$$

The cuttlefish cells are divided into four groups in which new patterns and colors are obtained, which in computational intelligence is associated with the acquisition of new solutions. These groups are defined as follows.

- Group I, movement strategy relying only the current solution:

$$reflection = R * x_c, \tag{10.13}$$

$$visibility = V * (x_{best} - x_c). \tag{10.14}$$

Algorithm 2 Flower pollination algorithm pseudocode.

Initialization;
$k \leftarrow 1;$
$f(x^*(0)) \leftarrow \infty;$
for $p = 1$ *to* P **do**
    | GenerateSolution $(x_p(k));$
find best;
for $p = 1$ *to* P **do**
    | $f(x_p(k)) \leftarrow$ EvaluateQuality $(x_p(k));$
    | **if** $f(x_p(k)) < f(x^*(k-1))$ **then**
    | | $x^*(k) \leftarrow x_p(k);$
    |
    | **else**
    | | $x^*(k) \leftarrow x^*(k-1);$
Main loop starts here;
repeat
    | **for** $p = 1$ *to* P **do**
    | | **if** *Real_Rand_in_*$(0,1) < prob$ **then**
    | | | *Global pollination;*
    | | | $s \leftarrow Levy(s_0, \gamma);$
    | | | $x_{trial} \leftarrow x_p(k) + s(x^*(k) - x_p(k));$
    | | **else**
    | | | *Local pollination;*
    | | | $\epsilon \leftarrow Real\_Rand\_in\_(0,1);$
    | | | $r, q \leftarrow Integer\_Rand\_in(1, M);$
    | | | $x_{trial} \leftarrow x_p(k) + \epsilon(x_q(k) - x_r(k));$
    | | *Check if new solution better;*
    | | $f(x_{trial}) \leftarrow$ EvaluateQuality $(x_{trial});$
    | | **if** $f(x_{trial}) < f(x_p(k))$ **then**
    | | | $x_p(k) \leftarrow x_{trial};$
    | | | $f(x_p(k)) \leftarrow f(x_{trial});$
    | *Find best and copy population;*
    | **for** $p = 1$ *to* P **do**
    | | **if** $f(x_p(k)) < f(x^*(k-1))$ **then**
    | | | $x^*(k) \leftarrow x_p(k);$
    | | **else**
    | | | $x^*(p) \leftarrow x^*(k-1);$
    | | $f(x(k+1)) \leftarrow f(x_k);$
    | | $x(k+1) \leftarrow x(k);$
    | $f(x^*(k+1)) \leftarrow f(x^*k);$
    | $x^*(k+1) \leftarrow x^*(k);$
    | $k \leftarrow k+1;$
until *stop_condition = false;*
return $f(x^*(k)), x^*(k), k$

- Group II, movement strategy relying only on the current and the best solution:

$$reflection = R * x_{best}, \tag{10.15}$$

$$visibility = V * (x_{best} - x_c). \tag{10.16}$$

- Group III, movement strategy relying only on the best solution:

$$reflection = R * x_{best}, \tag{10.17}$$

$$visibility = V * (x_{best} - x_{avg}). \tag{10.18}$$

- Group IV, movement strategy relying only on a random search within the investigated space:

$$reflection = \xi, \tag{10.19}$$

$$visibility = 0 \tag{10.20}$$

where x_{new} denotes a proposition of a new solution, x_{cur} is the current solution, x_{best} indicates the best solution, ξ is a random position of the investigated state space, R designates a random value generated from the uniform distribution within the interval $[r_1, r_2]$, V denotes a random value generated from the uniform distribution within the interval $[v_1, v_2]$, and x_{avg} is an average value of the vectors representing the best solution.

The population is divided into groups in order to increase the possibility of defining procedures that will lead to finding the best solution. Group I is aimed at exploiting the search space in order to find a global extreme, using the best solution found so far. Group IV has a similar role; here, a search of the space in a completely random way is carried out. However, a different role is played by Groups II and III. They are focused on exploring the local extreme in the hope that it is a global extreme.

The operation of the algorithm begins via the initialization of data structures, i.e., describing individuals, as well as the whole population. Initializing the data structure representing a single cuttlefish means situating it in a certain place of the feature space. For this purpose, it is recommended to employ random number generation following a uniform distribution. Like other algorithms inspired by nature, each individual represents one possible solution of the problem under consideration. After the initialization phase of the algorithm, it continues into a series of iterations.

Algorithm 3 presents the high-level pseudocode of the discussed algorithm. At the beginning, the population is initialized with random values of the N-dimensional space and the evaluation of individual solutions. In the next step, the population is divided into the four groups discussed above. If the current solution (in each of the four groups) is better than the best solution so far, then the best solution is overwritten with the new solution. These steps are repeated until the stop criterion is fulfilled.

The algorithm depends on the selection of four parameters: r_1, r_2, v_1, and v_2. For more information on the parameter values and their impact on the algorithm performance interested readers can refer to Kowalski et al. (2020).

Algorithm 3 High-level COA pseudocode.

Define and populate the algorithm's data structures;
Initialize random initial population, i.e., position of the swarm of cuttlefish;
Evaluate fitness of each individual cuttlefish on the basis of its position;
Find the best global solution;
Divide cuttlefish population into four subgroups G_I, G_{II}, G_{III}, and G_{IV};
while *stop criteria is not reached* **do**
    Calculate the average value of the best global solution X_{avg};
    for *i=1 to P* **do**
        Calculate the reflection based on Eq. (10.13);
        Calculate the visibility based on Eq. (10.14);
        Generate the new position of ith individual based on Eq. (10.12);
        Update current position (if the achieved cost is lower);
    for *i=1 to population size in G_{II}* **do**
        Calculate the reflection based on Eq. (10.15);
        Calculate the visibility based on Eq. (10.16);
        Generate the new position of ith individual based on Eq. (10.12);
        Update current position (if the achieved cost is lower)
    for *i=1:population size in G_{III}* **do**
        Calculate the reflection based on Eq. (10.17);
        Calculate the visibility based on Eq. (10.18);
        Generate the new position of ith individual based on Eq. (10.12);
        Update current position (if the achieved cost is lower);
    for *i=1:population size in G_{IV}* **do**
        Generate the new position of ith individual based on Eqs. (10.20) and (10.12);
        Update current position (if the achieved cost is lower);
    Evaluate fitness of each individual on the basis of its position;
    Update the best global solution (if the achieved cost is lower);
return x^*, k_{max}

10.3.3 Experimental results

To study the performance of aforementioned algorithms in the clustering tasks, we have performed experiments for selected benchmark datasets. Their properties are presented in Table 10.1. We have used four artificial datasets ($s1$, $s2$, $s3$, and $s4$), characterized by increasing cluster overlap. In addition, eight real-world benchmark instances taken from the UCI Machine Learning Repository (Lichman, 2013) were also included in our study.

For all algorithms we have used the same population size $P = 20$ and an identical number of iterations $k_{max} = 200$. The values of individual algorithms parameters were $N^{max} = 0.01$, $\omega_n = 0.5$, $V_f = 0.02$, $D^{max} = 0.01$, and $C^t = 0.5$ for KHA, $r_1 = -0.5$, $r_2 = -1.0$, $v_1 = -2.0$, and $v_2 = 2.0$ for COA, and $prob = 0.8$ for FPA, as established using preliminary pilot runs.

Table 10.1 Datasets used in the experiments.

Dataset	Sample size	Dimensionality	Clusters	Reference
s1	5000	2	15	Fränti and Virmajoki (2006)
s2	5000	2	6	Fränti and Virmajoki (2006)
s3	5000	2	3	Fränti and Virmajoki (2006)
s4	5000	2	6	Fränti and Virmajoki (2006)
ionosphere	351	34	2	Sigillito et al. (1989)
iris	150	4	3	Kowalski and Kulczycki (2017)
seeds	210	7	3	Charytanowicz et al. (2010)
sonar	208	60	2	Gorman and Sejnowski (1988)
thyroid	7200	21	3	Quinlan (1986); Quinlan et al. (1987)
vehicle	846	18	4	Setiono and Leow (1987)
WBC	683	10	2	Zhang (1992)
wine	178	13	3	Aeberhard et al. (1992)

Table 10.2 Results of data clustering using investigated algorithms.

Dataset	k-means		KHA		FPA		COA	
	$\overline{R}$	σ_R	$\overline{R}$	σ_R	$\overline{R}$	σ_R	$\overline{R}$	σ_R
s1	0.9748	0.0093	0.9782	0.0078	0.9950	0.0018	0.9930	0.0018
s2	0.9760	0.0072	0.9839	0.0053	0.9837	0.0037	0.9842	0.0037
s3	0.9522	0.0072	0.9548	0.0053	0.9583	0.0026	0.9587	0.0026
s4	0.9454	0.0056	0.9484	0.0048	0.9487	0.0023	0.9432	0.0023
iris	0.8458	0.0614	0.8872	0.0145	0.8931	0.0000	0.8991	0.0022
ionosphere	0.5945	0.0004	0.5573	0.0124	0.5946	0.0000	0.5949	0.0003
seeds	0.8573	0.0572	0.8709	0.0156	0.8839	0.0000	0.8731	0.0042
sonar	0.5116	0.0016	0.5145	0.0078	0.5128	0.0000	0.5191	0.0001
vehicle	0.5843	0.0359	0.6076	0.0194	0.6101	0.0006	0.6161	0.0006
WBC	0.5448	0.0040	0.5456	0.0000	0.5456	0.0000	0.5491	0.0056
wine	0.7167	0.0135	0.7257	0.0073	0.7299	0.0000	0.7452	0.0003
thyroid	0.5844	0.0982	0.4535	0.0339	0.5128	0.0000	0.5218	0.0032

Table 10.2 provides the results of experiments for investigated algorithms. As a point of reference the k-means clustering algorithm was also tested. Both mean $\overline{R}$ and standard deviation σ_R of Rand index values (Achtert et al., 2012) – calculated versus class labels for real-world benchmarks – are reported. All computational experiments were repeated 30 times.

To gain more insight into the performance of individual algorithms, a ranking based on the average value of the Rand index obtained for each dataset was also created. The results are presented in Table 10.3.

It can be seen that COA was found to be superior, with other nature-inspired techniques also performing much better than the classic k-means approach.

More details on the use of algorithms discussed here in the clustering tasks can be found in our papers Łukasik et al. (2016) and Kowalski et al. (2020), upon which this section is based.

Table 10.3 Algorithms' ranking based on average performance for each benchmark instance.

Dataset/algorithm	k-means	KHA	FPA	COA
s1	4	3	1	2
s2	4	2	3	1
s3	4	3	2	1
s4	3	2	1	4
iris	4	3	2	1
ionosphere	3	4	2	1
seeds	4	3	1	2
sonar	4	2	3	1
vehicle	4	3	2	1
WBC	4	2	2	1
wine	4	3	2	1
thyroid	1	4	3	2
Average rank	3.58	2.83	2.00	1.50
SD rank	0.90	0.72	0.74	0.90

10.4 Conclusion

This contribution provided a brief outlook on the applications of modern nature-inspired techniques in clustering tasks. Besides the description of existing strategies of metaheuristic clustering we have also demonstrated the application of three commonly used nature-inspired algorithms, namely, KHA, FPA, and COA, in the aforementioned problem. We found that all analyzed techniques outperform the standard k-means algorithm, in terms of both mean quality of obtained solutions and stability of final results. COA proved to be the most competitive one, offering the highest average accuracy in 8 out of 12 benchmark instances.

Follow-up studies could concern adding additional shape variables. As all considered techniques have also been modified to tackle multiobjective optimization tasks, additional clustering predeterminants could also be taken into account.

Acknowledgments

This work was partially financed (supported) by the Faculty of Physics and Applied Computer Science AGH UST statutory tasks with subsidy of the Ministry of Science and Higher Education.

The study was also supported in part by PL-Grid Infrastructure.

The authors would like to express their gratitude to Prof. Piotr Kulczycki and Prof. Małgorzata Charytanowicz, who provided insight and expertise that greatly assisted the research.

References

Achtert, E., Goldhofer, S., Kriegel, H.P., Schubert, E., Zimek, A., 2012. Evaluation of clusterings – metrics and visual support. In: 2012 IEEE 28th International Conference on Data Engineering, pp. 1285–1288.

Aeberhard, S., Coomans, D., De Vel, O., 1992. Comparison of classifiers in high dimensional settings. Tech. Rep. 92 (02). Dept. Math. Statist., James Cook Univ., North Queensland, Australia.

Arbelaitz, O., Gurrutxaga, I., Muguerza, J., Pérez, J.M., Perona, I., 2013. An extensive comparative study of cluster validity indices. Pattern Recognition (ISSN 0031-3203) 46 (1), 243–256. https://doi.org/10.1016/j.patcog.2012.07.021.

Azzalini, A., Scarpa, B., 2012. Data Analysis and Data Mining: An Introduction. Oxford University Press. ISBN 9780199942718.

Bandyopadhyay, S., Maulik, U., 2002. Genetic clustering for automatic evolution of clusters and application to image classification. Pattern Recognition (ISSN 0031-3203) 35 (6), 1197–1208. https://doi.org/10.1016/S0031-3203(01)00108-X.

Caliński, T., Harabasz, J., 1974. A dendrite method for cluster analysis. Communications in Statistics 3 (1), 1–27. https://doi.org/10.1080/03610927408827101.

Charytanowicz, M., Niewczas, J., Kulczycki, P., Kowalski, P.A., Łukasik, S., Żak, S., 2010. Complete gradient clustering algorithm for features analysis of x-ray images. In: Pietka, E., Kawa, J. (Eds.), Information Technologies in Biomedicine. In: Advances in Intelligent and Soft Computing, vol. 69. Springer, Berlin, Heidelberg, pp. 15–24.

Das, S., Abraham, A., Konar, A., 2009. Metaheuristic Pattern Clustering – An Overview. Springer Berlin Heidelberg, Berlin, Heidelberg. ISBN 978-3-540-93964-1, pp. 1–62.

Davies, D.L., Bouldin, D.W., 1979. A cluster separation measure. IEEE Transactions on Pattern Analysis and Machine Intelligence PAMI-1 (2), 224–227. https://doi.org/10.1109/TPAMI.1979.4766909.

Dietrich, T., Rundle-Thiele, S., Kubacki, K., 2016. Segmentation in Social Marketing: Process, Methods and Application. Springer, Singapore. ISBN 9789811018336.

Dunn, J.C., 1974. Well-separated clusters and optimal fuzzy partitions. Journal of Cybernetics 4 (1), 95–104. https://doi.org/10.1080/01969727408546059.

Eesa, A.S., Brifcani, A.M.A., Orman, Z., 2013. Cuttlefish algorithm-a novel bio-inspired optimization algorithm. International Journal of Scientific & Engineering Research 4 (9), 1978–1986.

Ester, M., Kriegel, H.P., Sander, J., Xu, X., 1996. A density-based algorithm for discovering clusters a density-based algorithm for discovering clusters in large spatial databases with noise. In: Proceedings of the Second International Conference on Knowledge Discovery and Data Mining. KDD'96. AAAI Press, pp. 226–231. http://dl.acm.org/citation.cfm?id=3001460.3001507.

Fahad, A., Alshatri, N., Tari, Z., Alamri, A., Khalil, I., Zomaya, A.Y., Foufou, S., Bouras, A., 2014. A survey of clustering algorithms for big data: taxonomy and empirical analysis. IEEE Transactions on Emerging Topics in Computing 2 (3), 267–279. https://doi.org/10.1109/TETC.2014.2330519.

Fränti, P., Virmajoki, O., 2006. Iterative shrinking method for clustering problems. Pattern Recognition (ISSN 0031-3203) 39 (5), 761–775. https://doi.org/10.1016/j.patcog.2005.09.012. http://www.sciencedirect.com/science/article/pii/S0031320305003778.

Gandomi, A.H., Alavi, A.H., 2012. Krill herd: a new bio-inspired optimization algorithm. Communications in Nonlinear Science and Numerical Simulation 17 (12), 4831–4845. https://doi.org/10.1016/j.cnsns.2012.05.010.

Gorman, R.P., Sejnowski, T.J., 1988. Analysis of hidden units in a layered network trained to classify sonar targets. Neural Networks 1 (1), 75–89.

José-García, A., Gómez-Flores, W., 2016. Automatic clustering using nature-inspired metaheuristics: a survey. Applied Soft Computing (ISSN 1568-4946) 41, 192–213. https://doi.org/10.1016/j.asoc.2015.12.001.

Kowalski, P.A., Kulczycki, P., 2017. Interval probabilistic neural network. Neural Computing & Applications (ISSN 1433-3058) 28 (4), 817–834. https://doi.org/10.1007/s00521-015-2109-3.

Kowalski, P.A., Łukasik, S., 2015. Experimental study of selected parameters of the krill herd algorithm. In: Intelligent Systems 2014. Springer Science Business Media, pp. 473–485.

Kowalski, P.A., Łukasik, S., Charytanowicz, M., Kulczycki, P., 2020. Optimizing clustering with cuttlefish algorithm. In: Kulczycki, P., Kacprzyk, J., Kóczy, L.T., Mesiar, R., Wisniewski, R. (Eds.), Information Technology, Systems Research, and Computational Physics. Springer International Publishing, Cham, pp. 34–43.

Kulczycki, P., Charytanowicz, M., 2010. A complete gradient clustering algorithm formed with kernel estimators. International Journal of Applied Mathematics and Computer Science (ISSN 1641-876X) 20 (1), 123–134.

Lichman, M., 2013. UCI machine learning repository. http://archive.ics.uci.edu/ml.

Łukasik, S., Kowalski, P.A., 2015. Study of flower pollination algorithm for continuous optimization. In: Angelov, P., Atanassov, K., Doukovska, L., Hadjiski, M., Jotsov, V., Kacprzyk, J., Kasabov, N., Sotirov, S., Szmidt, E., Zadrożny, S. (Eds.), Intelligent Systems'2014. In: Advances in Intelligent Systems and Computing, vol. 322. Springer International Publishing, pp. 451–459.

MacQueen, J., 1967. Some methods for classification and analysis of multivariate observations. In: Proc. 5th Berkeley Symp. Math. Stat. Probab.. Univ. Calif. 1965/1966, pp. 281–297.

Murthy, C.A., Chowdhury, N., 1996. In search of optimal clusters using genetic algorithms. Pattern Recognition Letters (ISSN 0167-8655) 17 (8), 825–832. https://doi.org/10.1016/0167-8655(96)00043-8.

Nanda, S.J., Panda, G., 2014. A survey on nature inspired metaheuristic algorithms for partitional clustering. Swarm and Evolutionary Computation (ISSN 2210-6502) 16, 1–18. https://doi.org/10.1016/j.swevo.2013.11.003.

Quinlan, J.R., 1986. Induction of decision trees. Machine Learning 1 (1), 81–106.

Quinlan, J.R., Compton, P.J., Horn, K., Lazarus, L., 1987. Inductive knowledge acquisition: a case study. In: Proceedings of the Second Australian Conference on Applications of Expert Systems, pp. 137–156.

Radionov, A., Evdokimov, S., Sarlybaev, A., Karandaeva, O., 2015. Application of subtractive clustering for power transformer fault diagnostics. In: International Conference on Industrial Engineering (ICIE-2015). Procedia Engineering (ISSN 1877-7058) 129, 22–28.

Setiono, R., Leow, W., 1987. Vehicle recognition using rule based methods. Turing Institute Research Memorandum TIRM-87-018 121.

Sheikholeslami, G., Chatterjee, S., Zhang, A., 2000. Wavecluster: a wavelet-based clustering approach for spatial data in very large databases. The VLDB Journal (ISSN 0949-877X) 8 (3), 289–304.

Sigillito, V.G., Wing, S.P., Hutton, L.V., Baker, K.B., 1989. Classification of radar returns from the ionosphere using neural networks. Johns Hopkins APL Technical Digest 10 (3), 262–266.

Łukasik, S., Kowalski, P.A., Charytanowicz, M., Kulczycki, P., 2016. Clustering using flower pollination algorithm and Calinski–Harabasz index. In: 2016 IEEE Congress on Evolutionary Computation (CEC), pp. 2724–2728.

Yang, X., 2014. Nature-Inspired Optimization Algorithms. Elsevier, London.

Yang, X.S., Karamanoglu, M., He, X., 2013. Multi-objective flower algorithm for optimization. Procedia Computer Science 18, 861–868.

Zhang, J., 1992. Selecting typical instances in instance-based learning. In: Proceedings of the Ninth International Conference on Machine Learning, pp. 470–479.

Bat-inspired algorithm for feature selection and white blood cell classification

11

Deepak Gupta, Utkarsh Agrawal, Jatin Arora, Ashish Khanna

Maharaja Agrasen Institute of Technology, Delhi, India

CONTENTS

11.1 Introduction

With the upcoming advancements in technology, every field is being explored nowadays with the intention of developing systems and machines which are self-driven and autonomous, i.e., having the capacity of decision making or executing a mechanical job. Such developments are actively pursued even in the medical department,

Nature-Inspired Computation and Swarm Intelligence. https://doi.org/10.1016/B978-0-12-819714-1.00022-1

in the hope of swift diagnosis of diseases, automated surgical mechanisms, auto-generated drug prescriptions, etc. The systems being developed in the biomedical field are extremely specific, and thus confined to predicting a singular or a limited set of symptoms. The work presented in this chapter deals with classification of white blood cells (WBCs). The scope for automating the classification and counting of WBCs can be used for the detection of a range of blood-related ailments and comprehensive pathological analysis.

WBCs, more commonly referred to as leukocytes in the medical field, are one of the special types of cells found in the human blood which aim to support the human immune system. During medical diagnosis, analyzing the patient's blood is often the first phase towards identifying the root cause of an ailment. This blood analysis mainly involves two types of counting mechanisms, i.e., complete blood count and differential blood count. The complete blood count, though a fast and automated mechanism, is not regarded as a definitive diagnostic test in medical sciences. The differential blood count, on the other hand, which gives the existing proportion of different types of WBCs, is still largely laborious and time consuming due to its complex classification. The hematological analysis of a blood sample classifies leukocytes into the following five types: monocytes, basophils, neutrophils, lymphocytes, and eosinophils. An unusual count in the variety of WBCs hints at an underlying ailment and requires further analysis.

Researchers have studied and applied various approaches to resolve this current issue of automation of classification of WBCs. The recent advances include applications of neural networks and deep learning. Sharma et al. (2019) developed a fully automated classification mechanism using deep learning methodology, thereby eliminating the need to extract features from images of blood smear; Jiang et al. (2018) developed a novel convolution neural network model, WBCNet, for the purpose of image feature extraction and recognition/classification; Shu et al. (2019) combined deep learning with quantitative phase imaging for classification of WBCs. Other techniques attempting to classify leukocytes include the technique of Fourier ptychographic microscopy for counting WBCs from a smear of blood, developed and studied by Chung et al. (2015). Ghosh et al. (2016) developed a hybrid technique of classifying WBCs using different fuzzy and nonfuzzy techniques on a variety of features extracted from the images. Rezatofighi and Soltanian-Zadeh (2011) demonstrated a detailed image processing-based algorithm used for automatic recognition of the five types of WBCs by implementing two well-known classifiers: artificial neural network and support vector machine. Honda et al. (2016) studied in detail the trend in WBC count as an indication for bacterial infection.

With the development of several automated and semiautomated techniques for the differential leukocyte recognition and count, the first step of feature extraction from the blood smear images remains the same. This has been leading to a huge pile-up of data, with extensive datasets leading to a nonaffordable space and time complexities, at the same time increasing the cost of processing. Another major drawback with bulk data is its susceptibility to noise and complications in data handling/processing. A favorable solution to this problem is feature selection, whereby a given feature set is reduced to a set of more discriminative features. This feature selection, also known as dimensionality reduction, is done in a manner that it adheres to the aim of lowering the overfitting problem which is dependent directly on the number of features.

Nature-inspired algorithms are a class of optimization algorithms which are closely based on the various self-sustaining natural systems found in the environment. This includes living/survival and hunting lifestyle of various plant and animal species. A majority of the nature-inspired, or bio-inspired, algorithms are metaheuristic algorithms, i.e., their prime focus is a quick solution, even though it may not be the optimal one. The genetic algorithm (GA) is one such popular algorithm which is inspired by the concept of natural selection in the evolution of living organisms. Similar to the changes occurring in a chromosome during the reproduction process, only favorable traits/features are passed on from the parent to the next generation. The basic operations in GA are reproduction, crossover, and mutation.

The bat algorithm (BA) is a nature-inspired algorithms researched by Yang (2010b) which simulates the echolocation technique employed by microbats for solving optimization problems. Other popular nature-inspired optimizers include particle swarm optimization (PSO), the cuttlefish algorithm, the whale optimization algorithm, ant colony optimization, and cuckoo search, as mentioned by Yang (2010a) in his book. BA has been successfully implemented to solve various optimization problems, thus proving its superior performance amidst most well-known evolutionary algorithms. For application to binary problems, which is quite often, the binary BA (BBA) was developed by Mirjalili et al. (2013).

Since feature selection is also an optimization problem, the idea of applying nature-inspired algorithms for solving it was considered, and it proved to be successful. This method has been employed previously by multiple authors. Gupta and Ahlawat (2017) employed feature selection on software usability attributes; Nakamura et al. (2012) implemented BBA and compared it with binary GA and PSO to conclude its preeminence. Yang et al. (2017) developed a novel approach, multiband averaging frequency offset synchronization, and the results were compared with some popular binary PSO- and BA-based algorithms over multiple datasets. Our previous work (Gupta et al., 2019) has laid the foundations for the contents of this chapter.

11.2 Background

11.2.1 Bat algorithm

BA is a well-known metaheuristic nature-inspired algorithm researched and developed by Dr. Xin-She Yang. The algorithm works by imitating the technique of echolocation used by bats for perceiving their immediate surrounding environment and locating their prey and/or an obstacle, as depicted. The intrinsic properties of the bat (loudness and pulse emission rate) vary, in accordance with its proximity to the prey or an object. A bat continuously emits pulses of sonar sound waves, which reflect off the surface of the obstacle or the prey, and this reflected wave is detected by the bat. Based on the time taken by the sound wave to return, the bat is able to calculate the distance and create a visual map of the surroundings. As the bat approaches the prey, the emission pulse rate increases, while the loudness of the emission decreases. At any instance, a population of bats tracks the common prey, and this mechanism is modeled by BA to evaluate the position of the bats for the next iteration around the local optima of the previous iteration(s). With the successful application of BA in a wide array of problems, its superior performance is quite evident among the other well-known evolutionary algorithms.

In a given optimization problem, the properties velocity (vel_i), position (x_i), frequency Q_i, loudness $loud_i$, and pulse rate $rate_i$ are initialized for each bat. The properties of each bat are updated iteratively until a prespecified value is reached for the variable $Iter_{max}$ by the following equations:

$$Freq_i = Freq_{min} + (Freq_{min} - Freq_{max})\beta, \tag{11.1}$$

$$vel^i_j = vel^i_j(t-1) + Pos_{Global} - Pos_{Global}(t-1)Q_i, \tag{11.2}$$

$$Pos^i_j = Pos^i_j(t-1) + vel^i_j(t). \tag{11.3}$$

In Eq. (11.1), $Freq_{min}$ and $Freq_{max}$ denote the minimum and maximum frequencies, respectively, and β is a number selected randomly in the interval [0,1]. For the ith bat with the jth dimension, Pos^i_j represents the position component and vel^i_j represents the velocity component. The variable Pos_{Global} stores the current global best solution, and we have

$$Pos_{new} = Pos_{old} + \epsilon loud(t), \tag{11.4}$$

$$loud_i(t+1) = \alpha loud_i(t), \tag{11.5}$$

$$rate_i(t+1) = rate_i(0)[1 - \exp^{-\gamma t}]. \tag{11.6}$$

Algorithm 1 Bat algorithm.

Initialize the population of bats with their corresponding properties, position Pos_i, velocity vel_i, and pulse frequency $Freq_i$.
Initialize characteristics including the loudness $loud_i$, maximum number of iterations $Iter_{Max}$, and pulse rate $rate_i$.
while *Iteration* < *Iter*$_{Max}$ **do**
   for *each bat bat$_i$* **do**
      Evaluate Eqs. (11.1), (11.2), and (11.3) to generate a new set of solutions;
      if *random* > r_i **then**
         Generate a local solution around one of the selected best solutions;
      if *(random* < *loud$_i$)&(fitness(x_i)* < *fitness(Pos_{Global}))* **then**
         Update Pos_{Global} and $fitness_{Best}$;
         Increase $rate_i$ and reduce $loud_i$ as per Eqs. (11.5) and (11.6);
   Rank the bats and find $Global_{Best}$

Both loudness and pulse emission rate at time $t = 0$, i.e., $loud_i(0) \in [1, 2]$ and $rate_i(0) \in [0, 1]$, are randomly chosen at the first step of the algorithm.

11.2.2 Binary bat algorithm

BA, as discussed above, provides an apt simulation for problems which are continuous-valued, such that the position of each bat represents a possible solution (optimal or not). Nakamura et al. extended the work with Xin-She Yang, and developed the BBA to exploit its application to binary-valued problems. The algorithm works in such a way that the search space of the problem is represented as an n-dimensional hypercube with the lattice location of the nodes being represented as a Boolean array. The value 0 or 1 is a corresponding indication of excluding or including the implying attribute. The sigmoid function as in Eq. (11.7) is used to evaluate the position of the bats given by Eq. (11.8), where

$$S(v_j^i) = \frac{1}{1 + \exp^{-v_j^i}}, \tag{11.7}$$

$$X = \begin{cases} 1, & \text{if } S(v_j^i) > \rho, \\ 0, & \text{otherwise}. \end{cases} \tag{11.8}$$

11.3 Methods

In this section the proposed methodology, experimental setup, input parameters, creation of dataset, feature extraction, and machine learning models are discussed.

11.3.1 Optimized binary bat algorithm (for feature selection and classification)

BBA has been optimized for feature selection. It has been applied on the WBC dataset. It is a kind of multiclassification problem in which each cell is classified as one of five major cell types, namely, monocytes, lymphocytes, eosinophils, basophils, and neutrophils. The fitness function for the suggested optimized BBA (OBBA) is defined as the accuracy of the machine learning model. K nearest neighbors (KNN), decision tree, logistic regression, and random forest have been used as machine learning models for the given optimization problem.

In the preprocessing stage the WBC dataset is divided into two sets, namely, the test set and the training set. After carrying out various tests, a split ratio of 80:20 was found to be the optimal one. Stratified split has been employed to ensure that the training and test sets have roughly the same proportion of samples of each target class. Both sets are further normalized to reduce any bias and ensure that all features weigh equally. Firstly, the complete feature set was given as an input to each of the four machine learning models to obtain initial accuracy values. These values were later used

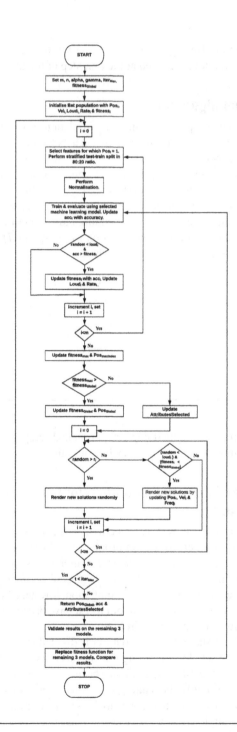

FIGURE 11.1

Flowchart of OBBA.

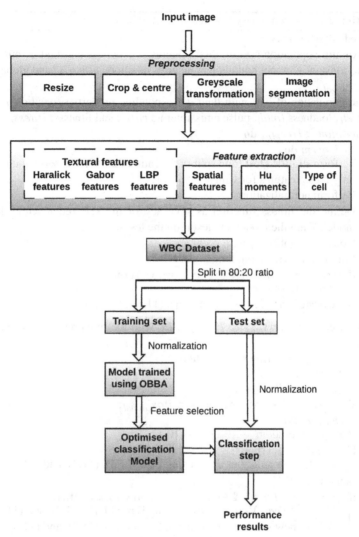

FIGURE 11.2

Implementational model.

for validation of our results. Secondly, accuracy of one of the four machine learning models (selected randomly) was defined as the fitness function of OBBA. Thirdly, the proposed methodology was verified by giving the feature subset as obtained from the previous stage to the remaining three unselected machine learning models. This process was repeated for each classifier to ensure that a robust fitness function can be obtained. Finally, a comparison was carried out between all the results. Fig. 11.1 illustrates the flowchart and Fig. 11.2 illustrates the model architecture.

Algorithm 2 Optimized binary bat algorithm (OBBA).

Required: $WBCDataset$

Set maximum population size m, total features n, loudness constant α, pulse emission rate constant γ, upper bound on number of iterations $Iter_{Max}$, and global fitness $fitness_{Global}$

Initialize population of bats with their corresponding properties, position Pos_i, velocity Vel_i, loudness $loud_i$, pulse emission rate $rate_i$, and fitness $fitness_i$

while *iteration* $< Iter_{Max}$ **do**

    for *each agent* **do**

        Perform stratified test train split in the ratio of 20:80. The sets should contain only those features for which $Pos_i = 1$;

        Perform normalization;

        Define the fitness function as accuracy of the selected machine learning model. Train the model and test it on the test set;

        Update variable acc_i with accuracy;

        Initialize a variable random $\leftarrow [0, 1]$;

        if $((random \leq loud_i)\&(acc_i \geq fitness_i))$ **then**

            $fitness_i \leftarrow acc_i$;

            Update $loud_i$ & $rate_i$ using Eqs. (11.5) & (11.6);

    Set variables $fitness_{max}$ & $Pos_{maxIndex}$ with the maximum fitness and index value;

    if $(fitness_{max} > fitness_{Global})$ **then**

        $fitness_{Global} \leftarrow fitness_{max}$;

        $Pos_{Global} \leftarrow Pos_{maxIndex}$;

    AttributesSelected $\leftarrow$ Number of 1s in Pos_{Global};

    for *each agent* **do**

        $\beta \leftarrow [0, 1]$, random $\leftarrow [0,1]$, e $\leftarrow [0,1]$;

        if $(random > r_i)$ **then**

            Render new solutions Pos_i using Eqs. (11.4), (11.7), and (11.8);

        random $\leftarrow [0,1]$;

        if $((random < loud_i) \& (fitness_i < fitness_{Global}))$ **then**

            Update $Freq_i$, Vel_i, and Pos_i using Eqs. (11.1), (11.2), and (11.3);

            Render new solutions Pos_i using Eqs. (11.4), (11.7), and (11.8);

Return Pos_{Global}, acc & AttributesSelected;

Validate results on the remaining three models;

Repeat all steps by replacing the model in fitness function

The modifications between the proposed OBBA and the original BBA are as follows:

- Normalization of each set has been carried out to ensure that each feature is treated equally.

- A stratified test–train split has been employed to ensure that each set has roughly the same proportion of samples of each target class.
- The suggested approach has been devised for multiple features while BBA was used for a singular feature.
- KNN, decision tree, logistic regression, and random forest have been chosen as the four machine learning models.
- The fitness of the algorithm is defined as the value of accuracy of the chosen machine learning model.

11.3.2 Dataset

The Leukocyte Images for Segmentation and Classification (LISC) database is comprised of 237 hematological pictures obtained from peripheral blood samples of eight normal and fit subjects. The blood slides were stained and smeared using the Gismo-Right technique, and they were further developed by a light Axioskope 40 microscope by means of an achromatic lens with a magnification level of 100. To capture these images, a camera (Sony Model No. SSCDC50AP) was used. The pictures have been gathered from the Hematology-Oncology and BMT Research Center of Imam Khomeini hospital in Tehran, Iran. All of them are colored, have an image size of 720×576, and are bitmap images. The images have also been classified into five WBC categories, namely, basophils, eosinophils, lymphocytes, monocytes, and neutrophils, by an expert. See Table 11.1.

Table 11.1 Original images from the LISC dataset.

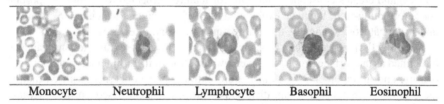

| Monocyte | Neutrophil | Lymphocyte | Basophil | Eosinophil |

11.3.3 Implementation details

11.3.3.1 Experimental setup

The system on which all the testing was carried out has the following configuration:

- macOS Mojave 10.14.4,
- 3.3 GHz Intel Core i7,
- 16 GB 2133 MHz LPDDR3,
- Intel Iris Graphics 550 1536 MB.

Python 3.7 and its libraries were used for implementation and testing purposes.

11.3.3.2 Input parameters

Table 11.2 depicts all the parameters that are initialized at the start of the algorithm.

Table 11.2 Initial values of parameters and their descriptions.

Parameter	Initial value	Description
m	35	Size of population
n	35	Total features
$Iter_{Max}$	20	Total iterations
Pos_j^i	[0,1]	Position of bats
Vel_j^i	0	Velocity of bats
α	0.9	Loudness constant
β	0.9	Pulse rate constant
γ	[0,1]	Frequency constant
e	[0,1]	Variable for rendering Pos_{New}
Rate	[0,1]	Pulse rate
Loud	[1,2]	Loudness
fitness	−1	Fitness
$fitness_{Global}$	−1	Global fitness

11.3.4 Feature extraction

Feature extraction plays a crucial role when it comes to machine learning. The process begins with an original set of data that can be of any form (numbers, letters, or images) and creates consequent attributes or features expected to be more important and information-rich. We have extracted a total of 35 features which can be grouped under two categories, namely, textural features and spatial features.

An object's appearance can be typically referred to as its texture; texture can also be defined as the surface properties given by a fraction of its rudimentary components, density, and arrangement. Textural feature extraction is concerned with extracting these kinds of features that can distinguish various objects based on their texture. A total of 17 texture features have been extracted for the purpose of our study. The breakdown is as follows (see Table 11.3):

- 13 Haralick features,
- 2 Gabor filter features,
- 2 local binary pattern features.

11.3.4.1 Haralick features

Haralick et al. (1973) proposed a new methodology to extract texture features from an image using the GLCM matrix. These features were referred to as Haralick features. The aim of extracting these features is to find out the most distinct and significant features that can categorize the five kinds of WBCs with high accuracy. Haralick features are texture-based features extracted using a gray-level cooccurrence matrix

Table 11.3 Processed images used for textural feature extraction.

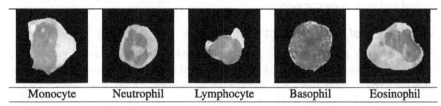

| Monocyte | Neutrophil | Lymphocyte | Basophil | Eosinophil |

(GLCM). GLCM features are put into operation to capture textural similarity in patterns of WBC images. Each type of WBC has clearly a distinct texture because of its unique cytoplasm–nucleus configuration. Haralick features are statistically global in nature, i.e., they are not concerned with a particular portion of the image but consider the image as a whole. It is a four-step process as shown below:

1. The first step is to use a quantization function to map the given image of size (P × Q) to a quantized image of size (1 × L), where L is the number of gray levels.
2. The second step is to create a nonnormalized GLCM matrix.
3. The third step is to normalize the GLCM matrix obtained from the previous step.
4. The fourth and final step is to extract the Haralick texture features. A total of 13 features were extracted for the purpose of our study.

The thirteen features are the following: correlation, contrast, entropy, angular second moment, sum variance, sum of squares variance, sum entropy, sum average, difference variance, difference entropy variance, difference inverse moment, and information measure correlation 1 and 2.

11.3.4.2 Gabor filter features

The Gabor filter feature extraction process starts with the application of a two-dimensional Gabor filter which is applied to each image individually. The process is guided by Gabor's uncertainty principle, which states that the product of frequency resolutions and time must be greater than a constant. The principle aids in better selection of orientations and frequencies. The principle also implies that frequency and spatial measurements must define a rectangular shape in Fourier space having an $area \geq 1/4\pi$. The Gabor filter function as given in the following equation has been applied to each image:

$$G_{p,q,f,\theta}(x, y) = \exp(-\frac{(x-p)^2}{\alpha^2} - \frac{(y-q)^2}{\beta^2})$$
$$\exp[j2\pi f((x-p)\cos\theta + (y-q)\sin\theta)]. \tag{11.9}$$

Here, f is spatial frequency, θ is orientation, (x, y) are spatial coordinates, (p, q) are central Gaussian envelope spatial coordinates, and (α, β) are spreads of the function in two dimensions. The result of applying the filter is an image. From these images two features, namely, Gabor filter entropy and Gabor filter energy, are extracted using the real and imaginary components.

11.3.4.3 Local binary pattern features

The local binary pattern (LBP) was first proposed in 1994. It is a rudimentary yet extremely efficient technique used in texture-based analysis of an image. The various steps involved in extracting LBP features are as follows:

- Initially starting with a grayscale image, consider a window of, say, $(m \times n)$ pixels, or in other words an $(m \times n)$ matrix, in which each value indicates the intensity of the pixel (0–255).
- The window is divided into cells.
- Now for each pixel in a given cell, the pixel is compared to its eight neighbors along a circle in either clockwise or counterclockwise direction.
- If the central pixel has a greater value than its neighbor, then set 0; otherwise, set 1. This results in an eight-digit binary number.
- An $(m \times n)$-dimensional histogram is computed over the cell for each combination.
- The histogram obtained in the previous step is normalized.
- All the normalized histograms are concatenated, resulting in a feature vector for the entire window.
- Finally, two features (LBP entropy and LBP energy) are evaluated using feature vector obtained.

11.3.4.4 Spatial features

The shape of an entity can be referred to as its physical form, peripheral structure, or outer surface appearance. Spatial features are useful in measuring, searching, and matching an object under consideration. A total of ten spatial features have been extracted for our study, namely, nucleus area, perimeter, roundness, and eccentricity, cell area, perimeter, roundness, and eccentricity, the nucleus to cell area ratio, and the cytoplasm area. See Table 11.4.

Table 11.4 Processed images used for spatial feature extraction.

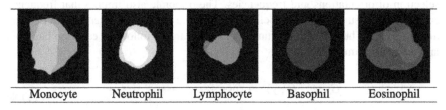

| Monocyte | Neutrophil | Lymphocyte | Basophil | Eosinophil |

11.3.4.5 Hu moments

We have also extracted seven Hu moments as suggested by Hu (1962). These Hu moments have some special characteristics, such as translation invariance, i.e., central moments are translation-invariant by default. They are also scale-invariant, achieved by dividing through a zero central moment which must be properly scaled, as shown by the following equation:

$$\phi_{pq} = \frac{m_{pq}}{m_{00}^{\left(1+\frac{p+q}{2}\right)}}, \tag{11.10}$$

where m denotes the central moment. Therefore a total of seven Hu moments invariant to translation, scale, and rotation are obtained, as illustrated in the following equations:

$$Hu_1 = \phi_{20} + \phi_{02}, \tag{11.11}$$

$$Hu_2 = (\phi_{20} - \phi_{02})^2 + 4\phi_{11}^2, \tag{11.12}$$

$$Hu_3 = (\phi_{30} - 3\phi_{12})^2 + (3\phi_{21} - \phi_{03})^2, \tag{11.13}$$

$$Hu_4 = (\phi_{30} + \phi_{12})^2 + (\phi_{21} + \phi_{03})^2, \tag{11.14}$$

$$Hu_5 = (\phi_{30} - 3\phi_{12})(\phi_{30} + \phi_{12})[(\phi_{30} + \phi_{12})^2 - 3(\phi_{21} + \phi_{03})^2] +$$
$$(3\phi_{21} - \phi_{03})(\phi_{21} + \phi_{03})[3(\phi_{30} + \phi_{12})^2 - (\phi_{21} + \phi_{03})^2], \tag{11.15}$$

$$Hu_6 = (\phi_{20} - \phi_{02})[(\phi_{30} + \phi_{12})^2 - (\phi_{21} + \phi_{03})^2] + 4\phi_{11}(\phi_{30} + \phi_{12})(\phi_{21} + \phi_{03}), \tag{11.16}$$

$$Hu_7 = (3\phi_{21} - \phi_{03})(\phi_{30} + \phi_{12})[(\phi_{30} + \phi_{12})^2 - 3(\phi_{21} + \phi_{03})^2] -$$
$$(\phi_{30} - 3\phi_{12})(\phi_{21} + \phi_{03})[3(\phi_{30} + \phi_{12})^2 - (\phi_{21} + \phi_{03})^2]. \tag{11.17}$$

In addition to the abovementioned features, an additional binary-valued feature is computed, referred to as "type." It represents whether the cell is mononuclear (type set to 0) or polynuclear (type set to 1). A matrix with 237 rows and 35 columns, referred to as the feature matrix, was finally constructed. The dataset characteristics are depicted in Table 11.5 and in Fig. 11.3. The objective of the model is to classify the cells as one of the five types based on the value present in the "Outcome" column.

Table 11.5 Characteristics of the dataset.

Characteristic	Value
Type of features	Real
Total attributes	35
Total samples	237
Total missing values	0

11.3.5 Machine learning models

In our study, four typical and popular machine learning models have been used. They are the following:

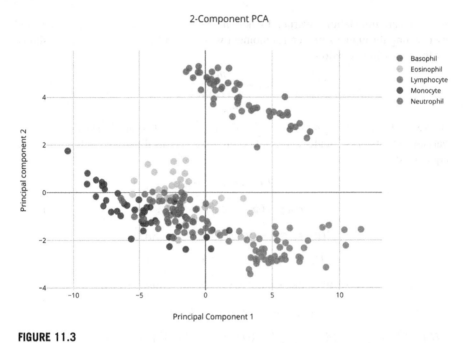

FIGURE 11.3

Spatial distribution of the dataset.

- Logistic regression: This model is used for classification problems. It aims to predict the class to which a particular sample belongs. The outcome is a discrete binary value, a probability between 0 and 1. The model uses a function known as logistic function or sigmoid function and measures the relationship between dependent (outcome) and independent variables (features). The function uses a threshold classifier to map the values between 0 and 1 to either 0 or 1.
- KNN: This algorithm is simple yet widely used. It can be used for both classification and regression problems. It is nonparametric, i.e., no assumptions are made on the underlying data distribution. It is also instance-based, i.e., KNN memorizes the training instances. To classify a sample, the distance between the sample and its K nearest neighbors or K nearest labeled samples is computed. The sample is classified based on the majority of nearest neighbors. The value of K has to be chosen carefully, and it is recommended to run test several values of K to find out its optimal value.
- Decision tree: This is a kind of supervised learning algorithm and is found to be mostly used in classification-based problems. It performs well with both continuous and categorical attributes. The decision tree can be thought of as a tree in which the leaf nodes symbolize a decision while branch nodes symbolize a choice between a number of alternatives. They have a simple if-then-else structure, which makes them easy to design and understand.

- Random forest: It is kind of supervised learning model used for both regression and classification. It uses ensemble learning to construct several decision trees and then combines them to get a prediction which is much more stable and has a higher accuracy. The aggregated outputs of individual decision trees ensure a more robust and stronger ensemble. It consists of two major phases, namely, the construction phase and the prediction phase.

11.4 Results

The section presents the results obtained by implementing the proposed algorithm on the abovementioned WBC dataset. The different models applied (decision tree, logistic regression, random forest, KNN) have been compared and contrasted for both accuracy and execution time.

Table 11.6 distinctly marks the improvement in the accuracy obtained after feature selection for each individual fitness function. Figure in part (A) of Table 11.13 corresponds to the graphical representation of the same data.

The feature set reduced from the original 35 attributes extracted has been mentioned in Table 11.7 with its parallel representation in Figure in part (C) of Table 11.13. Similarly, Table 11.8 and Figure in part (E) of Table 11.13 exemplify the run time obtained for 20 iterations, for each machine learning model as the fitness function.

As mentioned in Section 11.3.1, in the proposed algorithm, each of the four machine learning models was individually taken as the fitness function, and then evaluated on the remaining models. A detailed analysis of this approach can be understood from Table 11.9 and Figure in part (C) of Table 11.13, where each model is evaluated on all other models.

In the rest of the section, the abovementioned algorithm has been briefly juxtaposed with two other recently developed nature-inspired algorithms: the optimized crow search algorithm and the optimized cuttlefish algorithm. As can be seen in Tables 11.10, 11.11, and 11.12, OBBA is superior to Optimised Crow Search Algorithm (OCSA) and the Optimised Cuttlefish Algorithm (OCFA), with greater accuracy, reduced execution time, and also the smallest feature set. See Figure in part (D) and (F) of Table 11.13.

Table 11.6 Accuracy for different models.

Fitness function	Accuracy using all features (%)	Accuracy using selected features (%)
Decision tree	89.30	98.20
Logistic regression	96.83	98.20
Random forest	88.90	96.16
KNN	91.57	98.20

Table 11.7 Features selected.

Fitness function	Number of features selected (out of 35)
Decision tree	11
Logistic regression	18
Random forest	13
KNN	12

Table 11.8 Execution time for different models.

Fitness function	Run time for 20 iterations (s)
Decision tree	16.03
Logistic regression	18.42
Random forest	24.69
KNN	17.04

Table 11.9 Accuracy with evaluation and fitness function.

Fitness function	Decision tree	Logistic regression	Random forest	KNN
Decision tree	–	0.961	0.908	0.932
Logistic regression	0.855	–	0.918	0.941
Random forest	0.855	0.980	–	0.962
KNN	0.920	0.949	0.920	–

Table 11.10 Comparison with recently developed algorithms (accuracy, %).

Algorithm/model used	Decision tree	Random forest	KNN
OBBA	98.20	96.16	98.20
OCSA	92.23	92.89	92.89
OCFA	94.10	95.62	95.62

Table 11.11 Comparison with recently developed algorithms (feature selection).

Algorithm	Number of features selected (out of 35)
OBBA	11
OCSA	20
OCFA	15

Table 11.12 Comparison with recently developed algorithms (run time).

Fitness function	Average time (s)
OBBA	17.20
OCSA	20.98
OCFA	18.73

Table 11.13 Results.

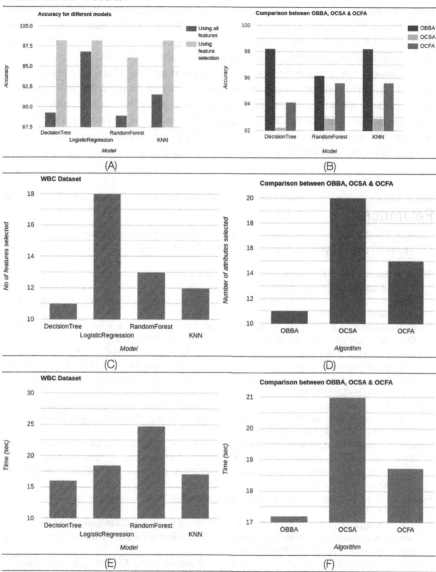

11.5 Conclusion and discussion

This study proposes a novel Optimized Binary Bat Algorithm (OBBA), a nature-inspired optimization algorithm for feature selection. The introduced algorithm provides much more accurate results along with reduced computational costs. The suggested approach has been applied to the White Blood Cells dataset. The results have been promising as OBBA selects 11 features out of 35 features with an accuracy of 97.69% as compared to an accuracy of 92.67% of OCSA (20 features selected)and 95.11% of OCFA (15 features selected). The proposed algorithm has demonstrated that it is fit to use in hematological laboratories by the physicians.

Additional research can be done in the arena of analyzing several hematological ailments & differentiating between the kinds of white blood cells, by developing practices to combine nature-inspired optimization algorithms along with machine learning models. Further methods of detection can also be investigated,scrutinized and corroborated with the algorithm to verify the results for much more efficient and accurate recognition. OBBA can also be implemented for datasets other than WBC.

References

Chung, J., Ou, X., Kulkarni, R.P., Yang, C., 2015. Counting white blood cells from a blood smear using Fourier ptychographic microscopy. PLoS ONE 10 (7), 1–10. https://doi.org/10.1371/journal.pone.0133489.

Ghosh, P., Bhattacharjee, D., Nasipuri, M., 2016. Blood smear analyzer for white blood cell counting: a hybrid microscopic image analyzing technique. Applied Soft Computing (ISSN 1568-4946) 46, 629–638. https://doi.org/10.1016/j.asoc.2015.12.038. http://www.sciencedirect.com/science/article/pii/S1568494615008170.

Gupta, D., Ahlawat, A.K., 2017. Usability feature selection via MBBAT: a novel approach. Journal of Computer Science 23, 195–203. https://doi.org/10.1016/j.jocs.2017.06.005.

Gupta, D., Arora, J., Agrawal, U., Khanna, A., de Albuquerque, V.H.C., 2019. Optimized binary bat algorithm for classification of white blood cells. Measurement (ISSN 0263-2241) 143, 180–190. https://doi.org/10.1016/j.measurement.2019.01.002. http://www.sciencedirect.com/science/article/pii/S0263224119300028.

Haralick, R.M., Shanmugam, K., Dinstein, I., 1973. Textural features for image classification. IEEE Transactions on Systems, Man and Cybernetics (ISSN 0018-9472) SMC-3 (6), 610–621. https://doi.org/10.1109/TSMC.1973.4309314.

Honda, T., Uehara, T., Matsumoto, G., Arai, S., Sugano, M., 2016. Neutrophil left shift and white blood cell count as markers of bacterial infection. Clinica Chimica Acta (ISSN 0009-8981) 457, 46–53. https://doi.org/10.1016/j.cca.2016.03.017. http://www.sciencedirect.com/science/article/pii/S0009898116301024.

Hu, M., 1962. Visual pattern recognition by moment invariants. I.R.E. Transactions on Information Theory 8, 179–187.

Jiang, M., Cheng, L., Qin, F., Du, L., Zhang, M., 2018. White blood cells classification with deep convolutional neural networks. International Journal of Pattern Recognition and Artificial Intelligence 32. https://doi.org/10.1142/S0218001418570069.

Mirjalili, S., Mirjalili, S., Yang, X.S., 2013. Binary bat algorithm. Neural Computing & Applications 25, 663–681. https://doi.org/10.1007/s00521-013-1525-5.

Nakamura, R., Pereira, L., Costa, K., Rodrigues, D., Papa, J., Yang, X.S., 2012. BBA: a binary bat algorithm for feature selection. In: 2012 25th SIBGRAPI Conference on Graphics, Patterns, and Image. IEEE, pp. 291–297.

Rezatofighi, H., Soltanian-Zadeh, H., 2011. Automatic recognition of five types of white blood cells in peripheral blood. Computerized Medical Imaging and Graphics: the Official Journal the Computerized Medical Imaging Society 35, 333–343. https://doi.org/10.1016/j.compmedimag.2011.01.003.

Sharma, M., Bhave, A., Janghel, R., 2019. White Blood Cell Classification Using Convolutional Neural Network: Methods and Protocols. ISBN 978-1-4939-8936-2, pp. 135–143.

Shu, X., Sansare, S., Jin, D., Tong, K.Y., Pandey, R., Zhou, R., 2019. White blood cell classification using quantitative phase microscopy based deep learning. In: Biophotonics Congress: Optics in the Life Sciences Congress 2019 (BODA,BRAIN,NTM,OMA,OMP). Optical Society of America. DT3B.3. http://www.osapublishing.org/abstract.cfm?URI=BODA-2019-DT3B.3.

Yang, X.S., 2010a. Nature-Inspired Metaheuristic Algorithms. Luniver Press, UK.

Yang, X.S., 2010b. A new metaheuristic bat-inspired algorithm. Studies in Computational Intelligence 284, 65–74. https://doi.org/10.1007/978-3-642-12538-6_6.

Yang, B., Lu, Y., Zhu, K., Yang, G., Liu, J., Yin, H., 2017. Feature selection based on modified bat algorithm. IEICE Transactions on Information and Systems E100.D, 1860–1869. https://doi.org/10.1587/transinf.2016EDP7471.

Modular granular neural network optimization using the firefly algorithm applied to time series prediction

12

Daniela Sánchez, Patricia Melin, Oscar Castillo

Tijuana Institute of Technology, Division of Graduate Studies and Research, Tijuana, Mexico

CONTENTS

12.1 Introduction

Intelligence techniques have become very useful tools applied to new technologies because techniques allow to perform real-world applications. These techniques include artificial neural networks (ANNs), fuzzy logic (FL), data mining, robotics, and others. Another important area is the bio-inspired optimization algorithm, where there are a lot of algorithms that allow to optimize parameters, architectures, or structures, such as the genetic algorithm (GA) (Holland, 1975; Man et al., 1975), ant colony optimization (ACO) (Dorigo, 1992), particle swarm optimization (PSO)

Nature-Inspired Computation and Swarm Intelligence. https://doi.org/10.1016/B978-0-12-819714-1.00023-3

(Eberhart and Kennedy, 1995), bat algorithm (BA) (Yang, 2010), cuckoo search (CS) (Yang and Deb, 2010), the firefly algorithm (FA) (Yang, 2009; Yang and He, 2013), and many others (Yang, 2014).

These optimization algorithms allow to achieve excellent results when they are combined with one or more intelligence techniques previously mentioned. Some of these works combine ANN with PSO (Carvalho and Ludermir, 2007), ACO (Salama and Abdelbar, 2014), or CS (Ong and Zainuddin, 2019). FL has been combined with BA (Talbi, 2019; Arora et al., 2016) and FA (Talib and Darus, 2017). Data mining has been combined with PSO (Durga and Lalitha, 2015) and ACO (Michelakos et al., 2011). In robotics, PSO (Taylan et al., 2013) and BA (Suárez and Iglesias, 2017) have been implemented, among others.

In this chapter the combination of modular neural networks, granular computing, and FA is shown, where time series prediction is performed to prove the effectiveness of this combination. The main contribution of this method is the optimization using FA for modular granular neural network architectures and the granulation of time series. The optimized parameters of the modular granular neural network are the number of submodules (granules), the error goal, and the number of hidden layers with their number of neurons.

12.2 Intelligent techniques

In this section, the intelligence techniques used in this chapter are described to facilitate the reader's understanding.

12.2.1 Modular neural networks

An ANN is a computational representation of the human nervous system. The main capability of ANNs is their learning capability, used successfully to solve nonlinear problems such as human recognition, clustering, function approximation, time series prediction, and control problems. ANNs can have different topologies, including perceptron, multilayer perceptron, feed-forward, radial basis network, and deep feed-forward, to mention some (Taheri et al., 2019; Gurevich and Stuke, 2019).

A modular neural network (MNN) is formed by several ANNs (called modules), where each module performs a different subtask of a whole task. This kind of architecture provides better conditions to learn and generalize, because each module is a specialist of certain information. To obtain a final result there are integration methods that allow to combine results of each module, depending on the application, such as average integration, winner takes all, and FL, among others (Melin, 2012; Sánchez and Melin, 2016; Sotirov et al., 2018; Goltsev and Gritsenko, 2015).

12.2.2 Granular computing

The granular computing basic ideas are connected with other disciplines or areas, mainly with fuzzy sets (Zadeh, 1965). Granular computing has emerged as an important and attractive area for researches because this area has become an umbrella which covers theories, techniques, and tools that uses granules to solve problems successfully. A granule is a collection of objects or information grouped together due to similarity and proximity of functionality (Bargiela and Pedrycz, 2003; Butenkov et al., 2017; Qian et al., 2014).

Granulation is very important for this area because granular computing involves the decomposition of a whole by the creation of granules. These granules have similar interpretation as sets, categories, clusters, or groups. Zadeh (1996) defined two operations for working with granules: decomposition of a whole into parts and integration of parts into a whole (Yao, 2018). Granular computing resolves problems and information processing using multiple levels of granularity (Yao and Zhao, 2012).

12.2.3 Firefly algorithm

FA was proposed in Yang (2009) and Yang and He (2013) and is based on the flashing behavior of fireflies. Three basic rules are established for this algorithm (Yang, 2009; KumarSrivastava and Singh, 2016):

1. The fireflies are unisex, and each firefly can be attracted to each other firefly.
2. The attractiveness and brightness are proportional, and their values decrease as their distance increases. For a couple of fireflies, the firefly with less brightness moves toward the other firefly; if they both have the same brightness, then their movement will be random.
3. The brightness of a firefly is obtained by the objective function.

The variation of attractiveness β with the distance r is defined by the following equation (Yang and He, 2013):

$$\beta = \beta_0 e^{-\gamma r^2}, \tag{12.1}$$

where the attractiveness at $r = 0$ is defined by β_0. The movement of firefly i to another brighter one j is defined as

$$x_i^{t+1} = x_i^t + \beta_0 e^{-\gamma r_{ij}^2}(x_j^t - x_i^t) + \alpha_t \epsilon_i^t, \tag{12.2}$$

where x_i is the position of a firefly in the iteration t, $\beta_0 e^{-\gamma r_{ij}^2}(x_j^t - x_i^t)$ defines the attraction between firefly j and firefly i, and ϵ_i^t is a vector of random numbers with a randomization parameter defined by α_t. This parameter is the initial randomness scaling factor, defined as

$$\alpha_t = \alpha_t \delta^t, \tag{12.3}$$

Table 12.1 Table of parameters for FA.

Parameters	Value
Fireflies (n)	10
Maximum number of iterations (t)	30
α	0.015
β	1
δ	0.95

where δ has a value between 0 and 1. The values for α, β, and δ applied in this work are shown in Table 12.1.

12.3 General descriptions

The proposed method optimizes modular granular neural networks applied to time series prediction. In this section, description and architectures are described.

12.3.1 Description of the granular modular neural networks

A dataset used in a neural network is mainly divided into a subdataset for the training phase and a subdataset for the testing phase. In Fig. 12.1, an example of this division when a time series is used is shown.

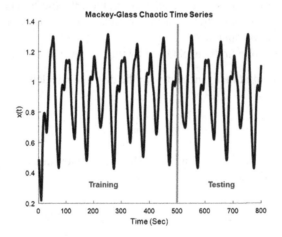

FIGURE 12.1

Division of a time series.

The dataset used in the training phase for the modular granular neural network is divided into subdatasets (subgranules) of information, where each of these sub-datasets is trained by a module of a modular neural network. This kind of neural networks can be applied to human recognition or, as in this chapter, time series prediction. In Sánchez and Melin (2014) and Sánchez et al. (2017) it was applied to human recognition, but with other optimization techniques, and in Sánchez and Melin (2015) time series prediction using FL as integration method was proposed. For the testing phase, the dataset used for testing is simulated in each module of the modular granular neural network. Fig. 12.2 shows an example of the proposed architecture, where each module trains a different number of data points but all of them simulate the same information of the testing phase.

12.3.2 Description of the firefly algorithm

As previously mentioned, the brightness of a firefly is obtained by the objective function. When each submodule simulates the testing dataset a vector of results is obtained. An average using the results of all the submodules is calculated and compared with the real data used in the testing phase. To calculate the prediction error the following equation is used:

$$error = \frac{\left(\sum_{i=1}^{D} |a_i - x_i|\right)}{D}, \tag{12.4}$$

where a represents the average obtained of the outputs achieved by submodules, x represents real data of the testing phase, and D is the number of data points used for the testing phase. Each dimension of a firefly represents a modular granular neural network parameter, and the number of dimensions of each firefly is calculated as

$$dimensions = 1 + \left(2 * m\right) + \left(m * h\right), \tag{12.5}$$

where the maximum number of modules used by FA is represented by m and h is the maximum number of hidden layers per module. Both these variables (m and h) can be freely adjusted depending on the application or the dataset. In Fig. 12.3, a representation of dimensions for each firefly is shown. In Table 12.1, the parameters used for FA are shown. The firefly number and the maximum number of iterations are based on previous works (Sánchez and Melin, 2014; Sánchez et al., 2017), and the other parameters are based on values recommended in Yang (2009) and Yang and He (2013).

To perform the training phase, the learning algorithm used was the Levenberg–Marquardt (LM) backpropagation, because in other works (Pulido et al., 2014; Melin and Pulido, 2014) this backpropagation algorithm provided better results than others when neural networks are applied to time series prediction. The search space to FA is shown in Table 12.2, where the minimum and maximum values for each parameter are shown. There are two stopping conditions for this FA:

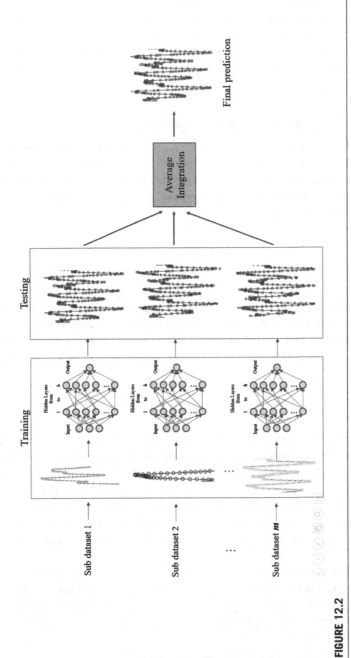

FIGURE 12.2

The architecture of the modular granular neural networks applied to time series prediction.

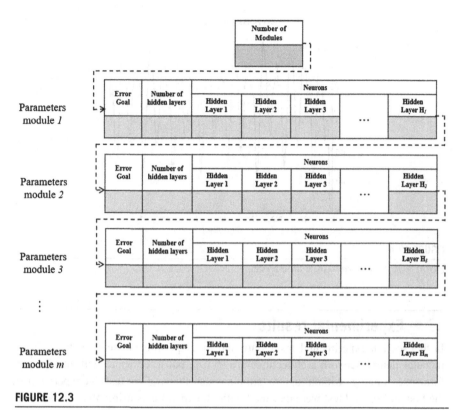

FIGURE 12.3

Dimensions of each firefly.

Table 12.2 Parameters for neural network architectures.

Parameters	Value
Submodules (m)	1–5
Neurons	5–20
Hidden layers	1–5
Error goal	0.000001–0.001

1. The maximum number of iterations is achieved.
2. The best firefly has an error value equal to zero.

12.3.3 Description of Mackey–Glass time series

The proposed method was tested using 800 data points of the Mackey–Glass time series (Mackey and Glass, 1997). These points correspond to a period from 9 November, 2005 to 15 January, 2009. An example of 800 data points of the Mackey–Glass time series is shown in Fig. 12.4.

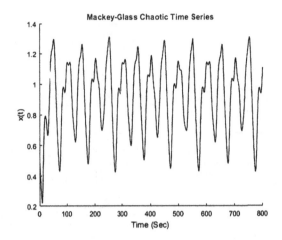

FIGURE 12.4

Example of the Mackey–Glass time series.

12.4 Experimental results

In this section experimental results are shown. Experiments using FA to optimize modular neural network architectures are shown, but to compare with the FA performance, trainings without optimization are also shown. Four tests were performed; the first and second test were executed with 30 experiments using 500 and 300 data points for the training phase (300 and 500 for testing) without optimization, respectively, while the third and fourth tests were executed with 30 experiments using 500 and 300 data points for the training phase (300 and 500 for testing), respectively, using FA for modular granular neural network architecture optimization.

12.4.1 Nonoptimized results

For these experiments, the modular granular neural network parameters were randomly established (the number of submodules, the number of neurons of two hidden layers for each submodule, and the error goal) using the ranges shown in Table 12.2. Only two hidden layers in each module of the modular neural network were used because in other works best results have been achieved (Pulido et al., 2014; Melin and Pulido, 2014) where neural networks for time series prediction were applied. For the first test, 30 nonoptimized trainings were performed using 500 data points for the training phase; of these nonoptimized trainings, the best architecture was obtained by the nonoptimized training #17, where a prediction error of 0.034371448 was obtained. In Fig. 12.5, results obtained by each module are shown. Each module shows the data points used for training and its corresponding prediction using the data points of the testing phase.

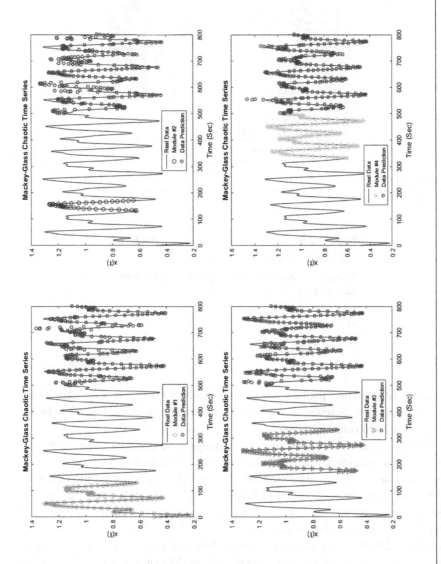

FIGURE 12.5

Nonoptimized training #17.

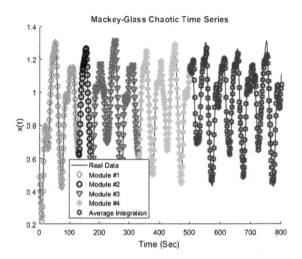

FIGURE 12.6

Nonoptimized training #17 result.

Table 12.3 Best nonoptimized architecture (500 data points).

Module	Neurons	Data points per module
Module #1	21, 11	1 to 130
Module #2	26, 24	131 to 170
Module #3	23, 28	171 to 330
Module #4	17, 15	331 to 500

In Fig. 12.6, the final prediction is shown. This final prediction is calculated with average integration using the prediction of each module shown in Fig. 12.5. In Table 12.3, the architecture is shown.

For the second test, 30 nonoptimized trainings were performed using 300 data points for the training phase; of these nonoptimized trainings, the best architecture was obtained by the nonoptimized training #10, where a prediction error of 0.046735933 was obtained. In Fig. 12.7, results obtained by each module are shown. Each module shows the data points used for training and its corresponding prediction using the data points of the testing phase.

In Fig. 12.8, the final prediction is shown. This final prediction is calculated with average integration using the prediction of each module shown in Fig. 12.7. In Table 12.4, the architecture is shown.

In Table 12.5, a summary of results of nonoptimized trainings is shown. It can be observed that better results were achieved when 500 data points were used for the training phase.

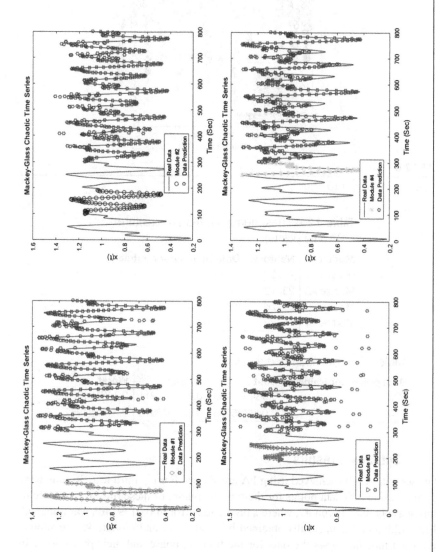

FIGURE 12.7

Nonoptimized training #10.

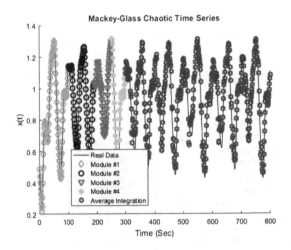

FIGURE 12.8

Nonoptimized training #10 result.

Table 12.4 Best nonoptimized architecture (300 data points).

Module	Neurons	Data points per module
Module #1	26, 20	1–100
Module #2	23, 12	101–190
Module #3	25, 26	191–250
Module #4	14, 29	251–300

Table 12.5 Nonoptimized results.

Data points for training	Best	Average	Worst
500	0.034371448	0.052881911	0.113121298
300	0.046735933	0.096949616	0.311284982

12.4.2 Optimized results

In this section, results achieved using FA are shown. For the third test, 30 runs were performed using 500 data points for the training phase; of these runs, the best architecture was obtained in run #7, where a prediction error of 0.01391408 was obtained. In Figs. 12.9 and 12.10, results obtained by module #1 and module #2 are shown, where each module shows the data for the training phase and their predictions. In Fig. 12.11, the final prediction is shown. In Table 12.6, the architecture of the best run is shown.

For the fourth test, 30 runs were performed using 300 data points for the training phase; of these runs, the best architecture was obtained by run #18, where a prediction error of 0.0194607 was obtained. In Figs. 12.12 and 12.13, results obtained

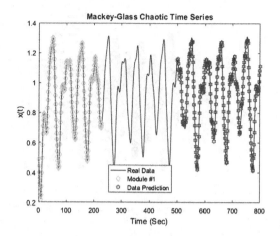

FIGURE 12.9

Run #7, module #1.

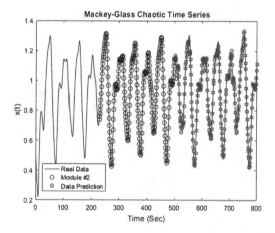

FIGURE 12.10

Run #7, module #2.

Table 12.6 Optimized architecture (500 data points).

Module	Neurons	Data points per module
Module #1	11	1–228
Module #2	13	229–500

by module #1 and module #2 are shown, where each module shows the data for the training phase and their predictions. In Fig. 12.14, the final prediction is shown. In Table 12.7, the architecture of the best run is shown.

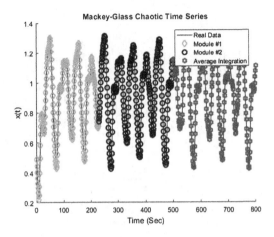

FIGURE 12.11

Final prediction, run #7.

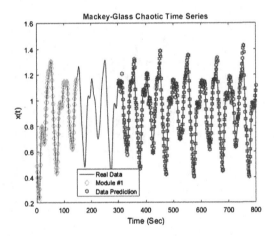

FIGURE 12.12

Run #18, module #1.

Table 12.7   Optimized architecture (300 data points).

Module	Neurons	Data points per module
Module #1	10	1–146
Module #2	14	147–300

In Table 12.8, a summary of the results of optimized trainings is shown. It can be observed that better results were achieved when 500 data points were used for the training phase.

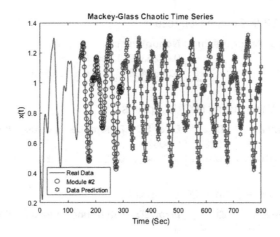

FIGURE 12.13

Run #18, module #2.

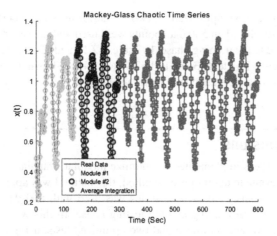

FIGURE 12.14

Final prediction, run #18.

Table 12.8 Optimized results.

Data points for training	Best	Average	Worst
500	0.01391408	0.01726544	0.02436075
300	0.0194607	0.02534833	0.03397797

12.4.3 Statistical comparison

In Table 12.9, the average result of each test is shown. It can be observed that the averages of error prediction using 500 or 300 data points are smaller when FA is used. To test if the optimization shown in this chapter is a good solution for modular

Table 12.9 Comparison results.

Data points for training	Nonoptimized	Optimized
500	0.052881911 (test #1)	0.01726544 (test #3)
300	0.096949616 (test #2)	0.02534833 (test #4)

Table 12.10 Values using 500 data points.

Test	N	Mean	SD	Error SD	Difference	t value	P value	DF
Test #1	30	0.0529	0.0202	0.0037				
Test #3	30	0.01727	0.00229	0.00042	0.03562	9.61	1.13567e−10	29

Table 12.11 Values using 300 data points.

Test	N	Mean	SD	Error SD	Difference	t value	P value	DF
Test #2	30	0.0969	0.0658	0.012				
Test #4	30	0.02535	0.00378	0.00069	0.0716	5.95	1.80842e−06	29

granular neural network optimization, two statistical tests were performed, the first comparing the results when 500 data points were used and the second when 300 data points were used to the training phase. In Tables 12.10 and 12.11, the values achieved by comparing test #1 vs test #3 and test #2 vs test #4 are shown, respectively. In both comparisons, results are statistically better when FA is applied to the optimization.

12.5 Conclusions

In this chapter, FA is implemented to optimize modular granular neural network architectures; to evaluate the proposed method, this method was applied to time series prediction. The Mackey–Glass time series was used with 800 data points, divided into data points to the training and the testing phase. In this work, four tests were performed, i.e., two nonoptimized and two optimized tests, using FA. In these tests the number of data points used for the training phase was also changed to compare performance with different numbers of data points.

The best architectures optimized by FA are smaller than without optimization (lower number of modules and hidden layers). As future work, other time series will be used and other parameters for the granular neural network will be added to be optimized, such as the number of data points used for the training phase and how many data points will be used for training by each module.

References

Arora, U., Lodhi, M.E.A., Saxena, T.K., 2016. PID parameter tuning using modified bat algorithm. Journal of Automation and Control Engineering 4 (5), 347–352.

Bargiela, A., Pedrycz, W., 2003. Granular Computing: An Introduction. Kluwer Academic.

Butenkov, S., Zhukov, A., Nagorov, A., Krivsha, N., 2017. Granular computing models and methods based on the spatial granulation. Procedia Computer Science 103, 295–302.

Carvalho, M., Ludermir, T.B., 2007. Particle swarm optimization of neural network architectures and weights. In: 7th International Conference on Hybrid Intelligent Systems (HIS 2007), pp. 336–339.

Dorigo, M., 1992. Optimization, Learning, and Natural Algorithms. Ph.D. Thesis. Politecnico di Milano, Milan, Italy.

Durga, M.L., Lalitha, P., 2015. Data mining by adopting particle swarm optimization. International Journal for Research in Science Engineering and Technology 2 (12), 32–36.

Eberhart, R.C., Kennedy, J., 1995. A new optimizer using particle swarm. In: Sixth International Symposium on Micro Machine and Human Science, pp. 39–43.

Goltsev, A., Gritsenko, V., 2015. Modular neural networks with radial neural columnar architecture. Biologically Inspired Cognitive Architectures 13, 63–74.

Gurevich, P., Stuke, H., 2019. Pairing an arbitrary regressor with an artificial neural network estimating aleatoric uncertainty. Neurocomputing 350, 291–306.

Holland, J., 1975. Adaptation in Nature and Artificial Systems. University of Michigan Press, Ann Arbor, MI, USA.

KumarSrivastava, A., Singh, H., 2016. An enhance firefly algorithm for flexible job shop scheduling. International Journal of Computer Applications 6 (5), 1–17.

Mackey, M.C., Glass, L., 1997. Oscillation and chaos in physiological control systems. Science 197, 287–289.

Man, K.F., Tang, K.S., Kwong, S., 1975. Genetic Algorithms: Concepts and Designs. Springer-Verlag, London.

Melin, P., 2012. Modular Neural Networks and Type-2 Fuzzy Systems for Pattern Recognition. Springer, Berlin.

Melin, P., Pulido, M., 2014. Optimization of ensemble neural networks with type-2 fuzzy integration of responses for the dow jones time series prediction. Intelligent Automation & Soft Computing 20 (3), 403–418.

Michelakos, I., Mallios, N., Papageorgiou, E., Vassilakopoulos, M., 2011. Ant Colony Optimization and Data Mining. Springer, Heidelberg.

Ong, P., Zainuddin, Z., 2019. Optimizing wavelet neural networks using modified cuckoo search for multistep ahead chaotic time series prediction. Applied Soft Computing 80, 374–386.

Pulido, M., Melin, P., Castillo, O., 2014. Particle swarm optimization of ensemble neural networks with fuzzy aggregation for time series prediction of the Mexican stock exchange. Information Sciences 280, 188–204.

Qian, Y., Zhang, H., Li, F., Hu, Q., Lianga, J., 2014. Set-based granular computing: a lattice model. International Journal of Approximate Reasoning 55 (3), 834–852.

Salama, K., Abdelbar, A.M., 2014. A novel ant colony algorithm for building neural network topologies. In: International Conference on Swarm Intelligence (ANTS 2014), pp. 1–12.

Sánchez, D., Melin, P., 2014. Optimization of modular granular neural networks using hierarchical genetic algorithms for human recognition using the ear biometric measure. Engineering Applications of Artificial Intelligence 27, 41–56.

Sánchez, D., Melin, P., 2015. Modular neural networks for time series prediction using type-1 fuzzy logic integration.

Sánchez, D., Melin, P., 2016. Hierarchical Modular Granular Neural Networks with Fuzzy Aggregation. Springer, Berlin.

Sánchez, D., Melin, P., Castillo, O., 2017. Optimization of modular granular neural networks using a firefly algorithm for human recognition. Engineering Applications of Artificial Intelligence 64, 172–186.

Sotirov, S., Sotirova, E., Atanassova, V., Atanassov, K., Castillo, O., Melin, P., Petkov, T., Surchev, S., 2018. A hybrid approach for modular neural network design using intercriteria analysis and intuitionistic fuzzy logic. Complexity 2018, 1–11.

Suárez, P., Iglesias, A., 2017. Bat algorithm for coordinated exploration in swarm robotics. In: International Conference on Harmony Search Algorithm (ICHSA 2017), pp. 134–144.

Taheri, M.H., Abbasi, M., Jamei, M.K., 2019. Using artificial neural network for computing the development length of MHD channel flows. Mechanics Research Communications 99, 8–14.

Talbi, N., 2019. Design of fuzzy controller rule base using bat algorithm. Energy Procedia 162, 241–250.

Talib, H., Darus, I.Z.M., 2017. Intelligent fuzzy logic with firefly algorithm and particle swarm optimization for semi-active suspension system using magneto-rheological damper. Journal of Vibration and Control 23 (3), 501–514.

Taylan, M., Dülger, L.C., Daş, G.S., 2013. Robotic applications with particle swarm optimization. In: CoDIT'13, pp. 160–165.

Yang, X.S., 2009. Firefly algorithms for multimodal optimization. In: Watanabe, O., Zeugmann, T. (Eds.), Stochastic Algorithms: Foundations and Applications. SAGA 2009. In: Lecture Notes in Computer Science, vol. 5792. Springer, Berlin, pp. 169–178.

Yang, X.S., 2010. A new metaheuristic bat-inspired algorithm. In: Cruz, C., González, J.R., Pelta, D.A., Terrazas, G. (Eds.), Nature Inspired Cooperative Strategies for Optimization (NISCO 2010). In: Studies in Computational Intelligence, vol. 284. Springer, Berlin, pp. 65–74.

Yang, X.S., 2014. Nature-Inspired Optimization Algorithms. Elsevier Insight, London.

Yang, X.S., Deb, S., 2010. Engineering optimisation by cuckoo search. International Journal of Mathematical Modelling and Numerical Optimisation 1 (4), 330–343.

Yang, X.S., He, X., 2013. Firefly algorithm: recent advances and applications. International Journal of Swarm Intelligence 1 (1), 36–50.

Yao, Y., 2018. Three-way decision and granular computing. International Journal of Approximate Reasoning 103, 107–123.

Yao, Y., Zhao, L., 2012. A measurement theory view on the granularity of partitions. Information Sciences 213, 1–13.

Zadeh, L.A., 1965. Fuzzy sets. Journal of Information and Control 8, 338–353.

Zadeh, L.A., 1996. Key roles of information granulation and fuzzy logic in human reasoning, concept formulation and computing with words. In: Proceedings of IEEE 5th International Fuzzy Systems, p. 1.

Artificial intelligence methods for music generation: a review and future perspectives

13

Maximos Kaliakatsos-Papakostas[a], Andreas Floros[b], Michael N. Vrahatis[c]

[a]*Athena Research and Innovation Centre,*
Institute for Language and Speech Processing, Athens, Greece
[b]*Ionian University, Department of Audio and Vidual Arts, Corfu, Greece*
[c]*University of Patras, Department of Mathematics, Patras, Greece*

CONTENTS

13.1 Introduction

Many diverse artificial intelligence (AI) methods have been proposed for music generation over many decades. From the rule-based and Markov approaches of the Illiac Suite (Hiller and Isaacson, 1979) to more recent deep learning approaches that allow interactive piano performance tools (Donahue et al., 2018) and score filling (Huang et al., 2019b,a), researchers find it intriguing to test AI methodologies for music generation. Among the many reasons that the application of AI for generating music is interesting and important, we find the fact that music is organized on many levels of abstraction, where even complex rules may not be enough to capture deeper structures. Even in the case of the Bach chorales, which is a style of music that is highly organized music with apparently strict rules, attempts to develop generative models for the Bach chorales with rule-based approaches are efficient up to a certain level.

Nature-Inspired Computation and Swarm Intelligence. https://doi.org/10.1016/B978-0-12-819714-1.00024-5

On the other hand, there is strong research interest about if or to what extent AI methods are creative. According to Boden (2004), there are three types of creativity: (i) explorational, (ii) transformational, and (iii) combinational. Certain AI methods potentially allow the exploration of musical styles, the transformation of rules for achieving novel musical results, and the combination of conceptual spaces for forming altogether new ones. There is, however, still debate on whether the creativity of such systems, some of which can arguably be categorized to one of the aforementioned creativity categories, is a reflection of the human agent's creativity. In other words, are the methods themselves "creative" or is the engineering of generative algorithms an essential creativity component that is a prerequisite for achieving computational creativity? Evaluation methods for answering such questions have been developed, e.g., with the FACE/IDEA models (Colton et al., 2011), where not only the creative output (e.g., generated music) is examined, but also the processes (e.g., the level of intervention of the agent constructing the algorithm) are examined to determine how creative a system is. Such evaluation models are still theoretical and they would potentially have very diverse implementations in different settings, e.g., for systems that generate music from scratch, assist composers, interact with musicians, etc.

This chapter presents AI methods that have been proposed for music generation over a wide variety of algorithmic approaches, attempting to predict which of the research directions in AI methods will be more promising for music generation. The concept of abstraction is highlighted and the chapter begins in Section 13.2 with presenting an information-based approach to how musical abstraction can provide deep musical meaning through simple geometric/computational modeling. Nonadaptive methods are presented in Section 13.3, which rely on human modeling for achieving interesting results. Section 13.4 presents methods that are adaptive and learn from data, and these methods are categorized as learning explicit or implicit features. Evolutionary approaches are presented in Section 13.5, which allow for intuitive interaction with users, highlighting the importance of transparent feature modeling. Finally, Section 13.6 gathers all the positive aspects of the aforementioned method, in an attempt to present what recent advancements are more promising towards developing systems that allow intuitive user involvement in generating novel music that interpolates learned styles or even extrapolates from them.

13.2 Information, abstraction, and music cognition

Music is a stream of information that can be comprehended by humans as having structure in the form of parts with beginnings and endings, conveys feelings, and presents meaning on different levels of abstraction, while the mechanisms that elicit emotions and make music interesting to humans are related to expectation and its fulfillment or violation (Huron, 2006). Several factors come into play when it comes to how humans understand, process, and value musical elements, ranging from low-level perceptual characteristics of the human anatomy (e.g., perception of harmonics due to the cochlear shape) to higher-level cognition that relates to memory capacity and statistical learning (Huron, 2006).

The importance of statistical learning in music is supported by studies where perception of music structures in Western listeners correlates with statistical findings in corpora, e.g., tonal center and mode (Krumhansl, 2001; Temperley, 2004). Listeners who are exposed to different musical environments have different norms of expectation to such an important extent as to allow scientists to support the *cultural distance* hypothesis: "the degree to which the musics of any two cultures differ in the statistical patterns of pitch and rhythm will predict how well a person from one of the cultures can process the music of the other" (Demorest and Morrison, 2016). Statistical learning, however, is evident on higher levels of information where musical information is abstracted from the "musical surface," i.e., the layer of discrete notes, and their vertical organization in chords and melodies.

In cognitive science, research has shown that humans employ some common basic mechanisms on an ultimately abstract level for understanding and categorizing concepts in their environment; those mechanisms have been called "schemata" (Gick and Holyoak, 1983; Hedblom et al., 2016). An example of a schema is the concept of the "container," where an object acts as a container to other objects, regardless of what those objects are. In music, the idea of schemata is mainly associated with tools that create abstractions from the musical surface and facilitate the acquisition of a mental knowledge structure (Leman, 2012). Examples of such abstractions that humans unconsciously extract when exposed to musical stimuli as studied, among other works, in Leman (2012), are the concepts of tonal center and mode (Krumhansl, 2001). Those abstractions allow listeners to relate and compare musical excerpts on more abstract levels, e.g., two pieces might be similar in that they sound "happy" because they both utilize elements of a major scale similarly.

From an information perspective, it is important to note that on higher levels of abstraction, the cognitive mechanisms function under geometric principles. For instance, the tonality of a tonal piece can be accurately predicted through the correlation of its pitch class profile with the pitch class profile templates extracted from the empirical experiment conducted by Krumhansl (2001). This also means that on abstract levels of information, the perception of similarity and therefore the notion of categories in music can be approached accurately by geometric relations.

As an example that shows the powerful interpretations that are offered by geometrical information reduction techniques, a set of 35 Bach chorales is considered, obtained from the COINVENT harmonic training dataset.[1] In this datasets two steps of abstraction are performed:

1. Pitch class abstraction: Each pitch in the Bach chorales is represented by its pitch class, i.e., its value modulo 12.
2. Tonality abstraction: Each phrase in each Bach chorale, being annotated according to its tonality, is shifted to a neutral tonality, making each pitch class from the

---

[1] https://github.com/maximoskp/COINVENT_HTD.git.

above mentioned step a *relative* pitch class rather than an *absolute* pitch class. For instance, for a piece in C major, the note G corresponds to relative pitch class 7, while for a piece in D major, the G pitch corresponds to relative pitch class 5.

The result of this process leads to the *relative pitch class* matrix representation, denoted as R, of a musical piece; an example where only note onsets (i.e., beginning times) are considered is illustrated in Fig. 13.1.

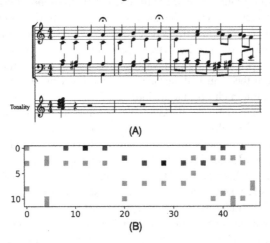

FIGURE 13.1

(A) A musical segment from a Bach chorale and its tonality annotation. (B) Illustration of its abstract relative pitch class representation in the form of a matrix, R, considering only note onsets and disregarding duration information. Colors in (B) correspond to occurrences, where darker colors represent higher values.

Each column in R represents a time instance (corresponding to a 16th of a measure). Even from the small example in Fig. 13.1, it is obvious that information can be compressed extensively and describe matrix R with its most often used components. By defining R as a matrix in $\mathbb{N}^{12 \times t}$, where t is the total number of 16th time steps in all Bach chorales, we can apply nonnegative matrix factorization (NMF) (Lee and Seung, 1999) and obtain a compressed representation of R, namely, $\hat{R}$, defined as

$$\hat{R} = W\,H, \text{ where } W \in \mathbb{R}^{12 \times 3} \text{ and } H \in \mathbb{R}^{3 \times t},$$

effectively reducing the dimension of R through a product of a basis vector (W) and the activation of each basis vector in time (H). The patterns that come out as basis vectors (columns of W) are shown in Fig. 13.2A, their activations (H) in (B), and the achieved reconstruction of R ($\hat{R}$) in (C).

Fig. 13.2A shows that when a geometry-based method (NMF) is used for compressing the information in three bases for the abstract representations of a set of 35 Bach chorales, each base has important musical meaning. The first column of W has large values in the locations $\{0, 3, 4, 7\}$, as indicated by the darker colors, the second in $\{0, 2, 5, 9\}$, and the third in $\{2, 7, 11\}$. The musical explanation of each column is:

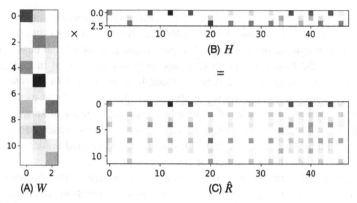

FIGURE 13.2

(A) A musical segment from a Bach chorale. (B) Illustration of its abstract relative pitch class representation in the form of a matrix, R, considering only note onsets and disregarding duration information. Colors in (A), (B) and (C) correspond to occurrences, where darker colors represent higher values.

1. $\{0, 3, 4, 7\}$ is the tonic scale degree, corresponding to either major ($\{0, 4, 7\}$) or minor ($\{0, 3, 7\}$);
2. $\{0, 2, 5, 9\}$ corresponds to the subdominant scale degree, either through IIm ($\{2, 5, 9\}$) or IV ($\{5, 9, 0\}$);
3. $\{2, 7, 11\}$ corresponds to the dominant ($\{7, 11, 2\}$).

Therefore, by simply representing the data after two steps of abstraction, a geometric method can clearly infer the "Schenkerian" basis of tonal music (W) and perform Schenkerian analysis (Cook, 1994) (H) on a set of Bach chorales.

This simple example shows the immense impact of proper abstractions from the musical surface; the NMF-inferred Schenkerian analysis simply demonstrated that there is a geometric/mathematical basis of music when considering proper abstractions. For generating music, however, abstractions on many levels and under many perspectives need to be taken into account. How can rhythmic abstractions be made? How about harmony/chords, melodies, or textural patterns (i.e., concerning voicing layouts)? The following sections, and especially Sections 13.4 and 13.5, focus on how methods decode, encode, and leverage abstractions obtained from data and/or inherited to generated musical surfaces.

13.3 Composing music with nonadaptive AI

In a "classical" AI approach, algorithms can be used that employ rules for defining the appropriateness of the output. As with Schenkerian analysis itself, musicologists have devised several abstract rules for defining musical style and for expressing what is "allowed" in specific types of compositions. These rules, however, are not sufficiently clear for producing new compositions. In fact, those rules describe very specific constraints that need to be satisfied while allowing absolute freedom on other aspects. For instance, harmonic motion (i.e., what chords are employed) and bass voice leading (i.e., how the bass voice is moving) are very strictly defined in cadences (i.e., phrase or piece endings), while they are more loosely defined in all other parts of a piece, allowing for many alternative approaches to be considered as valid for the Western tonal idiom. Examples of approaches to encoding musical knowledge of the Bach chorale style through expert-designed rules have been presented by Ebcioglu (1988) and Phon-Amnuaisuk et al. (2006), while the interested reader is referred to Pachet and Roy (2001) for a review of such methods.

Generative AI methods that are not adaptive incorporate proper mappings from the data they process to musical surfaces for music generation. Examples of such methods that are discussed in this section are *cellular automata* (CA), *L-systems*, and nonadaptive/autonomous *swarm intelligence*. The musical rules of expert systems mentioned in the previous paragraph are parts of the mapping, i.e., the numerical output of the generative algorithms was explicitly mapped to a musical entity (pitch, rhythm, intensity value, and/or information related with structure). The design and development of the mapping is what actually makes the musical output of such methods musically meaningful. In terms of the creativity reflected by such systems, the important task of the human designer/programmer in coming up with proper mappings plays an important role. Therefore, in a possible implementation of the FACE/IDEA (Colton et al., 2011) models, aiming to evaluate the creativity of such systems, the contribution of the human agent would be crucial to the creativity capabilities of the system.

Cellular automata

In CA, simple rules of interactions between neighboring units result in complex emergent behavior with structural characteristics that extend well beyond the radius of interaction between neighbors. CA can be implemented in any number of dimensions, but since their visual materialization is interesting and informative about the evolution of the CA society, one-, two-, and three-dimensional CA have been employed for music generation. The general setup of CA includes values for each unit, usually discrete or binary, that update iteratively in each "generation" according to rules that employ information about the values of neighboring units. The rules typically simulate physical phenomena as "extinction," "domination," or alternations between those extremes of units with specific values within the "universe" of the CA. The emergent behavior may have the following characteristics: (i) patterns disappear and units with a fixed value dominate; (ii) patterns repeat periodically, creating units that periodically change values; and (iii) patterns evolve chaotically/nonperiodically.

The diversity in the behavior of such systems has attracted the interest of many artists and researchers; detailed reviews of several approaches and mappings can be found in Burraston and Edmonds (2005), Burraston and Martin (2006), and Miranda and Al Biles (2007). Symbolic music mapping from CA to notes has been attempted in many contexts, with some diverse and notable examples including the piece *Horos* by Iannis Xanakis Solomos (2005), where areas of a CA universe were mapped to notes on a musical score; an application that employs similar principles for mapping to MIDI notes was presented in Millen (2005). CA have also been used for synthesizing audio through granular synthesis (Miranda, 2001), where the location of each active unit indicated which granule of sound would be active at any given time, or from given spectrograms (Serquera and Miranda, 2010) where each unit was acting as an amplitude envelope for the respective frequency area of an input spectrogram.

L-systems

L-systems generate fractal structures that resemble plants in their visual appearance (Prusinkiewicz and Lindenmayer, 2012). L-systems generate string sequences of symbols belonging to a given alphabet, based on substitution rules over an initial sequence. The typical and simplest form of L-systems belongs to the category of *deterministic context-free* grammars (DOL-systems), even though there are nondeterministic variants. As a form of formal presentation, L-systems incorporate an alphabet V of all possible symbols, a set of rules P that associate symbols in the alphabet with a string. Starting from an initial sequence of symbols $\omega \in V^+$,[2] denoted as x_0, the rules are applied for each symbol in x_0, resulting in a new sequence of symbols, denoted as x_1. Recursively, the sequence x_{n+1} is formed by applying the rules in P on each symbol in x_n. After a number of k steps, the sequence x_k will constitute a symbol sequence that potentially exhibits interesting structural characteristics on many levels, i.e., not only neighboring but also remote symbols.

As with CA, the visual interpretation of L-systems makes it evident that the higher-level structures that emerge can potentially provide a sense of structural hierarchy when properly mapped to sound/music. The first study of transforming L-systems to music via direct interpretation of generated symbols to notes was presented by Prusinkiewicz (1986); later, McCormack (1996) presented probabilistic L-systems that include probabilities about several possible rules associated with a symbol, mapping the resulting symbols to melodic notes. Further exploration of mappings between various forms of L-systems and musical notes can be found in Worth and Stepney (2005), while approaches to generating sound with the output of L-systems can be found in Manousakis (2006). A variation of the L-systems, namely, the *finite L-systems* (FL-systems), has been proposed by Kaliakatsos-Papakostas et al. (2012b), where the produced strings at each next step were truncated to a fixed length, producing strings that had quasiperiodic characteristics at different levels. Methodologies that evolve the rules of musical L-systems (de la Puente et al., 2002a; Lourenc and Brand, 2009) and FL-systems (Kaliakatsos-Papakostas et al., 2012d) through grammatical evolution have been also explored, which offer gradual alterations of the output, and adaptive behavior.

---

[2] V^+ is the set of nonempty words in V, i.e., nonempty symbol sequences comprising symbols of V.

Swarm intelligence

Swarm intelligence leads to the emergence of collective spatial behavior through the individual readjustment of the location of unique individuals, based on the application of simple rules that update the velocity of each agent according to the location and velocity of other neighboring agents. Variations of such emerging behavior have proved successful in optimization, through particle swarm optimization (Kennedy, 2010). Artistic expression has been examined in a specific setting of swarm intelligence that simulates the flocking behavior of real-world animals. The algorithm introduced by Reynolds (1987) has gained attention in computer and game graphics for simulating the motion of swarms, herds, and flocks (e.g., in the Batman Returns movie from 1992). This algorithm defines the motion of agent based on three components: *shoaling*, where each agent moves towards the center of mass of its neighboring agents, *collision avoidance*, where each agent moves away from the agents that are too close, and *schooling*, where the velocity of each agent gets aligned with the mean velocity of the neighboring agents.

Several parts of the aforementioned social characteristics have been embodied to interactive agents, leading to music and sound output. A swarm that was able to improvise with symbolic music output with the guidance of a human singing voice was presented by Blackwell and Bentley (2002b). The latter work was also enhanced with the addition of collision avoidance skills to the agents (Blackwell and Bentley, 2002a). Symbolic music has also been composed by agents that were specialized in certain musical tasks (Blackwell, 2003). A thorough review of these systems can be found in Blackwell (2007). Such intelligent societies have also been used for additive synthesis (Apergis et al., 2018) and granular synthesis (Blackwell and Young, 2004; Blackwell, 2008) as well as granular synthesis with spatial characteristics (Wilson, 2008). Finally, an interactive system has been proposed that receives feedback from the user to create audio and visual material using swarm intelligence and genetic algorithms (Jones, 2008). The sonification of the swarm intelligence agents behavior has been integrated into the "Swarmlake" (Kaliakatsos-Papakostas et al., 2014a) game, which expanded the social behavior with user-controlled commands and attributed different agents with different sound properties, according to specific conditions of the game.

Section summary

This section presented AI methods for music generation that are not adaptive and thus rely on human expertise in generating structured output. The creativity of such methods relies heavily on the creativity of the developer and thus the generative strengths of such models rely heavily on musical abstractions in the human agent's mind and how well the human agent can communicate those abstractions to method-related variables.

13.4 **Learning from data**

Trying to generate music in a specific style with nonadaptive systems is a difficult task. This is evident even for the structured style of Western tonal music, not to mention the most structured subset of this style, the Bach chorales. Regarding rule-based modeling of the tonal harmony, leaving aside voicing layout and rhythm, grammatical structures have been proposed by Rohrmeier (2011) and Koops et al. (2013), while attempts have been made to expand tonal grammars to the jazz style by Granroth-Wilding and Steedman (2014). This modeling is, however, incomplete, since it disregards all other aspects except harmony, which is by itself already complex to describe formally. In contrast to nonadaptive AI methods, methods that statistically adapt to given data have been proposed. Such methods either capture the statistical behavior of "explicitly" defined features (e.g., chord transition probabilities) or learn "implicit" representations from the musical surface into latent feature spaces. This section presents work on explicit and implicit AI modeling for music generation.

13.4.1 **Explicit modeling**

Probabilistic generative models can capture probabilities of occurrence of specific elements in a dataset. Regarding music, probabilities of explicitly defined features from a musical score can be captured, e.g., note or chord occurrences, note or chord transitions, and conditional probabilities of chords over given notes. Capturing such statistical information allows the development of trained models that reflect specific characteristics of the musical style in the training data. In contrast to rule-based, non-adaptive modeling, sampling from such models can potentially generate new music that reflects the characteristics of a given style.

Capturing combined and conditional probabilities of elements on a musical surface allows for the generation of new music under different settings. For methods that learn explicit representations, the review will focus on two specific test cases common to Western compositional practice, namely, *four-part harmonization* and *melodic harmonization*. In four-part harmonization, the goal is to compose a piece with a soprano, alto, tenor, and bass voice layout, where those voices are combined properly to form both concise harmonic, i.e., proper vertical positioning, and melodic streams, i.e., each voice should be a well-formed melodic part. Melodic harmonization is the composition of concise harmony over a given melody, without necessarily implementing compositions with specific voicing layout, i.e., some studies go as far as to simply assign proper chord symbols without any voicing information.

The most popular probabilistic AI techniques employed for solving such problems include hidden Markov models (HMMs) and more generalized Bayesian networks (BNs). Such models are suitable for modeling and generating music that incorporates relations between various elements, since these models allow the formulation of conditional probabilities across various aspects. For instance, the typical probability conditions for HMMs that model and generate melodic harmonizations learn two aspects of the musical surface from data: (i) chord transition (hidden state) relations through probabilities of the form $P(c_t|c_{t-1} = C)$, i.e., the probability density function of the current chord/state given the previous, and (ii) $P(c_t|m_t = \vec{M})$, i.e., the probability of the current chord given the current observed note sequence.

Regarding four-part harmonization, an approach that employed a dual HMM was proposed by Allan and Williams (2004). The role of the first HMM was to define a coarse harmonic layout given a melodic (soprano) line, modeling chords as unified symbols rather than independent voices. The second HMM was generating ornamentations, given the selected chords from the first HMM. A BN has been proposed in Suzuki et al. (2013) for four-part harmonization, which incorporated different nodes for each voice. Specifically, each voice was hierarchically conditioned on its higher voice, i.e., alto was conditioned on the soprano voice, tenor on alto, etc., while for each voice, current notes were conditioned on their previous ones. This method was able to generate the ATB voices given a melody/soprano voice. The employment of an additional node for conditioning chord symbols on the ATB voices was also examined, and the results were compared, showing the importance of the chord symbol for generating harmonically concise four-part harmonizations.

HMMs have been extensively studied for melodic harmonization, where the hidden states are chord symbols and observations are melodic notes. Microsoft has presented the then-called MySong (Simon et al., 2008) application, which allowed users to sing melodies, and after the melodic note fundamentals were extracted with digital signal processing, an HMM composed chord sequences on the given melody. The system was trained in two harmonic styles, classical and jazz. The study by Raczyński et al. (2013) examined the idea of incorporating additional information as the local tonality of the piece, therefore conditioning chord selection on tonality as well.

Among the main weaknesses of Markov-based models is their inability to capture structures on a larger timescale, since they are able to capture statistical information only to the extent that their order allows. Human-composed music incorporates meaning on many structural levels, with intermediate phrases and repetitions of large harmonic segments that cannot be captured by low-order HMMs. On the other hand, using high-order Markov models for modeling harmony leads to extremely specialized models that cannot capture style, but rather capture unaltered segments of pieces in the training dataset. Hierarchical Markov models have been proposed for capturing long-term structure (Thornton, 2009), preserving the generalization capabilities of lower-order Markov models. Those models rely on modeling repeating parts of hidden states in new hidden states, building hierarchically Markov models on top of each other for consecutively capturing patterns on different levels of time granularity. Graphical models for modeling chord progressions (Paiement et al., 2005) and melodic harmonization (Paiement et al., 2006) have been proposed, which are capable of capturing long-term relationships between chords through tree-like nodes that model conditional probabilities from top to bottom. Such methods, however, model only a fixed number of chords in a sequence – 16 chords in both aforementioned examples.

Higher-level structure in human composition is to a great extent evident in the presence *cadences*, both intermediate and final, which are parts of a piece that convey a feeling of closure, i.e., that a section closes or ends and a new section begins – in the case of a final cadence, that the entire piece ends. This fact has led many researchers to develop variations of Markov-based models that focus on generating proper cadences. For instance, Borrel-Jensen and Hjortgaard Danielsen (2010) and Yi and Goldsmith (2007) evaluated entire melodic harmonizations based on a cadence score that rated higher more typical cadential schemes of Western harmony. Other methods implemented a backwards propagation of the harmonization process (Allan and Williams, 2004; Hanlon and Ledlie, 2002), beginning from the end (cadence) to ensure that the ending part will be as concise as possible. Given the importance of cadences, a study presented by Yogev and Lerch (2008) studied the identification of possible intermediate cadence locations. If the positions of cadences are known, then Markov models with constraints (Pachet et al., 2011) could be employed, forcing the harmonization system to apply intermediate cadences at proper locations, therefore reflecting longer time structures. A first trivial approach towards this direction was presented by Kaliakatsos-Papakostas and Cambouropoulos (2014), where constraints were merely straightforwardly added in the trellis diagram, for selecting chord progressions that belong to learned cadential schemes. This method was incorporated in the CHAMELEON[3] melodic harmonization assistant (Kaliakatsos-Papakostas et al., 2017), but this method requires the user to annotate the location of intermediate cadences.

13.4.2 Implicit learning

The features to be captured by the methods in the previous paragraphs are defined "explicitly," meaning that their definition is transparent and they encompass concrete meaning; e.g., "the probability of appearance of a chord over a set of given melodic notes." Features can also be extracted "implicitly." A popular example for implicit feature computation is artificial neural networks (ANNs), which, in the case of music, process information from the musical surface and produce more abstract representations in each layer. Those representations, however, are not transparent, in the sense that there is no distinction on which aspects of the musical surface are represented at each computational unit of the ANN. Recently, deep learning has increased the attention of the research community on methods that incorporate vast amounts of computational units (i.e., neurons) organized in multiple layers, which learn implicitly from large amounts of data. Implicit learning with deep ANNs offers important possibilities for categorization and prediction, without still giving clear information about what aspects of the data are more important for taking decisions. There is significant research on alleviating this "trade-off" (lack of transparency in what the latent abstract features represent) with which those powerful methods come, towards mak-

---

[3] http://ccm.web.auth.gr/chameleonmain.html.

ing ANNs that are "explainable," "intuitive," or "interpretable"; Section 13.6 refers to such studies.

Early approaches on using ANNs for composing music employed basic recurrent neural network (RNN) architectures, where a hidden layer was used in between input and output layer. The hidden layer, also called "states" of the network, included recurrent connections from each unit to all other units of this layer, which reinjected a weighted sum of the states values in the previous step to each unit in the states layer. The recurrent connections allow such networks to learn local dynamics of data, developing a local memory. Even from the first implementations of such networks (Todd, 1989), research was focused not only on making RNNs that reproduce music in a specific style, but also allow the network to switch styles according to an input "plan," which was actually a separate input vector to the system with the binary code that corresponded to the piece name (plan) that was currently incorporated in training. This allowed the experimentation on interpolating and extrapolating from learned melodies, by properly manipulating the "plan" part of the input.

During the early days of studying RNNs for melodic generation, the effect of psychoacoustical modeling of the inputs was examined by Mozer and Soukup (1991). According to this approach, the representation of inputs and outputs was not a simple one-hot encoding of the note currently played, but a vector of coordinates that combined pitch height (one dimension), the (x, y) coordinates on the chromatic circle, and the (x, y) coordinates on the circle of fifths. The aforementioned method exhibited the ability to learn scales, the form of interspersed random walks, and to generate melodies in the style of Bach chorales. The authors have validated, however, that such architectures were poor in capturing long-term structure of the learned melodies. To this end, the system was improved by incorporating a blurred "bird's eye" view as an additional input, for getting information from further back in the past (Mozer, 1994). Even though the results were better, significant improvements on adaptation to long-term structure was exhibited by using the long short-term memory (LSTM) networks proposed by Eck and Schmidhuber (2002). Those networks include trainable gates that selectively forget information from the past or recall information from arbitrarily back in time. The first study on how those networks learned on musical data (Eck and Schmidhuber, 2002) showed that they are capable of learning long time dependencies that allowed them to learn and generate structures belonging to the style of 12-bar blues.

A more recent approach to using LSTM RNNs for generating melodies was presented by Sturm et al. (2015, 2016), where folk tunes (monophonic melodies) were modeled in the ABC format, a text and character-based representation that includes metadata, overview of musical setup (e.g., tempo and time signature), metric information (i.e., measure boundaries), and the music surface. In contrast to the directly numerical format of musical data representation, elements of characters and strings corresponding to elements of a melody were extracted into a one-hot dictionary representation (binary array with a single unit).

Many alternatives have been proposed in representing polyphonic music for efficient processing by ANNs. Many studies have examined the efficiency of proposed representations and how ANNs process them by learning and generating pieces in the style of Bach chorales, since this style of music has a very strictly defined form. A "quasimonophonic" approach to modeling polyphonic data was presented by Liang et al. (2017), where note symbols, bar limits, and fermata symbols are learned and generated sequentially, from top to bottom and from left to right. A similar representation approach was followed by Colombo et al. (2018) with more refined information about the duration and offset time of notes. In these approaches, the ANN was fed with a sequence of single notes, where simultaneous notes simply had the same onset (beginning) time. Both aforementioned studies incorporated learning and generating polyphonic music in the style of Bach chorales; in addition to the partly different representations they used, another difference was that Liang et al. (2017) used LSTM units while Colombo et al. (2018) used gated recurrent units (GRUs) for polyphonic symbolic music generation. GRUs, like LSTMs, include a gating mechanism, but only for selectively resetting and updating the content of information in the recursive connections. The GRU architecture is simpler than the LSMT architecture, thus GRUs are less computationally expensive, while at the same time being approximately equally efficient to the LSTMs (Chung et al., 2014).

An important concern in generative RNNs is not only to enable them to capture longer-time dependencies, which LSTMs and GRUs achieve quite efficiently, but also enable them to capture different modes of structures for long-time relations, e.g., to compose a piece in 4/4 or 7/8 time signature. To this end, constraints have been introduced in Hadjeres et al. (2017) which allow the networks representing each voice to have a more robust understanding about the overall metric structure and the activity in each voice. Typical bidirectional LSTM layers in the architecture were responsible for learning the motion in single voices, while other parts of the network were imposing constraints for the metric structure, allowing the network to learn to generate four-part harmonizations in specific time signatures defined in the input. Additionally, the network was generating music through sampling and therefore any voice could have any set of notes fixed as a priori constraints, allowing the network to fill in the remaining notes for completing a composition. A similar approach to imposing constraints was presented for drum rhythm generation by Makris et al. (2017, 2019). In the latter studies, indications were given that proper representation of the metric constraints could allow the network to compose rhythms in time signatures that were not encountered during training. For instance, the network was trained in pieces in 4/4 and 7/8 time signatures of a given style and could compose consistent rhythms in 5/4, 9/8 and 17/16 that were compatible with such rhythms in the learned style.

Except for RNNs, other types of networks have been explored for generating music. Convolutional neural networks (CNNs) are able to capture patterns in data through the employment of filters that adapt to specific regularities that appear often. In a study by Yang et al. (2017) a generative adversarial neural network (GAN) setting was presented that employed a generator and a discriminator based on CNNs for generating monophonic melodies. In GANs there are two networks "competing" with each other: the generator produces data that the discriminator tries to identify as "artificial" in comparison with given ground-truth data. The generator, therefore, gradually learns to generate data that more persuasively appear as being part of a given dataset, while the discriminator gradually becomes more sensitive in identifying artificial data produced by the generator, leading to a feedback loop that makes both parts of the network more effective in their task. The interesting aspect of GANs is that the generator is not necessarily straightforwardly trained from data, since it can start from generating random material with initially randomized internal parameters, that are gradually refined during training. Additionally, a lead-sheet music setting was employed where chord symbols were given and the network learned to generate musical surface that corresponded to a rhythmic and a melodic part (Liu et al., 2018). The aforementioned methodology has also expanded to incorporate multiple tracks (Dong et al., 2018). More studies on ANNs are discussed in Section 13.6, along with their potential to offer new possibilities.

Section summary

This section presented models for learning music, based on adaptation to training data. Explicit learning methods were first analyzed that have the advantage of being "transparent," i.e., it is clear what features the network learns. Next, implicit learning methods were discussed, which create abstractions from data that are not transparent, but can describe deep structures on many levels of information. Especially regarding deep implicit learning methods, some studies were discussed that allowed some form of control over the generated output, e.g., by defining the key signature. More work on ANNs is mentioned in Section 13.6, when discussing future perspectives of AI in generative music.

13.5 Evolutionary approaches

Evolutionary algorithms evolve generations of individuals by selecting and breeding individuals based on their fitness value, which interprets numerically some criteria for the goal to be achieved. Each individual is represented by a genotype (the genetic material that can be modified during the breeding stage) and a phenotype (the materialization of the genotype); the genotypes and phenotypes of individuals may coincide, depending on the formulation of the problem. Evolutionary algorithms have been studied for music generation under various setups regarding how the musical surface is represented (in terms of phenotype and genotype) and how a good musical surface should be formally described, i.e., what the fitness criteria should be. Especially regarding the fitness criteria, cognitive-based features extracted from data play an important role, making abstraction a necessary step towards assessing the fitness of musical individuals during evolution. This section focuses on two music generation approaches where evolutionary algorithms have given interesting results: feature-based composition and interactive composition.

13.5.1 **Feature-based composition**

Humans create abstractions from the musical surface that allow them to do content categorization and measure similarity (Cambouropoulos, 2001). As mentioned in Section 13.2, an example of this process is the categorization of a musical piece in the category of tonal music: if a piece employs some standard tonal harmonic devices, e.g., diatonic scales and cadences that resolve from highly dissonant to highly consonant harmonies (Huron, 2006), then this piece is included in the tonal category. Research has shown that such high-level features, e.g., how strong the presence of a diatonic scale in a musical excerpt is, can be directly computed from the musical surface, as with the pitch class profile of the excerpt, and can accurately predict the diatonicity of an excerpt based on the templates extracted by Krumhansl (2001). Numerous such examples have been shown in the literature, where features computed from the musical surface can indicate qualitative aspects of the data. Another example concerns the perception of rhythm, where empirical studies presented by Madison and Sioros (2014) and Sioros et al. (2014) have shown that there are strong correlations between the feature of syncopation and the sensation of groove in rhythms.

Widely used methodologies and software have been proposed and developed extracting symbolic music features (Eerola and Toiviainen, 2004; McKay and Fujinaga, 2006). A fact that makes the efficiency of such features more evident is that they are producing accurate results in various content categorization tasks, such as composer identification (Purwins et al., 2004; Wolkowicz et al., 2008; Kaliakatsos-Papakostas et al., 2010, 2011) and the style and genre classification (Kranenburg and Backer, 2004; Mckay and Fujinaga, 2004; Hillewaere et al., 2009b; Hillewaere and M, 2009a; Herremans et al., 2015a; Zheng et al., 2017). Furthermore, features that generate information-theoretic abstractions of data, e.g., Shannon information entropy or fractal dimension, have been studied for the characterization of "esthetic" quality in music, leading to models that examine relations between complexity and human perception in music (Shmulevich et al., 2001; Madsen and Widmer, 2007) and also to models of subjective preference (Manaris et al., 2002; Machado et al., 2003; Manaris et al., 2005; Hughes and Manaris, 2012).

On the one hand, such features can indicate the category, mood, or complexity of composed pieces. On the other hand, evolutionary methods can be used to generate novel excerpts that belong to a certain category, mood, or complexity, given proper fitness functions and representations of musical surfaces – mappings from "genotypes" to "phenotypes." The feature extraction methods discussed above are again "explicit," in the sense that their computation from the musical surface is transparent. Such features can be used in evolutionary generative methods during fitness evaluation to examine whether the generated material meets the criteria set from those features. Even from the early days of generative music systems, there were some exceptional studies on the evolutionary generation of melodies that employed "implicit" feature extraction methods, implemented with ANNs. The work of Spector and Alpern (1995) and the work of Pearce (2000) are such examples of using fitness functions in evolutionary methods for music generation that are based on implicit learning. Those so-called "artificial critics" are trained to give positive feedback to collected melodies that incorporate desired characteristics (e.g., Charlie Parker solos) and negative feedback to either random or empty melodies. They are therefore capable of providing "implicitly computed" fitness evaluation to melodies generated through evolutionary processes.

Other than the aforementioned approaches that employed implicit fitness evaluation through ANN critics, fitness evaluation by means of distance from targeted/desired explicit features has been the most usual approach with evolutionary music generation methods. Such methodologies incorporate a set of target features that either incorporate information-related target features or features related to music cognition and theory; the evolutionary component of those methods generates music that adapts to the target features as generations progress. Regarding information-related metrics, features that compute the fractal dimension in distributions of several diverse elements obtained from the musical surface (e.g., pitch, interval, or rhythm-related distributions) were presented in Manaris et al. (2007); similarly, the normalized compression distance in Alfonseca et al. (2007) was employed for generating music with specific complexity characteristics. Cognitive and music-theoretic target features have been developed in other studies that quantified approaches to describe rules for counterpoint (Herremans and Sörensen, 2012), four-part harmonizations theory (Donnelly and Sheppard, 2011; Phon-Amnuaisuk and Wiggins, 1999), and melodic harmonization (Phon-Amnuaisuk et al., 2006), or quantities related to how humans perceive and process music (Wiggins and Papadopoulos, 1998; Özcan and Ercal, 2008; Matic, 2010; Hofmann, 2015; Herremans et al., 2015b). It should be noted that some approaches employed ANNs as "artificial critics" (Manaris et al., 2007; Machado et al., 2003), but therein, ANNs actually evaluated the similarity of the explicitly defined features related with fractal dimension and transition probabilities between the generated and a set of training data. The neural networks in this context receive many such features as input and their goal is to create abstract/latent representations of these features (instead of the musical surface, as discussed in the previous paragraph).

Another point of distinction between methods that have been employed for music generation is the genotypical and phenotypical representation. The methods mentioned so far evolve individuals that directly represent musical surface, i.e., the genotype comprises representations of notes. Other methodologies attempt to leverage the structural coherence that nonadaptive AI methods (discussed in Section 13.3) present. Examples of such methods include grammatical evolution (de la Puente et al., 2002b) and genetic evolution of CA rules (Lo, 2012) and of FL-systems (Kaliakatsos-Papakostas et al., 2012d). Additionally, the evolution of parameters of dynamical systems that present chaotic behavior was examined in Kaliakatsos-Papakostas et al. (2013a), where the parameters were tuned using differential evolution (Price et al., 2006).

13.5.2 Interactive composition

Evolutionary processes offer ways for human users to affect the generative process in different ways. In evolutionary schemes that employ interactive evolution, fitness evaluation is given directly by the human user and, therefore, the results are directly guided by the human agent. Due to the interpretable nature of explicitly defined features, it is possible for evolutionary processes to involve the human user in processes that "indirectly" affect the outcome. In this chapter two such methods are identified: "dissimilarity-based interaction" and "active musical interaction"; all approaches are discussed in the remainder of this section.

"Interactive evolution" comprises methods that require human evaluation for assigning fitness values for the evolved individuals in the population in the form of rating, ranking, or selection. Several studies have focused on music generation through interactive evolution with individual selection based on ratings (Unehara et al., 2005; Fortier and Van Dyne, 2011; Kaliakatsos-Papakostas et al., 2012a; MacCallum et al., 2012) or direct selection for reproduction (Sánchez et al., 2007). Rhythm generation (Horowitz, 1994; Johanson and Poli, 1998) was the first field of application and subsequently more musical aspects were included, where interactive evaluation incorporated partial rating of different aspects of music, e.g., rhythm, tonality, and style (Fortier and Van Dyne, 2011; Moroni et al., 2000). Such methodologies for music generation have the theoretical advantage that fitness evaluation is absolutely adaptive to the human user and that esthetic convergence is possible, given sufficient time; however, additional problems are practically introduced in comparison with methods that are noninteractive. The main problem of interactive evolution methods for music generation is the practical infeasibility to combine and alter large numbers of individuals within the course of many generations. Human users are not able to undergo vast amounts of listening and rating (or selecting) sessions, since *user fatigue* occurs during the very few first minutes of rating/selective sessions. As an even more negative result, the uncertainty in user ratings or selections is also increased, leading to inconsistent ratings that eventually "detune" the evolutionary effectiveness. Early approaches (Tokui, 2000) involved intermediate steps in between generations, where many individuals were generated and only a small part of them were shown to the user, based on an intermediate evaluation offered by an ANN.

"Dissimilarity-based interaction" accepts a user-given musical segment and a user-defined dissimilarity value; genetically modified musical segments are then evolved towards generating segments that are desirably dissimilar to the user-given segment. The dissimilarity value is computed according to some features that offer a layer of human–machine communications, where information is interpretable both by human and machine. Some studies have focused on generating novel rhythms based on an input rhythm provided by the user and a value of dissimilarity for the new rhythms (Kaliakatsos-Papakostas et al., 2013b; Nuanáin et al., 2015; Vogl et al., 2016).

During "active musical interaction," the user generates musical objects and expects relevant musical responses from the system. Therefore, the human performance affects the AI performance and vice versa, leading to a "creative loop" between the human and the artificial agent. The first approaches that employed evolutionary algorithms for active musical interaction were presented by Biles (2002), Thom (2001), and, more recently, Manaris et al. (2011), where the human and the artificial agents were exchanging phrases, i.e., the artificial agent needed to record and encode the human phrase, analyze its features, and playback a proper response; such an implementation was presented by Weinberg et al. (2008), but the responses of the artificial agent were performed by a hardware robotic musician. In Kaliakatsos-Papakostas et al. (2012c), another approach was presented, which incorporated concurrent performance from the human and the artificial agent in the form of intelligent accompaniment. Therein, however, the system was able only to identify the current playing status of the human musician, failing to predict possible structures and therefore leading to "constraint-free" improvisation.

Section summary

Evolutionary computation methods for music generation have been mainly studied as methods that evolve musical individuals to capture explicitly defined feature. Having interpretable/explicit features, on the one hand, allows such methods to model and reproduce specific aspects of musical styles and, on the other hand, allows interactive applications, where human and artificial agents interact towards formulating a result that is interesting to the user. Even though there are inherent limitations in interactive evolutionary systems (i.e., user fatigue), other modes of interaction (dissimilarity-based and active musical interaction) potentially allow for better human–machine collaboration results.

13.6 Towards intuitive musical extrapolation with AI

Section 13.4 discussed the importance of capturing deep structures in music implicitly, without the necessity to employ human expertise for describing all the necessary information for representing an entire musical style. Section 13.5 presented evolutionary algorithms, which are based on feature design for capturing desired characteristics of generated pieces. Describing style can be achieved either by explicit or by implicit feature extraction methods. The questions that this paragraph tries to answer are the following:

1. How is it possible to cross the borders of musical style?
2. How can methods provide an intuitive, interpretable layer, for how stylistic crossing is perceived?

Answers to these questions are sought by examining recent work in three generative AI methods: (i) evolutionary computation, (ii) conceptual blending, and (iii) deep learning.

Evolutionary computation

Regarding evolutionary algorithms, an approach to interactively extrapolating and crossing stylistic boarders in a consistent way was presented by Kaliakatsos-Papakostas et al. (2016). In this approach, a human user was listening to quadruples of generated polyphonic melodies according to rhythmic and pitch characteristics.

Based on those ratings, two evolutionary processes were utilized: an "upper-level" evolution of rhythmic and pitch features using the particle swarm optimization (PSO) algorithm (Kennedy, 2010) and a "lower-level" evolution of melodies with genetic algorithms with fitness evaluation targeting the features generated on the upper level. The goal of the upper level is to converge to rhythm and pitch features that the user prefers, while the goal of the lower level is to materialize those features to actual musical excerpts.

Even though this study incorporates some assumptions that need to be examined more thoroughly, it nonetheless offers a possible way to exploring generatively new musical areas by traversing potentially unforeseen areas of musical feature spaces. The cognitive advantage of this study, in comparison with studies that simply apply interactive evolution on musical excepts, is that evolution and user ratings concern the cognitively informed layer of musical features rather than the musical surface. Even though human evaluation of this method has not been implemented, the idea behind it is that evolving features towards directions given by the PSO algorithm encompasses a cognitive coherence, making the evolutionary process more meaningful. Contrarily, traditional methods for evolving musical excerpts by mere mutation and crossover of their parts do not guarantee to generate new excerpts that also combine high-level features.

Conceptual blending

The conceptual blending (CB) theory (Fauconnier and Turner, 2003) describes the cognitive processes that humans undergo when generating new concepts, based on the experiences of already known conceptual spaces. In CB theory, two input conceptual spaces are considered, which incorporate properties and relations between elements, and a blended space is generated by consistently combining properties and elements of the inputs. Initially, CB theory was used as a theoretic tool for interpreting creative artifacts created by humans, i.e., the blended space of a creative outcome was considered (e.g., a musical piece) and the task was to identify the constituent parts as independent input spaces (e.g., the musical/conceptual tool combined by the artist). Algorithmic approaches that use CB theory generatively have more recently been developed, i.e., two input spaces are given and a blended space is algorithmically constructed that consistently and creatively combines properties and relations in the input. This approach to computational creativity is related with *combinational creativity*, which Boden (2004) maintains is the most difficult to describe formally.

In music, generative formulations of generative CB have produced interesting results. In Eppe et al. (2015), it was shown that proper encoding of conceptual spaces describing cadences (defined as the last pair of chords in a chord sequence) can lead to the generation of interesting cadences. An interesting example presented therein was the algorithmic construction of the tritone substitution cadence, which is omnipresent in jazz music after the 20th century, by using two input cadences that were employed in music centuries earlier, namely, the perfect and the Phrygian cadences. The algorithmic materialization of this example agrees with music-theoretic perspectives that indeed relate the characteristics of the tritone substitution with the most salient characteristics of the perfect and the Phrygian cadences. Describing the characteristics of cadences and weighing them by importance (in order to be included in blends that are rated as more successful) is, however, a task that requires extensive musical expertise. Therefore, the algorithmic backbone of generative CB by itself is not sufficient for describing what are the most important characteristics of the inputs that should be included in the blends; depending on the domain of application, extensive human expertise and knowledge engineering are required for acquiring meaningful results. There have been attempts for automating the process of computing the salience/importance of features in the inputs through statistical approaches (Kaliakatsos-Papakostas and Cambouropoulos, 2019), but further examination is necessary before verifying that salience computation can be achieved effectively directly from data.

In practical terms, the employment of generative CB can prove useful for developing systems that generate entire melodic harmonizations. The CHAMELEON melodic harmonization assistant (Kaliakatsos-Papakostas et al., 2017) is such an example. This system expands on the ideas developed for the example of cadence blending for blending chord progressions between the Markov transition tables of two learned musical styles and generates blended transition matrices of two learned idioms. The blended styles integrate the most salient, in terms of statistical frequency, characteristics of the inputs, a fact that leads to results that not only interpolate between the two input styles but also extrapolate from them. An evaluation study with students in a music department, who were well aware of tonal and jazz music, indicated that they would categorize blended harmonizations either as "in between" tonal and jazz, or oftentimes as belonging to altogether "other" styles (Zacharakis et al., 2018).

Deep learning

Even though both evolutionary and blending approaches potentially offer the mechanisms to cross stylistic boundaries and interpolate between or even extrapolate from known styles, they require explicit knowledge description and extensive knowledge engineering. These requirements are not by themselves necessarily a drawback; there is extensive, however, continuous extensive research on how to allow computational methods do by themselves implicit feature extraction that is adaptive to the style of the training data. As discussed in Section 13.4.2, implicit learning methods do not require extensive human knowledge, which is very time consuming, tedious, prone to errors, and nonadaptive, in the sense that not all styles behave under comparable statistical rules. For instance, the polyphonic songs of Epirus present a "horizontal" rather than "vertical" interpretation of harmony, i.e., streams of voices move rather independently from each other, forming hard dissonances that cannot be accurately captured by the consonant organization of Western harmony (Kaliakatsos-Papakostas et al., 2014b). Therefore, modeling polyphonic songs from Epirus with statistical learning (e.g., through Markovian processes) on explicitly defined chord structures is problematic in generating persuasive results.

The implicit learning methods discussed in Section 13.4.2 can effectively learn latent representations of features directly from musical surfaces, alleviating the necessity for extensive human engineering. Such methods are therefore effective in learning a style though deep representations; are they, however, capable of meaningfully interpolating between or extrapolating from learned styles? Examples in image generation have shown that leveraging the spatially interpretational capabilities of the latent space in the variational autoencoder (VAE) (Kingma and Welling, 2013) can lead to generating new images that combine characteristics of existing images (Gulrajani et al., 2016). This is possible due to the sampling process that occurs during training, in combination with the fact that the latent space is trained not only to reconstruct faithful representations of the input data (e.g., of a given image) but also to follow a Gaussian distribution. Sampling from (initially "detuned") Gaussian distributions at each training epoch leads to latent representations that interpret meaningful information throughout the entire extent of the Gaussian distribution – given enough data.

In music, interesting results have been presented with the utilization of VAE for music generation (Roberts et al., 2018). Learning the latent space of various musical excerpts allowed the system to both interpolate and extrapolate from two given excerpts in a meaningful way. For instance, if two melodies were given that were primarily different in one feature, e.g., one was polyphonic and the other monophonic, sampling from interpolated points between the latent representations of the inputs generated new melodies with intermediate characteristics. Specifically, sampling from latent points that were closer to, e.g., the polyphonic excerpt generated more polyphonic output than sampling closer to the monophonic end. This creates a "morphing continuum" between any two points in the latent space, which is musically meaningful, in the sense that the different features that are implicitly captured in the latent space are consistently mapped to the musical output, creating excerpts that morph between two extremes. Furthermore, extrapolating from the line that connects the latent representations of two input excerpts would generate excerpts that "exaggerate" the feature differences towards the extrapolation end. For instance, in the example of the polyphonic and monophonic input excerpts, extrapolating towards the end of the polyphonic excerpt would produce an excerpt that has even more polyphony than the input excerpt in the polyphonic extreme.

Section summary and discussion

This section discussed methods that potentially allow music generation that intuitively and meaningfully interpolates and extrapolates from learned styles. Such methods were presented that either follow explicit approaches to representing features or implicit learning from data. The intuitiveness offered by methods that employ explicit feature representations comes at the cost of extensive human design, which is not only inaccurate and tedious but also style-specific and error-prone.

Regarding implicit learning, VAEs constitute a promising example of how machines can learn interconnected and meaningful latent representations that not only homogeneously connect areas that represent samples of the training data, but are also able to infer connections with latent points that represent data beyond the borders of what has been learned. The typical VAE, even though effective, incorporates a latent space that has "entangled" representations of features, i.e., a single feature (e.g., polyphony) might be expressed by more than one latent variable with complex relations. A modification of the typical VAE has been proposed recently, namely, the *beta*-VAE (Higgins et al., 2017), which generates disentangled representations of the latent features, in the sense that unique features are represented by the fewest possible latent variables. For example, the polyphony feature can be represented only by a single latent feature; therefore, changing the value of this feature could simply change the polyphony of an excerpt without necessarily having to provide a second excerpt for computing the interpolation/extrapolation direction in the latent space for achieving the desired result in polyphony.

13.7 Conclusion

This chapter has presented a review on AI methods for music generation. Initially, the important role of abstraction in music was highlighted, giving an example with the application of NMF on abstracted information obtained from the music surface of a set of Bach chorales, leading to an information-based extraction of the Schenkerian analysis. Subsequently, AI methods for music generation were presented in three categories, according to how they implement abstraction from the musical surface: (i) nonadaptive, (i) probabilistic, and (iii) evolutionary methods. The role of a human design was highlighted in nonadaptive systems, where abstractions are taking the form of rules designed by the artist/researcher who creates the method. Probabilistic methods, in contrast to nonadaptive that employ fixed rules, adapt to data by learning statistical relations between musical elements. In probabilistic methods, two categories were defined according to what knowledge they acquire: explicit modeling of knowledge involves learning specific features/abstractions defined in a way that is computationally transparent to the human user, and implicit modeling allows the model itself to create proper abstractions and learn their statistical behavior. Finally, evolutionary methods were analyzed, which are based on explicit modeling but also allow interactive interventions by users, based on the intuitive modeling they offer.

The chapter then focused on work that will expectedly open new possibilities for the involvement of AI in music generation. Specifically, methods based on evolutionary computation, conceptual blending, and deep learning are analyzed, based on the fact that they attempt to offer genuinely new/creative results that cross stylistic borders, allowing interpolations between and extrapolations from musical styles. Especially for deep learning methods, it seems possible that recent advantages will allow intuitive interpretations on what they produce, allowing them to be employed in more interactive and useful settings in real-world applications. Those developments appear to be significant since deep learning methods learn by themselves abstractions that cover multiple aspects of the learned styles, covering deep relations that a human designer might fail to properly describe with explicit feature design.

References

Alfonseca, M., Cebrián, M., Ortega, A., 2007. A simple genetic algorithm for music generation by means of algorithmic information theory. In: IEEE Congress on Evolutionary Computation. 2007, CEC 2007, pp. 3035–3042.

Allan, M., Williams, C.K.I., 2004. Harmonising chorales by probabilistic inference. In: Advances in Neural Information Processing Systems, vol. 17. MIT Press, pp. 25–32.

Apergis, A., Floros, A., Kaliakatsos-Papakostas, M., 2018. Sonoids: interactive sound synthesis driven by emergent social behaviour in the sonic domain. In: Proceeding of the 15th Sound and Music Computing Conference (SMC). SMC 2018.

Biles, J.A., 2002. GenJam: evolution of a jazz improviser. In: Bentley, P.J., Corne, D.W. (Eds.), Creative Evolutionary Systems. Morgan Kaufmann Publishers Inc., San Francisco, CA, USA, pp. 165–187.

Blackwell, T.M., 2003. Swarm music: improvised music with multi-swarms. In: Proceedings of the 2003 AISB Symposium on Artificial Intelligence and Creativity in Arts and Science, pp. 41–49.

Blackwell, T.M., 2007. Swarming and music. In: Miranda, E.R., Biles, J.A. (Eds.), Evolutionary Computer Music. Springer, London, pp. 194–217.

Blackwell, T., 2008. Swarm granulation. In: Romero, J., Machado, P. (Eds.), The Art of Artificial Evolution. In: Natural Computing Series. Springer, Berlin, Heidelberg, pp. 103–122.

Blackwell, T.M., Bentley, P., 2002a. Don't push me! Collision-avoiding swarms. In: Proceedings of the 2002 Congress on Evolutionary Computation. 2002, CEC '02, vol. 2. IEEE, pp. 1691–1696.

Blackwell, T.M., Bentley, P., 2002b. Improvised music with swarms. In: Proceedings of the 2002 Congress on Evolutionary Computation. 2002, CEC '02, vol. 2. IEEE, pp. 1462–1467.

Blackwell, T., Young, M., 2004. Swarm granulator. In: Applications of Evolutionary Computing, Proceedings of the 2nd European Workshop on Evolutionary Music and Art. EVOMUSART 2004. Springer, Berlin, Heidelberg, pp. 399–408.

Boden, M.A., 2004. The Creative Mind: Myths and Mechanisms. Routledge.

Borrel-Jensen, N., Hjortgaard Danielsen, A., 2010. Computer-assisted music composition – a database-backed algorithmic composition system. B.S. Thesis. Department of Computer Science, University of Copenhagen, Copenhagen, Denmark.

Burraston, D., Edmonds, E., 2005. Cellular automata in generative electronic music and sonic art: a historical and technical review. Digital Creativity 16 (3), 165–185.

Burraston, D., Martin, A., 2006. Wild nature and the digital life. In: Special Issue. Leonardo Electronic Almanac 14 (7-8).

Cambouropoulos, E., 2001. Melodic cue abstraction, similarity, and category formation: a formal model. Music Perception: An Interdisciplinary Journal 18 (3), 347–370.

Chung, J., Gulcehre, C., Cho, K., Bengio, Y., 2014. Empirical evaluation of gated recurrent neural networks on sequence modeling. Preprint. arXiv:1412.3555.

Colombo, F., Brea, J., Gerstner, W., 2018. Learning to generate music with BachProp. Preprint. arXiv: 1812.06669.

Colton, S., Charnley, J.W., Pease, A., 2011. Computational creativity theory: the face and idea descriptive models. In: ICCC, pp. 90–95.

Cook, N., 1994. A Guide to Musical Analysis. Oxford University Press, USA.

de la Puente, A.O., Alfonso, R.S., Moreno, M.A., 2002a. Automatic composition of music by means of grammatical evolution. In: Proceedings of the 2002 Conference on APL: Array Processing Languages: Lore, Problems, and Applications. APL '02, pp. 148–155.

de la Puente, A.O., Alfonso, R.S., Moreno, M.A., 2002b. Automatic composition of music by means of grammatical evolution. In: ACM SIGAPL APL Quote Quad, vol. 32(4), pp. 148–155.

Demorest, S.M., Morrison, S.J., 2016. Quantifying culture: the cultural distance hypothesis of melodic expectancy. In: The Oxford Handbook of Cultural Neuroscience, p. 183.

Donahue, C., Simon, I., Dieleman, S., 2018. Piano genie. Preprint. arXiv:1810.05246.

Dong, H.W., Hsiao, W.Y., Yang, L.C., Yang, Y.H., 2018. Musegan: multi-track sequential generative adversarial networks for symbolic music generation and accompaniment. In: Thirty-Second AAAI Conference on Artificial Intelligence.

Donnelly, P., Sheppard, J., 2011. Evolving four-part harmony using genetic algorithms. In: European Conference on the Applications of Evolutionary Computation, pp. 273–282.

Ebcioglu, K., 1988. An expert system for harmonizing four-part chorales. Computer Music Journal 12 (3), 43–51. ISSN 01489267.

Eck, D., Schmidhuber, J., 2002. Learning the long-term structure of the blues. In: International Conference on Artificial Neural Networks, pp. 284–289.

Eerola, T., Toiviainen, P., 2004. Midi toolbox: Matlab tools for music research.

Eppe, M., Confalonier, R., Maclean, E., Kaliakatsos-Papakostas, M., Cambouropoulos, E., Schorlemmer, M., Codescu, M., Kühnberger, K.U., 2015. Computational invention of cadences and chord progressions by conceptual chord-blending. In: International Joint Conference on Artificial Intelligence (IJCAI). 2015.

Fauconnier, G., Turner, M., 2003. The Way We Think: Conceptual Blending and the Mind's Hidden Complexities, reprint edition ed. Basic Books, New York.

Fortier, N., Van Dyne, M., 2011. A genetic algorithm approach to improve automated music composition. International Journal of Computers 5 (4), 525–532.

Gick, M.L., Holyoak, K.J., 1983. Schema induction and analogical transfer. Cognitive Psychology 15 (1), 1–38.

Granroth-Wilding, M., Steedman, M., 2014. A robust parser-interpreter for jazz chord sequences. Journal of New Music Research 43 (4), 355–374.

Gulrajani, I., Kumar, K., Ahmed, F., Taiga, A.A., Visin, F., Vazquez, D., Courville, A., 2016. Pixelvae: a latent variable model for natural images. Preprint. arXiv:1611.05013.

Hadjeres, G., Pachet, F., Nielsen, F., 2017. DeepBach: a steerable model for Bach chorales generation. In: Proceedings of the 34th International Conference on Machine Learning, vol. 70, pp. 1362–1371.

Hanlon, M., Ledlie, T., 2002. CPU Bach: an automatic chorale harmonization system. http://www.timledlie.org/cs/CPUBach.pdf.

Hedblom, M.M., Kutz, O., Neuhaus, F., 2016. Image schemas in computational conceptual blending. Cognitive Systems Research 39, 42–57.

Herremans, D., Sörensen, K., 2012. Composing first species counterpoint with a variable neighbourhood search algorithm. Journal of Mathematics and the Arts 6 (4), 169–189.

Herremans, D., Sörensen, K., Martens, D., 2015a. Classification and generation of composer-specific music using global feature models and variable neighborhood search. Computer Music Journal 39 (3), 71–91.

Herremans, D., Weisser, S., Sörensen, K., Conklin, D., 2015b. Generating structured music for bagana using quality metrics based on Markov models. Expert Systems with Applications 42 (21), 7424–7435.

Higgins, I., Matthey, L., Pal, A., Burgess, C., Glorot, X., Botvinick, M., Mohamed, S., Lerchner, A., 2017. beta-VAE: learning basic visual concepts with a constrained variational framework. In: International Conference on Learning Representations, vol. 2(5), p. 6.

Hiller, L.A., Isaacson, L.M., 1979. Experimental Music; Composition with an Electronic Computer. Greenwood Publishing Group Inc.

Hillewaere, R., Manderick, B., Conklin, D., 2009a. Global feature versus event models for folk song classification. In: Proceedings of the 10th International Society for Music Information Retrieval Conference (ISMIR 2009). Kobe, Japan, 26–30 October, pp. 729–733.

Hillewaere, R., Manderick, B., Conklin, D., 2009b. Melodic models for polyphonic music classification. In: Proceedings of the 2nd International Workshop on Machine Learning and Music (MML 2009). Held in Conjunction with The European Conference on Machine Learning and Principles and Practice of Knowledge Discovery in Databases (ECML-PKDD 2009), Bled, Slovenia, 7 September, pp. 19–24.

Hofmann, D.M., 2015. A genetic programming approach to generating musical compositions. In: Evolutionary and Biologically Inspired Music, Sound, Art and Design. Springer, pp. 89–100.

Horowitz, D., 1994. Generating rhythms with genetic algorithms. In: Proceedings of the Twelfth National Conference on Artificial Intelligence. AAAI'94, vol. 2. American Association for Artificial Intelligence, Menlo Park, CA, USA, pp. 1459–1460.

Huang, C.Z.A., Cooijmans, T., Roberts, A., Courville, A., Eck, D., 2019a. Counterpoint by convolution. Preprint. arXiv:1903.07227.

Huang, C.Z.A., Hawthorne, C., Roberts, A., Dinculescu, M., Wexler, J., Hong, L., Howcroft, J., 2019b. The Bach Doodle: approachable music composition with machine learning at scale. Preprint. arXiv: 1907.06637.

Hughes, D., Manaris, B., 2012. Fractal dimensions of music and automatic playlist generation: similarity search via MP3 song uploads. In: 2012 Eighth International Conference on Intelligent Information Hiding and Multimedia Signal Processing (IIH-MSP). Piraeus, Greece, pp. 436–440.

Huron, D.B., 2006. Sweet Anticipation: Music and the Psychology of Expectation. MIT Press.

Johanson, B.E., Poli, R., 1998. GP-Music: an interactive genetic programming system for music generation with automated fitness raters. Technical Report CSRP-98-13. University of Birmingham, School of Computer Science.

Jones, D., 2008. AtomSwarm: a framework for swarm improvisation. In: Proceedings of the 2008 Conference on Applications of Evolutionary Computing. Evo'08. Springer-Verlag, Berlin, Heidelberg, pp. 423–432.

Kaliakatsos-Papakostas, M., Cambouropoulos, E., 2014. Probabilistic harmonisation with fixed intermediate chord constraints. In: Proceeding of the Joint 11th Sound and Music Computing Conference (SMC) and 40th International Computer Music Conference (ICMC). ICMC–SMC 2014.

Kaliakatsos-Papakostas, M., Cambouropoulos, E., 2019. Conceptual blending of high-level features and data-driven salience computation in melodic generation. Cognitive Systems Research 58, 55–70.

Kaliakatsos-Papakostas, M.A., Epitropakis, M.G., Vrahatis, M.N., 2010. Musical composer identification through probabilistic and feedforward neural networks. In: Applications of Evolutionary Computation. In: LNCS, vol. 6025. Springer, Berlin/Heidelberg, pp. 411–420.

Kaliakatsos-Papakostas, M.A., Epitropakis, M.G., Vrahatis, M.N., 2011. Feature extraction using pitch class profile information entropy. In: Mathematics and Computation in Music. In: Lecture Notes in Artificial Intelligence, vol. 6726. Springer, Berlin/Heidelberg, pp. 354–357.

Kaliakatsos-Papakostas, M.A., Epitropakis, M.G., Floros, A., Vrahatis, M.N., 2012a. Interactive evolution of 8-bit melodies with genetic programming towards finding aesthetic measures for sound. In: Proceedings of the 1st International Conference on Evolutionary and Biologically Inspired Music, Sound, Art and Design. EvoMUSART 2012, Malaga, Spain. In: LNCS, vol. 7247. Springer Verlag, pp. 140–151.

Kaliakatsos-Papakostas, M.A., Floros, A., Vrahatis, M.N., 2012b. Intelligent generation of rhythmic sequences using Finite L-systems. In: Proceedings of the Eighth International Conference on Intelligent Information Hiding and Multimedia Signal Processing (IIHMSP 2012). Piraeus, Athens, Greece, 18–20 July, pp. 424–427.

Kaliakatsos-Papakostas, M.A., Floros, A., Vrahatis, M.N., 2012c. Intelligent real-time music accompaniment for constraint-free improvisation. In: Proceedings of the 24th IEEE International Conference on Tools with Artificial Intelligence (ICTAI 2012). Piraeus, Athens, Greece, 5–7 November, pp. 444–451.

Kaliakatsos-Papakostas, M.A., Floros, A., Vrahatis, M.N., Kanellopoulos, N., 2012d. Genetic evolution of L and FL–systems for the production of rhythmic sequences. In: Proceedings of the 2nd Workshop in Evolutionary Music Held During the 21st International Conference on Genetic Algorithms and the 17th Annual Genetic Programming Conference (GP). GECCO 2012, Philadelphia, USA, 7–11 July, pp. 461–468.

Kaliakatsos-Papakostas, M., Epitropakis, M., Floros, A., Vrahatis, M., 2013a. Chaos and music: a study of tonal time series and evolutionary music composition. International Journal of Bifurcation and Chaos.

Kaliakatsos-Papakostas, M., Floros, A., Vrahatis, M.N., 2013b. evoDrummer: deriving rhythmic patterns through interactive genetic algorithms. In: Evolutionary and Biologically Inspired Music, Sound, Art and Design. In: Lecture Notes in Computer Science, vol. 7834. Springer, Berlin, Heidelberg, pp. 25–36.

Kaliakatsos-Papakostas, M., Floros, A., Drossos, K., Koukoudis, K., Kyzalas, M., Kalantzis, A., 2014a. Swarm lake: a game of swarm intelligence, human interaction and collaborative music composition. In: ICMC.

Kaliakatsos-Papakostas, M., Katsiavalos, A., Tsougras, C., Cambouropoulos, E., 2014b. Harmony in the polyphonic songs of Epirus: representation, statistical analysis and generation. In: 4th International Workshop on Folk Music Analysis (FMA).

Kaliakatsos-Papakostas, M., Floros, A., Vrahatis, M.N., 2016. Interactive music composition driven by feature evolution. SpringerOpen 5 (1), 826.

Kaliakatsos-Papakostas, M., Queiroz, M., Tsougras, C., Cambouropoulos, E., 2017. Conceptual blending of harmonic spaces for creative melodic harmonisation. Journal of New Music Research 46 (4), 305–328.

Kennedy, J., 2010. Particle swarm optimization. In: Encyclopedia of Machine Learning, pp. 760–766.

Kingma, D.P., Welling, M., 2013. Auto-encoding variational Bayes. Preprint. arXiv:1312.6114.

Koops, H.V., Magalhães, J.P., de Haas, W.B., 2013. A functional approach to automatic melody harmonisation. In: Proceedings of the First ACM SIGPLAN Workshop on Functional Art, Music, Modeling & Design. FARM '13. ACM, New York, NY, USA, pp. 47–58.

Kranenburg, P.V., Backer, E., 2004. Musical style recognition – a quantitative approach. In: Parncutt, R., Kessler, A., Zimmer, F. (Eds.), Proceedings of the Conference on Interdisciplinary Musicology (CIM04). Graz, Austria, 15–18 April, pp. 1–10.

Krumhansl, C.L., 2001. Cognitive Foundations of Musical Pitch. Oxford University Press.

Lee, D.D., Seung, H.S., 1999. Learning the parts of objects by non-negative matrix factorization. Nature 401 (6755), 788.

Leman, M., 2012. Music and Schema Theory: Cognitive Foundations of Systematic Musicology, vol. 31. Springer Science & Business Media.

Liang, F.T., Gotham, M., Johnson, M., Shotton, J., 2017. Automatic stylistic composition of Bach chorales with deep LSTM. In: ISMIR, pp. 449–456.

Liu, H.M., Wu, M.H., Yang, Y.H., 2018. Lead sheet generation and arrangement via a hybrid generative model. In: Proc. Int. Soc. Music Information Retrieval Conf., Late Breaking and Demo Papers.

Lo, M.Y., 2012. Evolving cellular automata for music composition with trainable fitness functions. Ph.D. thesis. University of Essex.

Lourenc, B.F., Brand, C.P., 2009. L–systems, scores, and evolutionary techniques. In: Proceedings of the SMC 2009–6th Sound and Music Computing Conference, pp. 113–118.

MacCallum, R.M., Mauch, M., Burt, A., Leroi, A.M., 2012. Evolution of music by public choice. Proceedings of the National Academy of Sciences 109 (30), 12081–12086.

Machado, P., Romero, J., Manaris, B., Santos, A., Cardoso, A., 2003. Power to the critics – a framework for the development of artificial critics. In: Proceedings of 3rd Workshop on Creative Systems, 18 th International Joint Conference on Artificial Intelligence (IJCAI 2003). Acapulco, Mexico, pp. 55–64.

Madison, G., Sioros, G., 2014. What musicians do to induce the sensation of groove in simple and complex melodies, and how listeners perceive it. Frontiers in Psychology 5.

Madsen, S.T., Widmer, G., 2007. A complexity-based approach to melody track identification in MIDI files. In: Proceedings of the International Workshop on Artificial Intelligence and Music (MUSIC-AI 2007). Held in Conjunction with the 20th International Joint Conference on Artificial Intelligence (IJCAI 2007), Hyderabad, India, 6–12 January, pp. 1–9.

Makris, D., Kaliakatsos-Papakostas, M., Karydis, I., Kermanidis, K.L., 2017. Combining LSTM and feed forward neural networks for conditional rhythm composition. In: International Conference on Engineering Applications of Neural Networks, pp. 570–582.

Makris, D., Kaliakatsos-Papakostas, M., Karydis, I., Kermanidis, K.L., 2019. Conditional neural sequence learners for generating drums' rhythms. Neural Computing & Applications 31 (6), 1793–1804.

Manaris, B., Purewal, T., McCormick, C., 2002. Progress towards recognizing and classifying beautiful music with computers – MIDI-encoded music and the Zipf-Mandelbrot law. In: Proceedings IEEE SoutheastCon. 2002. IEEE, pp. 52–57.

Manaris, B., Romero, J., Machado, P., Krehbiel, D., Hirzel, T., Pharr, W., Davis, R.B., 2005. Zipf's law, music classification, and aesthetics. Computer Music Journal (ISSN 0148-9267) 29 (1), 55–69. ArticleType: research-article / Full publication date: Spring, 2005 / Copyright © 2005 The MIT Press.

Manaris, B., Roos, P., Machado, P., Krehbiel, D., Pellicoro, L., Romero, J., 2007. A corpus-based hybrid approach to music analysis and composition. In: Proceedings of the 22nd National Conference on Artificial Intelligence, vol. 1. AAAI Press, Vancouver, British Columbia, Canada, pp. 839–845.

Manaris, B., Hughes, D., Vassilandonakis, Y., 2011. Monterey mirror: combining Markov models, genetic algorithms, and power laws. In: Proceedings of 1st Workshop in Evolutionary Music, 2011 IEEE Congress on Evolutionary Computation (CEC 2011), pp. 33–40.

Manousakis, S., 2006. Musical L-systems. Master's thesis. The Royal Conservatory, the Hague, the Netherlands.

Matic, D., 2010. A genetic algorithm for composing music. Yugoslav Journal of Operations Research 20 (1), 157–177.

McCormack, J., 1996. Grammar-based music composition. Complexity International 3.

Mckay, C., Fujinaga, I., 2004. Automatic genre classification using large high-level musical feature sets. In: Proceedings of the 5th International Society for Music Information Retrieval Conference (ISMIR 2004). Barcelona, Spain, 10–14 October, pp. 525–530.

McKay, C., Fujinaga, I., 2006. jSymbolic: a feature extractor for MIDI files. In: ICMC.

Millen, D., 2005. An interactive cellular automata music application in Cocoa. In: Proceedings of the 2004 International Computer Music Conference (ICMC 2004).

Miranda, E.R., 2001. Evolving cellular automata music: from sound synthesis to composition. In: Proceedings of the Workshop on Artificial Life Models for Musical Applications (ECAL 2001), pp. 87–98.

Miranda, E.R., Al Biles, J., 2007. Evolutionary Computer Music. Springer.

Moroni, A., Manzolli, J., Zuben, F.V., Gudwin, R., 2000. Vox populi: an interactive evolutionary system for algorithmic music composition. Leonardo Music Journal, 49–54.

Mozer, M.C., 1994. Neural network music composition by prediction: exploring the benefits of psychoacoustic constraints and multi-scale processing. Connection Science 6 (2–3), 247–280.

Mozer, M.C., Soukup, T., 1991. Connectionist music composition based on melodic and stylistic constraints. In: Advances in Neural Information Processing Systems, pp. 789–796.

Nuanáin, C.O., Herrera, P., Jorda, S., 2015. Target-based rhythmic pattern generation and variation with genetic algorithms. In: Sound and Music Computing Conference.

Özcan, E., Ercal, T., 2008. A genetic algorithm for generating improvised music. In: Monmarché, N., Talbi, E.G., Collet, P., Schoenauer, M., Lutton, E. (Eds.), Artificial Evolution. In: Lecture Notes in Computer Science, vol. 4926. Springer, Berlin/Heidelberg, pp. 266–277.

Pachet, F., Roy, P., 2001. Musical harmonization with constraints: a survey. Constraints 6 (1), 7–19.

Pachet, F., Roy, P., Barbieri, G., 2011. Finite-length Markov processes with constraints. In: Twenty-Second International Joint Conference on Artificial Intelligence.

Paiement, J.F., Eck, D., Bengio, S., 2005. A probabilistic model for chord progressions. In: Proceedings of the Sixth International Conference on Music Information Retrieval (ISMIR).

Paiement, J.F., Eck, D., Bengio, S., 2006. Probabilistic melodic harmonization. In: Conference of the Canadian Society for Computational Studies of Intelligence, pp. 218–229.

Pearce, M., 2000. Generating rhythmic patterns: a combined neural and evolutionary approach. Ph.D. thesis, Master's thesis. Department of Artificial Intelligence, University of Edinburgh, Scotland. http://www.soi.city.ac.uk/ek735/msc/msc.html, 2000.

Phon-Amnuaisuk, S., Wiggins, G.A., 1999. The four-part harmonisation problem: a comparison between genetic algorithms and a rule-based system. In: Proceedings of the AISB'99 Symposium on Musical Creativity, pp. 28–34.

Phon-Amnuaisuk, S., Smaill, A., Wiggins, G., 2006. Chorale harmonization: a view from a search control perspective. Journal of New Music Research 35 (4), 279–305.

Price, K., Storn, R.M., Lampinen, J.A., 2006. Differential Evolution: a Practical Approach to Global Optimization. Springer Science & Business Media.

Prusinkiewicz, P., 1986. Score generation with L–systems. Computer, 455–457.

Prusinkiewicz, P., Lindenmayer, A., 2012. The Algorithmic Beauty of Plants. Springer Science & Business Media.

Purwins, J., Graepel, T., Blankertz, B., Obermayer, C., 2004. Correspondence analysis for visualizing interplay of pitch class, key, and composer. In: Mazzola, G., Noll, T., Luis-Puebla, E. (Eds.), Perspectives in Mathematical and Computational Music Theory. Universities Press, pp. 432–454.

Raczyński, S.A., Fukayama, S., Vincent, E., 2013. Melody harmonization with interpolated probabilistic models. Journal of New Music Research 42 (3), 223–235.

Reynolds, C.W., 1987. Flocks, Herds and Schools: A Distributed Behavioral Model, vol. 21(4). ACM.

Roberts, A., Engel, J., Raffel, C., Hawthorne, C., Eck, D., 2018. A hierarchical latent vector model for learning long-term structure in music. Preprint. arXiv:1803.05428.

Rohrmeier, M., 2011. Towards a generative syntax of tonal harmony. Journal of Mathematics and Music 5 (1), 35–53.

Sánchez, A., Pantrigo, J.J., Virseda, J., Pérez, G., 2007. Spieldose: an interactive genetic software for assisting to music composition tasks. In: Proceedings of the 2nd International Work-Conference on The Interplay Between Natural and Artificial Computation, Part I: Bio-Inspired Modeling of Cognitive Tasks. IWINAC '07. Springer-Verlag, Berlin, Heidelberg, pp. 617–626.

Serquera, J., Miranda, E.R., 2010. Evolutionary sound synthesis: rendering spectrograms from cellular automata histograms. In: European Conference on the Applications of Evolutionary Computation, pp. 381–390.

Shmulevich, I., Yli-Harja, O., Coyle, E., Povel, D.J., Lemstram, K., 2001. Perceptual issues in music pattern recognition: complexity of rhythm and key finding. Computers and the Humanities 35, 23–35.

Simon, I., Morris, D., Basu, S., 2008. MySong: automatic accompaniment generation for vocal melodies. In: Proceedings of the SIGCHI Conference on Human Factors in Computing Systems, pp. 725–734.

Sioros, G., Miron, M., Davies, M., Gouyon, F., Madison, G., 2014. Syncopation creates the sensation of groove in synthesized music examples. Frontiers in Psychology 5.

Solomos, M., 2005. Cellular automata in Xenakis' music. Theory and practice. In: Proceedings of the International Symposium Iannis Xenakis, pp. 120–137.

Spector, L., Alpern, A., 1995. Induction and recapitulation of deep musical structure. In: Proceedings of International Joint Conference on Artificial Intelligence. IJCAI, vol. 95, pp. 20–25.

Sturm, B., Santos, J.F., Korshunova, I., 2015. Folk music style modelling by recurrent neural networks with long short term memory units. In: 16th International Society for Music Information Retrieval Conference.

Sturm, B.L., Santos, J.F., Ben-Tal, O., Korshunova, I., 2016. Music transcription modelling and composition using deep learning. Preprint. arXiv:1604.08723.

Suzuki, S., Kitahara, T., Univercity, N., 2013. Four-part harmonization using probabilistic models: comparison of models with and without chord nodes. In: Proceedings of the Sound and Music Computing Conference, pp. 628–633.

Temperley, D., 2004. The Cognition of Basic Musical Structures. MIT Press.

Thom, B., 2001. Machine learning techniques for real-time improvisational solo trading. In: Proceedings of the 2001 International Computer Music Conference (ICMC).

Thornton, C., 2009. Hierarchical Markov modeling for generative music. In: International Computer Music Conference.

Todd, P.M., 1989. A connectionist approach to algorithmic composition. Computer Music Journal 13 (4), 27–43.

Tokui, N., 2000. Music composition with interactive evolutionary computation. Communication 17 (2), 215–226.

Unehara, M., Onisawa, T., 2005. Music composition by interaction between human and computer. New Generation Computing 23 (2), 181–191. Tokyo, Japan.

Vogl, R., Leimeister, M., Nuanáin, C.Ó., Jordà, S., Hlatky, M., Knees, P., 2016. An intelligent interface for drum pattern variation and comparative evaluation of algorithms. Journal of the Audio Engineering Society 64 (7–8), 503–513.

Weinberg, G., Godfrey, M., Rae, A., Rhoads, J., 2008. A real-time genetic algorithm in human-robot musical improvisation. In: Computer Music Modeling and Retrieval. Sense of Sounds. Springer-Verlag, Berlin, Heidelberg, pp. 351–359.

Wiggins, G., Papadopoulos, G., 1998. A genetic algorithm for the generation of jazz melodies. In: Proceedings of the Finnish Conference on Artificial Intelligence (STeP'98). Jyväskylä, Finnland.

Wilson, S., 2008. Spatial swarm granulation. In: Proceedings of the 2008 International Computer Music Conference.

Wolkowicz, J., Kulka, Z., Keselj, V., 2008. n-gram based approach to composer recognition. Archives of Acoustics 33 (1), 43–55.

Worth, P., Stepney, S., 2005. Growing music: musical interpretations of L-systems. In: EvoWorkshops. In: Lecture Notes in Computer Science. Springer, pp. 545–550.

Yang, L.C., Chou, S.Y., Yang, Y.H., 2017. MidiNet: a convolutional generative adversarial network for symbolic-domain music generation. Preprint. arXiv:1703.10847.

Yi, L., Goldsmith, J., 2007. Automatic generation of four-part harmony. In: Laskey, K.B., Mahoney, S.M., Goldsmith, J. (Eds.), BMA. In: CEUR Workshop Proceedings, vol. 268. CEUR-WS.org.

Yogev, N., Lerch, A., 2008. A system for automatic audio harmonization.

Zacharakis, A., Kaliakatsos-Papakostas, M., Tsougras, C., Cambouropoulos, E., 2018. Musical blending and creativity: an empirical evaluation of the chameleon melodic harmonisation assistant. Musicae Scientiae 22 (1), 119–144.

Zheng, E., Moh, M., Moh, T.S., 2017. Music genre classification: a n-gram based musicological approach. In: Advance Computing Conference (IACC), 2017 IEEE 7th International, pp. 671–677.

Optimized controller design for islanded microgrid employing nondominated sorting firefly algorithm

14

Quazi Nafees Ul Islam, Ashik Ahmed
Islamic University of Technology (IUT),
Electrical and Electronic Engineering, Gazipur, Bangladesh

CONTENTS

14.1 Introduction

Global optimization algorithms have gained a lot of attention in recent years due to their ability to solve different real-life challenges. Their nonlinear and multimodal nature and the presence of multilocal optima make global optimization problems difficult to solve. Researchers are now moving towards stochastic algorithms

Nature-Inspired Computation and Swarm Intelligence. https://doi.org/10.1016/B978-0-12-819714-1.00025-7

which mostly include metaheuristics methods like the genetic algorithm (GA) (Goldberg and Holland, 1988), particle swarm optimization (PSO) (Kennedy and Eberhart, 1995), evolutionary programming (Fogel et al., 1966), ant colony optimization (ACO) (Dorigo and Birattari, 2010), modified PSO (Kennedy, 2010; Gao et al., 2015; Du et al., 2015), and artificial bee colony (ABC) (Karaboga and Basturk, 2007), which have shown great potential in solving complex global optimization problems. All these are swarm intelligence algorithms as they are based on the biological nature of breeding, food searching, and other biotic processes of different creatures. Due to its efficiency in dealing with nonlinear multimodal global optimization problems, the firefly algorithm (FA) is used in different sectors for optimization.

FA was first introduced in 2008 by Xin-She Yang, where flashing lights produced by bioluminescent fireflies along with general behavior of the tropical fireflies were taken into consideration for developing the algorithm (Yang, 2008, 2009). FA has been used in different engineering applications; in Apostolopoulos and Vlachos (2010) it was used to solve economic emissions of a load dispatch problem, in Chatterjee et al. (2012) to design digital controlled reconfigurable switched beam concentric ring array antennae, in Sayadi et al. (2010) for minimization in permutation flow shop scheduling problems, and in Horng et al. (2012) and Horng (2012) to minimize the computational time for digital image compression. It was shown in Basu and Mahanti (2011) and Zaman et al. (2012) that FA is capable of producing better global solutions for linear antenna array design compared with ABC and PSO. The pioneering FA algorithm works on the basis of light intensity variation and attraction capability of the fireflies which were further modified to improve the performance of the algorithm. Generation of random directions for the determination of the best direction of fireflies with highest brightness and modification of attractiveness was initiated to observe its effect on objective functions (Tilahun and Ong, 2012). Several binary FA algorithms were also developed as a potential solution for different problems; in Palit et al. (2011) binary FA was developed to decode cipher to plain text, and this algorithm outperformed GA in terms of efficiency. In order to speed up the convergence of solution of the algorithm, Gaussian distribution was introduced so that all fireflies move to the global best after each iteration and eventually this modified algorithm outperformed the classical FA (Farahani et al., 2011). In dos Santos Coelho et al. (2011), a combination of FA with chaotic maps was proposed, where the introduction of a chaotic sequence helped in escaping local optima efficiently. In Abdullah et al. (2012) a hybrid evolutionary FA was proposed where classical FA was incorporated with the evolutionary process of the differential evolution algorithm for better searching efficiency. All of these proposed algorithms efficiently handled single-objective optimization problems.

In single-objective optimization, the aim generally is to search for the best design or decision, which is usually the global solution of the optimization problem. But in the case of multiple objectives, there may exist one or more solutions which may be the best (global minimum or maximum) with respect to all objectives. Generally, a set of solutions is obtained for this type of problem which may be better than the rest of solutions in the search space. For multiobjective optimization problems, the nondominated sorting GA (NSGA), which was introduced by Srinivas and Deb (1994), gained much appreciation due to its ability to solve such problems with better search results and efficiency.

Renewable energy sources are gaining more popularity nowadays due to the production of clean energy with no emission of greenhouse gases and less dependency on limited fossil fuel resources. Microgrids (MGs) are a competent solution for power system management, control, and integration of renewables as distributed sources within utility grids (Katiraei et al., 2008). Microgrids have two modes of operation, i.e., grid-connected and islanded. In islanded mode, there is basically one or more reliable energy source or energy storage compensating for generated power fluctuation of renewables (Chen et al., 2013; Li et al., 2013). Microgrid performance depends upon the selection of microgrid sources, controller performance, and load scheduling. Voltage source converters are used to interface a majority of distributed generation (DG) units to network. Droop control is a well-established approach for autonomous microgrid operation. Due to the low-inertia nature of such converter-dominated systems, the stability of autonomous microgrids is a critical issue. In Mohamed and El-Saadany (2008) and Pogaku et al. (2007), small signal-based stability analysis has been reported for studying the stability of autonomous droop-controlled microgrid systems. Here, static (RL-type) loads are considered to simplify the modeling and analysis tasks addressing microgrid stability issues. In future power networks and recent advances in power converter ratings and topologies, medium-voltage multi-MW microgrid systems will be created with a wide pattern of both static and dynamic loads. The operation of dynamic induction motor loads in droop-controlled converter-based microgrids yields special characteristics due to direct and relatively fast frequency control (e.g., droop control) using load power. Modeling of the small signal dynamic equations is very complex because of the complex control strategies for DGs. Ahmed et al. (2017) characterized small-signal stability of a hybrid AC/DC microgrid with static and dynamic loads using state space and dynamic simulation models developed in MATLAB®, but it faced challenges in maintaining the voltage and frequency within permissible limits during the standalone mode. The impedance mismatch between inverter-interfaced distributed generation (IIDG) units and induction motor (IM) loads, addressed in Radwan and Mohamed (2013) and Kahrobaeian and Mohamed (2013), showed that the presence of IM load causes medium frequency instabilities in the range of tens to a few hundreds of Hz. Jain et al. (2018) investigated the impact of load dynamics and load sharing among IIDG units on the stability and dynamic performance of islanded AC microgrids but lacks a description of controller requirement for damping out unstable, inter-area, and low-frequency oscillations introduced by dynamics of active and IM loads.

For maintaining required control performance and power quality during severe conditions, an excessive trial-and-error-based repeated tuning process is required. The problem of tuning control parameters optimally based on control objectives has been mentioned in many researches. There are traditional methods for tuning PI controllers, such as the Ziegler–Nichols method and frequency domain methods considering gain and phase margins. Since change of one PI control parameter tuning is likely to affect other parameter tuning, applying these methods to multiple DG controllers tuning is critical. Chung et al. (2008) first presented control schemes for the coordination of multiple microgrid generators, so that they can work on both grid-connected and autonomous modes. They used PSO for control parameter tuning during islanded mode operation. Chung et al. (2011) proposed an effective control parameter tuning method for multiple distributed generators in a microgrid using the PSO algorithm and a gain scheduling method, which were found to be effective for weak power systems only. In Hassan and Abido (2010), PSO has been used to give optimal settings of the optimized control parameters in each mode. Moreover, they proposed eigenvalue-based objective functions for enhancing the damping characteristics and finally presented nonlinear time domain-based simulation for minimization of error in the measured power. In Yu et al. (2015) GA was introduced to optimize the control parameters to improve the dynamic performance of the microgrid under load variation. Similarly, in Tilahun and Ong (2012), an artificial fish swarm algorithm was used to optimize the gains of a PI controller to obtain better frequency output during islanded operation of a microgrid. All the aforementioned work gained decent performance by optimizing the control parameters, although they treated the controller parameter tuning as a single-objective optimization problem. In Wang et al. (2018) a multiobjective-based GA was used to optimize the control parameters for stable operation of microgrid but left much opportunity for improvement. Thus, in this work our aim is to obtain global optimum control parameters with better dynamic performance under load variation for islanded microgrids in the presence of both static and dynamic loads by applying the nondominated sorting technique with FA to form a nature-inspired hybrid NSFA algorithm.

14.2 Mathematical model of microgrid

An islanded microgrid consisting of both static and dynamic load has been used for our study. The microgrid model shown in Fig. 14.1 consists of two DG units where static load has been installed on one unit and an induction motor as dynamic load on the other unit. In this section the dynamic model of the aforementioned islanded microgrid will be discussed, where necessary static and dynamic load equations, line equations, and controller equations will be developed for completion of a total state space model of the studied microgrid. The equations for the complete microgrid model were developed in Pogaku et al. (2007) and Kahrobaeian and Mohamed (2013).

The modeling of a microgrid is basically comprised of three main segments: inverter, loads, and network. Inverters were designed in their own reference frame and a power sharing controller was used to fix their angular frequency. In this study, the reference frame of one of the inverters was taken as the common reference frame and the load and network state equations were represented in terms of that common reference frame including the other inverters. D-Q transformation was used for this purpose, as shown in Eqs. (14.1) and (14.2) (Uudrill, 1968). Here D-Q axes were

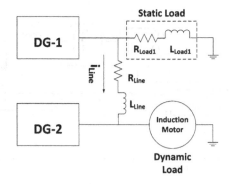

FIGURE 14.1

DGs with both static and dynamic load.

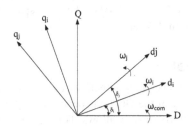

FIGURE 14.2

Transformation of reference frame.

taken as common global reference frame, where $(d - q)_i$ and $(d - q)_j$ were taken as the individual reference frame of the ith and jth inverters, respectively, and angle δ_i was taken as the reference angle for the ith inverter, as shown in Fig. 14.2. D-Q axes rotate at a frequency ω_{com}, which is taken as the common reference frame frequency, whereas $(d - q)_i$ and $(d - q)_j$ axes rotate with frequency ω_i and ω_j, respectively. Now angle δ denotes the difference of angle between system and inverter, which can be calculated with Eq. (14.3). We have

$$[f_{DQ}] = [T_i][f_{dq}], \tag{14.1}$$

$$[T_i] = \begin{bmatrix} \cos(\delta_i) & -\sin(\delta_i) \\ \sin(\delta_i) & \cos(\delta_i) \end{bmatrix}, \tag{14.2}$$

$$\delta = \int \omega - \omega_{com}. \tag{14.3}$$

A DG inverter block diagram is shown in Fig. 14.3 which is connected to the microgrid. There are three controller units of which power controller has been designed for determining the magnitude and frequency of the inverter output voltage following droop characteristics (Marwali et al., 2004). For obtaining characteristics of the output voltage, voltage control unit has been used and current control unit is used

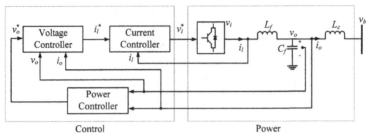

FIGURE 14.3

Block diagram of the DG inverter.

for controlling the LC filter current and damp out the high-frequency disturbances (Prodanovic, 2004; Marwali et al., 2004). The state space model for each of these controllers is represented in this section.

14.2.1 Power controller

The droop controller in power loop generates the output reference voltage magnitude and voltage frequency of the DG units depending upon their active and reactive power values and the controller also allows them to share their active and reactive power demand. But before this, first the instantaneous active power, $p(t)$, and instantaneous reactive power, $q(t)$, are calculated from the output voltage and output current, and then the average active power, P, and average reactive power, Q, are obtained by passing their instantaneous values through a low pass filter, as depicted in the following equations:

$$p(t) = 1.5(v_{od}i_{od} + v_{oq}i_{oq}), \quad q(t) = 1.5(v_{od}i_{oq} - v_{oq}i_{od}), \qquad (14.4)$$

$$P = \frac{\omega_c}{s + \omega_c}p(t), \quad Q = \frac{\omega_c}{s + \omega_c}q(t). \qquad (14.5)$$

Here, ω_c is the cutoff frequency of the low pass filter and v_{od}, v_{oq} are the output voltage and i_{od}, i_{oq} are the output current in the d-q reference frame. The power sharing between active and reactive power due to the droop controller is depicted in the following equations, where m_p and n_q are static active and reactive power droop gains, respectively:

$$\omega = \omega_{nl} - m_p P, \qquad (14.6)$$

$$v_{od}^* = V_n - n_q Q, \quad v_{oq}^* = 0. \qquad (14.7)$$

Here, ω represents the output angular frequency, ω_{nl} is the no-load frequency set point, V_n is the output voltage nominal set point, and superscript '*' represents a reference value. The static droop control gains m_p and n_q are calculated for a certain

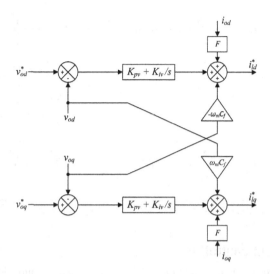

FIGURE 14.4

Voltage controller of the microgrid.

bandwidth of frequency and magnitude of voltage using the following equations:

$$m_p = \frac{\omega_{max} - \omega_{min}}{P_{max}}, \quad n_q = \frac{V_{odmax} - V_{odmin}}{Q_{max}}. \quad (14.8)$$

14.2.2 Voltage controller

A block diagram of output voltage controller consisting of the controller gains and feedback is shown in Fig. 14.4. A standard PI controller has been used here for controlling the output voltage. The main objective of voltage controller is to measure the reference value and actual value of voltage and frequency and then using the PI controller to generate inductor output current. The corresponding voltage controller equations are

$$\dot{\varphi}_d = v^*_{od} - v_{od}, \quad (14.9)$$

$$\dot{\varphi}_q = v^*_{oq} - v_{oq}, \quad (14.10)$$

$$i^*_{ld} = K_{pv}\dot{\varphi}_d + K_{iv}(\int \dot{\varphi}_d dt) - \omega_{nl}C_f v_{oq} + F i_{od}, \quad (14.11)$$

$$i^*_{lq} = K_{pv}\dot{\varphi}_q + K_{iv}(\int \dot{\varphi}_q dt) + \omega_{nl}C_f v_{od} + F i_{oq}. \quad (14.12)$$

Here, K_{pv} and K_{iv} are proportional and integral gains, respectively, for the voltage controller of the $d - q$ axes; φ_d, φ_q are the voltage controller's integrator states and F is the feed-forward gain, which is used to minimize the disturbance effect of output grid current components i_{od} and i_{oq}.

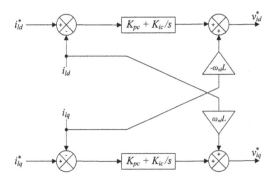

FIGURE 14.5

Current controller of the microgrid.

14.2.3 Current controller

The current controller is shown in Fig. 14.5, where a current controller basically compares the measured filter currents with the reference filter current (i^*_{ldq}). It controls the voltage across the inductor in such a way that the current error becomes minimum. The PI controller helps in controlling the output filter current. The corresponding current controller equations are given in Eqs. (14.13)–(14.15). Here, $\dot{\gamma}_d$ and $\dot{\gamma}_q$ denote the state of the current controller, and K_{pc} and K_{ic} are proportional and integral gains of the current controller, respectively. We have

$$\dot{\gamma}_d = i^*_{ld} - i_{ld}, \quad \dot{\gamma}_q = i^*_{lq} - i_{lq}, \tag{14.13}$$

$$v^*_{id} = K_{ic}\gamma_d + K_{pc}(i^*_{id} - i_{id}) - \omega_n L_f i_{lq}, \tag{14.14}$$

$$v^*_{iq} = K_{ic}\gamma_q + K_{pc}(i^*_{iq} - i_{iq}) + \omega_n L_f i_{ld}. \tag{14.15}$$

14.2.4 LCL filter

The LCL filter in this work mainly includes the inductor in the filter (L_f), the capacitor in the filter (C_f), and the coupling inductor (L_c) and coupling resistance R_c of the transformer; R_f is considered as the parasitic resistance and v_{bd}, v_{bq} are the d-q frame bus voltages on the grid side. A low pass filter is used in the microgrid model so that the switching frequency can be discarded as it has a dynamic effect on system stability. Here it is assumed that the voltage on the input and output side of the inverter is equal, i.e., $v_i = v^*_i$. The LCL filter dynamic states are shown in the following equations:

$$\dot{i}_{ld} = \omega i_{lq} - \frac{1}{L_f}(v_{od} - v_{id} + R_f i_{ld}), \tag{14.16}$$

$$\dot{i}_{lq} = -\omega i_{ld} - \frac{1}{L_f}(v_{oq} - v_{iq} + R_f i_{lq}), \tag{14.17}$$

$$i_{od} = \omega i_{oq} - \frac{1}{L_c}(v_{bd} - v_{od} + R_c i_{od}),\tag{14.18}$$

$$i_{oq} = -\omega i_{od} - \frac{1}{L_c}(v_{bq} - v_{oq} + R_c i_{oq}),\tag{14.19}$$

$$\dot{v}_{od} = \omega v_{oq} + \frac{1}{C_f}(i_{ld} - i_{od}),\tag{14.20}$$

$$\dot{v}_{oq} = -\omega v_{od} + \frac{1}{C_f}(i_{lq} - i_{oq}).\tag{14.21}$$

14.2.5 Complete inverter model

There are two DG units along with an induction motor as dynamic load in the microgrid model. For completion of the whole microgrid state space model, each of the states was transferred to a common global reference frame using Eqs. (14.1) and (14.2). Now linearizing Eqs. (14.4)–(14.7) and (14.9)–(14.21) and then transferring them in a common reference frame, the complete inverter state space model is given as follows:

$$\Delta\dot{x}_{INVi} = -A_{INVi}\Delta x_{INVi} + B_{INVi}\Delta v_{bDQi} + B_{com}\Delta\omega_{com},\tag{14.22}$$

$$\Delta i_{oDQi} = C_{INVi}\Delta x_{INVi},\tag{14.23}$$

$$\Delta x_{INVi} = [\Delta\delta_i\ \Delta P_i\ \Delta Q_i\ \Delta\varphi_{dqi}\ \Delta\gamma_{dqi}\ \Delta i_{ldqi}\ \Delta v_{odqi}\ \Delta i_{odqi}].\tag{14.24}$$

Here, $\Delta\varphi_{dqi}$ and $\Delta\gamma_{dqi}$ are the voltage and current controller states, respectively. An individual inverter model basically consists of 13 states, three inputs, and two inputs. The components of A_{INVi}, B_{INVi}, B_{com}, and C_{INVi} are given in matrix form in the following equations:

$$A_{INVi} = \begin{bmatrix} 0 & -m_p & 0 & 0 & 0 & 0 & 0 & 0 & 0 & 0 & 0 & 0 & 0 \\ 0 & -\omega_c & 0 & 0 & 0 & 0 & 0 & 0 & 0 & 1.5\omega_c I_{odi} & 1.5\omega_c I_{oqi} & 1.5\omega_c V_{odi} & 1.5\omega_c V_{oqi} \\ 0 & 0 & -\omega_c & 0 & 0 & 0 & 0 & 0 & 0 & 1.5\omega_c I_{oqi} & -1.5\omega_c I_{odi} & -1.5\omega_c V_{oqi} & 1.5\omega_c V_{odi} \\ 0 & 0 & -n_q & 0 & 0 & 0 & 0 & 0 & 0 & -1 & 0 & 0 & 0 \\ 0 & 0 & 0 & 0 & 0 & 0 & 0 & 0 & 0 & 0 & -1 & 0 & 0 \\ 0 & 0 & -n_i K_{pv} & K_{ic} & 0 & 0 & 0 & -1 & 0 & -K_{pv} & -\omega_{noload} & F & 0 \\ 0 & 0 & 0 & K_{ic} & 0 & 0 & 0 & 0 & -1 & \omega_{noload} & -K_{pv} & 0 & F \\ 0 & -m_p I_{lqi} & \frac{-n_q K_{pc} K_{pv}}{L_f} & \frac{K_{ic} K_{pc}}{L_f} & 0 & \frac{K_{ic}}{L_f} & 0 & \frac{-K_{pc}-R_f}{L_f} & \omega_o - \omega_{noload} & \frac{-K_{pc} K_{pv}-1}{L_f} & \frac{-\omega_{noload} C_f K_{pc}}{L_f} & \frac{K_{pc} F}{L_f} & 0 \\ 0 & m_p I_{ldi} & 0 & 0 & \frac{K_{ic} K_{pc}}{L_f} & 0 & \frac{K_{ic}}{L_f} & \omega_o - \omega_{noload} & \frac{-K_{pc}-R_f}{L_f} & \frac{\omega_{noload} C_f K_{pc}}{L_f} & \frac{-K_{pc} K_{pv}-1}{L_f} & 0 & \frac{K_{pc} F}{L_f} \\ 0 & -m_p V_{oqi} & 0 & 0 & 0 & 0 & 0 & \frac{1}{C_f} & 0 & 0 & \omega_o & \frac{-1}{C_f} & 0 \\ 0 & -m_p V_{odi} & 0 & 0 & 0 & 0 & 0 & 0 & \frac{1}{C_f} & \omega_o & 0 & 0 & \frac{-1}{C_f} \\ \frac{1}{L_c}\left(V_{bDi}\sin\delta_i - V_{bQi}\cos\delta_i\right) & -m_p I_{oqi} & 0 & 0 & 0 & 0 & 0 & 0 & 0 & \frac{1}{L_c} & 0 & \frac{-R_c}{L_c} & \omega_o \\ \frac{1}{L_c}\left(V_{bDi}\cos\delta_i + V_{bQi}\sin\delta_i\right) & m_p I_{odi} & 0 & 0 & 0 & 0 & 0 & 0 & 0 & 0 & \frac{1}{L_c} & -\omega_o & \frac{-R_c}{L_c} \end{bmatrix},\tag{14.25}$$

$$B_{INVi} = \begin{bmatrix} 0 & \cdots & 0 & \frac{-\cos\delta_i}{L_c} & \frac{-\sin\delta_i}{L_c} \\ 0 & \cdots & 0 & \frac{\sin\delta_i}{L_c} & \frac{-\cos\delta_i}{L_c} \end{bmatrix}^T_{13\times 2},$$

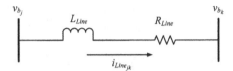

FIGURE 14.6

Line network of the microgrid.

$$B_{com} = \begin{bmatrix} -1 & 0 & 0 & 0 & \ldots & 0 \end{bmatrix}^{T}_{13 \times 1},$$ (14.26)

$$C_{INVi} = \begin{bmatrix} -I_{odi}\sin\delta_i - I_{oqi}\cos\delta_i & 0 & 0 & \ldots & 0 & 0 & \cos\delta_i & -\sin\delta_i \\ I_{odi}\cos\delta_i - I_{oqi}\sin\delta_i & 0 & 0 & \ldots & 0 & 0 & \sin\delta_i & \cos\delta_i \end{bmatrix}_{2 \times 13}.$$ (14.27)

14.2.6 Line network and static load model

Line current flows from one node to other node of the line connecting the buses. As the system has one single line network, if two nodes of a line are j and k, then the state equations of the line network of the microgrid can be shown as in Eqs. (14.28) and (14.29). Eq. (14.30) shows the state space model of the line network. A network model for a two-bus system is shown in Fig. 14.6, where one bus static load is connected and another bus dynamic load is connected. We have

$$\dot{i}_{LineD} = \frac{v_{bDj}}{L_{Line}} - \frac{v_{bDk}}{L_{Line}} - \frac{R_{Line}}{L_{Line}}i_{LineD} + \omega i_{LineQ},$$ (14.28)

$$\dot{i}_{LineQ} - \frac{v_{bQj}}{L_{Line}} - \frac{v_{bQk}}{L_{Line}} - \frac{R_{Line}}{L_{Line}}i_{LineQ} - \omega i_{LineD},$$ (14.29)

$$[\Delta \dot{i}_{LineDQ}] = A_{Line}[\Delta i_{LineDQ}] + B_{1Line}[\Delta v_{bDQ}] + B_{2Line}[\Delta \omega].$$ (14.30)

As the line network the load model of the microgrid is formulated in Eqs. (14.31)–(14.33). All state equations formulated for the line network and load model are considered on the common global reference frame. We have

$$\dot{i}_{LoadD} = \frac{v_{bD}}{L_{Load}} - \frac{R_{Load}}{L_{Load}}i_{LoadD} + \omega i_{LoadQ},$$ (14.31)

$$\dot{i}_{LoadQ} = \frac{v_{bQ}}{L_{Load}} - \frac{R_{Load}}{L_{Load}}i_{LoadQ} - \omega i_{LoadD},$$ (14.32)

$$[\Delta \dot{i}_{LoadDQ}] = A_{Load}[\Delta i_{LoadDQ}] + B_{1Load}[\Delta v_{bDQ}] + B_{2Load}[\Delta \omega].$$ (14.33)

Now the components of A_{Line}, B_{1Line}, B_{2Line}, A_{Load}, B_{1Load}, and B_{2Load} are given as follows:

$$A_{Line} = \begin{bmatrix} \frac{-R_{Line}}{L_{Line}} & \omega_0 \\ -\omega_0 & \frac{-R_{Line}}{L_{Line}} \end{bmatrix}, \quad B_{1Line} = \begin{bmatrix} \frac{1}{L_{Line}} & 0 & \frac{-1}{L_{Line}} & 0 \\ 0 & \frac{1}{L_{Line}} & 0 & \frac{-1}{L_{Line}} \end{bmatrix},$$

$$B_{2Line} = \begin{bmatrix} I_{LineQ} \\ -I_{LineD} \end{bmatrix}, \tag{14.34}$$

$$A_{Load} = \begin{bmatrix} \frac{-R_{Load}}{L_{Load}} & \omega_o \\ -\omega_o & \frac{-R_{Load}}{L_{Load}} \end{bmatrix}, \quad B_{1Load} = \begin{bmatrix} \frac{1}{L_{Load}} & 0 & \frac{-1}{L_{Load}} & 0 \\ 0 & \frac{1}{L_{Load}} & 0 & \frac{-1}{L_{Load}} \end{bmatrix},$$

$$B_{2Load} = \begin{bmatrix} I_{LoadQ} \\ -I_{LoadD} \end{bmatrix}. \tag{14.35}$$

Now node voltages v_{bDQ} are basically used as input in the above subsystems. For proper implementation of node voltages a sufficiently large virtual resistance R_n is considered between each node and ground. The expression of node voltages is given as follows:

$$v_{bD} = R_n(i_{oD} + i_{LineD} - i_{LoadD}), \tag{14.36}$$

$$v_{bQ} = R_n(i_{oQ} + i_{LineQ} - i_{LoadQ}). \tag{14.37}$$

Now i_{oDQ} can be transferred into their common global reference frame using Eqs. (14.1) and (14.2).

14.2.7 State space model of induction motor as dynamic load

In the system induction motor has been used as a dynamic load. The stator and rotor voltages of an induction motor developed in a common reference frame are given by Eqs. (14.38)–(14.41) following Horng (2012), where v_{DQs} and v_{DQr} are stator and rotor side voltage, respectively, i_{DQs} and i_{DQr} are stator and rotor current, respectively, L_s, R_s are stator inductance and resistance, respectively, L_r, R_r are rotor inductance and resistance, respectively, L_m is mutual inductance, ω is stator angular frequency, and s is the rotor slip. We have

$$v_{Ds} = R_s i_{Ds} + L_s \frac{d}{dt} i_{Ds} + L_m \frac{d}{dt} i_{Dr} - \omega L_s i_{Qs} - \omega L_m i_{Qr}, \tag{14.38}$$

$$v_{Qs} = R_s i_{Qs} + L_s \frac{d}{dt} i_{Qs} + L_m \frac{d}{dt} i_{Qr} + \omega L_s i_{Ds} + \omega L_m i_{Dr}, \tag{14.39}$$

$$v_{Dr} = R_r i_{Dr} + L_r \frac{d}{dt} i_{Dr} + L_m \frac{d}{dt} i_{Ds} - \omega s L_m i_{Qs} - \omega s L_r i_{Qr}, \tag{14.40}$$

$$v_{Qr} = R_r i_{Qr} + L_r \frac{d}{dt} i_{Qr} + L_m \frac{d}{dt} i_{Qs} + \omega s L_m i_{Ds} + \omega s L_r i_{Dr}. \tag{14.41}$$

The relationship between mechanical speed of the motor and electromechanical torque developed by the motor is shown in Eqs. (14.42) and (14.43), where T_E is electrical torque, T_L is load torque, J is the inertia of the motor and load combined, and ρ is the number of poles. We have

$$T_E = \frac{3\rho L_m}{4}(i_{Qs} i_{Dr} - i_{Ds} i_{Qr}), \tag{14.42}$$

$$T_E - T_L = J\frac{d}{dt}((1-s)\omega). \tag{14.43}$$

Now after linearizing the motor equations and rearranging them, the state space model of the induction motor is given in Eq. (14.44), where the components of ΔX_{IM}, ΔU_{IM}, A_{IM}, B_{IM}, C_{IM}, and D_{IM} are given in Eqs. (14.45)–(14.48). Here $\Delta\omega$ and $\Delta\dot{\omega}$ can be transferred and expressed according to the states of the microgrid. We have

$$\Delta\dot{X}_{IM} = (-A_{IM}^{-1}B_{IM})\Delta X_{IM} + A_{IM}^{-1}\Delta U_{IM}$$
$$+ (-A_{IM}^{-1}C_{IM})\Delta\omega + (-A_{IM}^{-1}D_{IM})\Delta\dot{\omega}, \tag{14.44}$$

$$\Delta X_{IM} = \begin{bmatrix} \Delta i_{Qs} \\ \Delta i_{Ds} \\ \Delta i_{Qr} \\ \Delta i_{Dr} \\ \Delta s \end{bmatrix}, \quad \Delta U_{IM} = \begin{bmatrix} \Delta v_{Qs} \\ \Delta v_{Ds} \\ \Delta v_{Qr} \\ \Delta v_{Dr} \\ \Delta T_L \end{bmatrix}, \tag{14.45}$$

$$A_{IM} = \begin{bmatrix} L_s & 0 & L_m & 0 & 0 \\ 0 & L_s & 0 & L_m & 0 \\ L_m & 0 & L_r & 0 & 0 \\ 0 & L_m & 0 & L_r & 0 \\ 0 & 0 & 0 & 0 & \frac{4J\omega_0}{3\rho} \end{bmatrix}, \tag{14.46}$$

$$B_{IM} = \begin{bmatrix} R_s & \omega_0 L_s & 0 & \omega_0 L_m & 0 \\ -\omega_0 L_s & R_s & -\omega_0 L_m & 0 & 0 \\ 0 & s_0\omega_0 L_m & R_r & s_0\omega_0 L_r & -\omega_0(L_m i_{Ds0} + L_r i_{Dr0}) \\ -s_0\omega_0 L_m & 0 & -s_0\omega_0 L_r & R_r & \omega_0(L_m i_{Qs0} + L_r i_{Qr0}) \\ L_m i_{Dr0} & -L_m i_{Qr0} & -L_m i_{Ds0} & L_m i_{Qs0} & 0 \end{bmatrix}, \tag{14.47}$$

$$C_{IM} = \begin{bmatrix} L_s i_{Ds0} + L_m i_{Dr0} \\ -L_s i_{Qs0} - L_m i_{Qr0} \\ s_0 L_m i_{Ds0} + s_0 L_r i_{Dr0} \\ -s_0 L_m i_{Qs0} - s_0 L_r i_{Qr0} \\ 0 \end{bmatrix}, \quad D_{IM} = \begin{bmatrix} 0 \\ 0 \\ 0 \\ 0 \\ \frac{-4J(1-s_0)}{3\rho} \end{bmatrix}. \tag{14.48}$$

For ease of calculation $(-A_{IM}^{-1}B_{IM})$, A_{IM}^{-1}, $(-A_{IM}^{-1}C_{IM})$, and $(-A_{IM}^{-1}D_{IM})$ were considered as E_{IM}, F_{IM}, G_{IM}, and H_{IM}, respectively.

14.2.8 Complete microgrid model

The complete microgrid model of our system was developed by combining the individual inverter, line network, load, and induction motor state space model from

Eqs. (14.22), (14.30), (14.33), and (14.44), respectively. The complete microgrid model is depicted as follows:

$$\Delta \dot{X}_{MG} = A_{MG} \Delta X_{MG}(t), \tag{14.49}$$

where ΔX_{MG} are the integrated states of the microgrid model. The components of ΔX_{MG} and A_{MG} are given as follows:

$$\Delta X_{MG} = [\Delta x_{INV1} \; \Delta x_{INV2} \; \Delta i_{LineDQ} \; \Delta i_{LoadDQ} \; \Delta X_{IM}]_{1\times35}, \tag{14.50}$$

$$A_{MG} = \begin{bmatrix} [A_{INV} + B_{INV} R_N M_{INV} C_{INV}]_{26\times26} & [B_{INV} R_N M_{NET}]_{26\times4} & [B_{INV} R_N M_{IM}]_{26\times5} \\ [B_{1Line} R_N M_{INV} C_{INV} + B_{2Line} C_{INV_\omega}]_{2\times26} & [A_{INV} + B_{1Line} R_N M_{NET}]_{2\times4} & [B_{1Line} R_N M_{LOAD}]_{2\times5} \\ [B_{1Load} R_N M_{INV} C_{INV} + B_{2Line} C_{INV_\omega}]_{2\times26} & [B_{1Load} R_N M_{NET}]_{2\times4} & [A_{Load} + B_{1Load} R_N M_{LOAD}]_{2\times5} \\ [[F_{IM} T][R_N M_{INV} C_{INV}] + G_{IM} C_{INV_\omega} + H_{IM} C_{INV_\omega} A_{INV}] & [F_{IM} T][R_N M_{NET}] & E_{IM} + [F_{IM} T][R_N M_{INV}] \end{bmatrix}_{35\times35}, \tag{14.51}$$

where M_{INV}, M_{NET}, M_{LOAD}, and M_{IM} are used for mapping the connection points between nodes of the network and line. The size of M_{INV} is (2 × nodes) × (2 × inverters). Similarly, the size of M_{LOAD} is (2 × nodes) × (2 × No. of load points), M_{IM} is 4 × 5, M_{NET} is (2 × nodes) × (2 × No. of lines), and R_N is (2 × nodes) × (2 × nodes). The diagonal elements of R_N are equal to R_n and components of M_{NET} are either +1 or −1, depending upon the direction of line current. For linking the stator voltage of the induction motor with the microgrid system, matrix T is used, whose size is 5 × 4. The components of C_{INV_ω} are given as follows:

$$C_{INV_\omega com} = [0 \; -m_p \; 0 \; .... \; 0]_{1\times13}, \; C_{INV_\omega} = [C_{INV_\omega com} \; 0 \; .... \; 0]_{1\times26}. \tag{14.52}$$

14.3 Problem formulation

Microgrid stability while operating in islanded mode is an important factor to be addressed. For the aforementioned microgrid model, which consists of both static and dynamic load, the prime concern is to obtain a stable optimized system with optimized controller parameters.

14.3.1 Objective functions

For the abovementioned microgrid system the controller parameters were optimized for both inverters, i.e., K_{pv1}, K_{iv1}, K_{pc1}, K_{ic1}, K_{pv2}, K_{iv2}, K_{pc2}, and K_{ic2}. The damping ratio and eigenvalue-based objective functions are given as follows:

$$J_1 = (-0.3 - Minimum(\sigma)), \; J_2 = (0.7 - Minimum(\lambda)). \tag{14.53}$$

Here σ and λ represent real eigenvalues and damping ratio, respectively. Here reference values −0.3 and 0.7 are used to limit the objective functions within our desired boundary. The problem constraints for the developed model are given as follows:

$$0 < K_{pv}, \; K_{pi}, \; K_{iv}, \; K_{ic} < 500. \tag{14.54}$$

14.3.2 Optimization process of nondominated sorting firefly algorithm

The technique used for optimization of the controller parameters in the proposed system is nondominated sorting, which is used mostly for multiobjective problems. In nondominated sorting optimization there are several conflicting objective functions from which a set of optimal solutions can be found, and they can be named as Pareto-optimal solutions. Pareto-optimality is calculated from the Pareto-optimal solutions to differentiate between the dominated and nondominated solutions. The Pareto technique followed in the proposed algorithm is nondominated ranking for moving the solutions towards the Pareto front. This nondominated sorting is hybridized with FA to form the nondominated sorting FA (NSFA). The detailed step-wise procedure of NSFA is explained through a flowchart shown in Fig. 14.7.

Step 1: At first the system parameters are defined and the initial population of fireflies is generated for the state variables as stated in Eq. (14.50). The initial population size was 50.

Step 2: The fitness function F(x) of the fireflies is evaluated, where basically the light intensity I of the fireflies is determined as attractiveness of a firefly depending on its brightness (Yang, 2009).

Step 3: Nondominated sorting of the initial fitness function is performed for sorting the fireflies of the initial generation according to their fitness function.

Step 4: In the next step the ranks of fireflies are determined according to their fitness function and the best fitness amongst them is noted. This step is repeated for maximum iteration.

Step 5: Now, two *for loops* are run as there are two fireflies i and j and their fitness values are compared. If the jth firefly is brighter than the ith firefly, then the ith firefly is moved towards the jth firefly using the following equation:

$$x_i(t+1) = x_i(t) + \beta \exp^{-\gamma r^2}(x_i - x_j) + \alpha \epsilon_i, \tag{14.55}$$

where β denotes the attractiveness of fireflies, α denotes the mutation coefficient, γ is the light absorption coefficient, and r represents the distance between the ith and jth firefly. The values of β, α, and γ were 2, 0.2, and 1, respectively, $\beta \exp^{-\gamma r^2}(x_i - x_j)$ is used for the attraction of the jth firefly, and $\alpha \epsilon_i$ is used as a randomization parameter.

Step 6: If the previous step is not satisfied, then the fitness function of the fireflies is evaluated again. Rank and the best fitness are updated accordingly.

Step 7: In this step old and new fitness of the fireflies are merged and nondominated sorting of the fitness function is carried out. If the condition given in step 5 is satisfied, then step 6 is skipped and the result is moved directly from step 5 to this step.

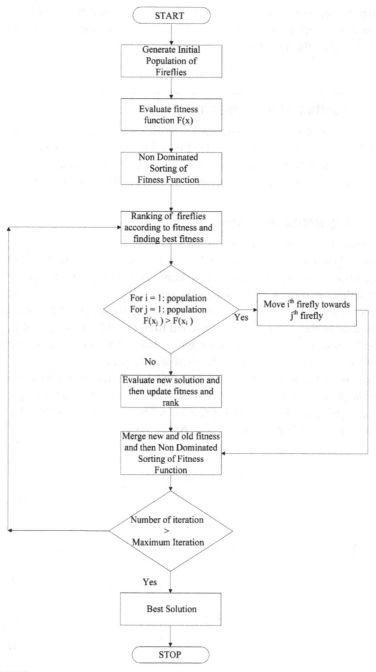

FIGURE 14.7

Flow chart for NSFA.

Step 8: The loop from step 4 is continued to run until the maximum number of iterations defined at the beginning and finally the best fitness is determined after completion of all the iterations.

14.4 Results and discussion

A mathematical model of the microgrid discussed above has been developed using MATLAB coding for analysis of the stability and performance of the microgrid using the NSFA optimization algorithm. The analysis is subdivided into three parts, namely, (i) eigenvalue analysis, (ii) time domain simulation, and (iii) statistical tests.

14.4.1 Eigenvalue analysis

First, eigenvalue analysis was carried out for the determination of the ability of the proposed algorithm in obtaining stability of the proposed system. Fig. 14.8 shows the condition of the eigenvalues of the system before optimization, where it can be seen that there exist some eigenvalues at the right half-plane, i.e., on the positive axis. This proves that the system is not stable. Now using the proposed NSFA algorithm, optimization of the system is performed and from Fig. 14.9 it is noticeable that the system has become stable as the unstable eigenvalues have moved from the positive axis to the negative axis, i.e., on the left half of the s plane.

This proves the ability of the proposed NSFA algorithm to stabilize an unstable system with optimized settings. The optimized values of the controller parameters are $K_{pv1} = 60.4500$, $K_{iv1} = 155.1168$, $K_{pc1} = 354.7484$, $K_{ic1} = 490.0740$, $K_{pv2} = 100.7187$, $K_{iv2} = 439.9541$, $K_{pc2} = 499.6512$, and $K_{ic2} = 436.8814$.

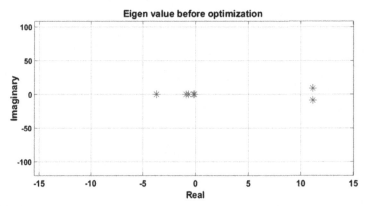

FIGURE 14.8

Eigenvalue of the system before optimization.

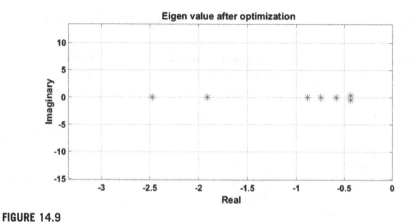

FIGURE 14.9

Eigenvalue of the system after optimization.

14.4.2 Time domain simulation analysis

The results obtained using NSFA optimization techniques are compared with the nondominated sorting genetic algorithm-II (NSGA-II) optimization technique to determine their differences in performance. The overshoot and oscillation frequency of real power, reactive power, inductor current, and output current of DG-1 and DG-2 for both NSFA and NSGA algorithms are presented in a tabulated form in Table 14.1. Along with the tabulated data, the step response of real power, reactive power, inductor current (d-q), and output current (d-q) are shown in Fig. 14.10, Fig. 14.11, Fig. 14.12, Fig. 14.13, Fig. 14.14, and Fig. 14.15, respectively. In Fig. 14.10 it is seen that in both DG-1 and DG-2, real power is more oscillatory and has overshoot for NSGA, compared with zero overshoot in NSFA.

In Fig. 14.11, the NSFA algorithm provides a lower oscillation frequency for reactive power of both DGs compared to NSGA. But for DG-1, NSFA provides more overshoot and in DG-2, NSGA has negligible overshoot whereas NSFA has zero overshoot. From Fig. 14.12 and tabulated data it is observable that for inductor current of DG-1 (d axis), NSFA has a higher oscillation frequency compared with NSGA and for DG-2 (d axis) NSFA has more overshoot. Fig. 14.13 and tabulated data show that for inductor current of DG-1 (q axis), NSFA's overshoot is more compared with NSGA, but vice versa for oscillation frequency, i.e., NSFA has a lower oscillation frequency compared with NSGA. Similar performance was found for inductor current of DG-2 (q axis). For output voltage it is clearly seen from Fig. 14.14, Fig. 14.15, and Table 14.1 that the performance of NSFA is better than that of NSGA as overshoot and oscillation frequency are higher with the NSGA algorithm. From the performance analysis it can be seen that the damping characteristics of the system have been improved using the NSFA algorithm.

Table 14.1 NSGA and NSFA comparison on the basis of overshoot and oscillation frequency.

Criterion	Parameters	NSGA	NSFA
Real power of DG-1	Overshoot (%)	0.658	0
	Oscillation frequency (Hz)	1943184.60945872	1943184.58052368
Real power of DG-2	Overshoot (%)	2.89e−7	0
	Oscillation frequency (Hz)	3530.53971016327	3012.13652730558
Reactive power of DG-1	Overshoot (%)	1.02e+11	4.06e+14
	Oscillation frequency (Hz)	1322438.80856331	1322438.70465282
Reactive power of DG-2	Overshoot (%)	1.06e−8	0
	Oscillation frequency (Hz)	3530.53971016327	3012.13652730558
Inductor current of DG-1 (d axis)	Overshoot (%)	0	0
	Oscillation frequency (Hz)	66757.6167391984	77408.4617388257
Inductor current of DG-2 (d axis)	Overshoot (%)	1.59e+12	9.01e+12
	Oscillation frequency (Hz)	0	0
Inductor current of DG-1 (q axis)	Overshoot (%)	2.97e+4	3.04e+04
	Oscillation frequency (Hz)	46067.9986622605	50339.8409294205
Inductor current of DG-2 (q axis)	Overshoot (%)	1.06e+12	2.98e+12
	Oscillation frequency (Hz)	8.62138639160466	7.91008749588845
Output voltage of DG-1 (d axis)	Overshoot (%)	0	0
	Oscillation frequency (Hz)	46067.9986622605	50339.8409294205
Output voltage of DG-2 (d axis)	Overshoot (%)	0.077	0
	Oscillation frequency (Hz)	8.62138639160466	7.91008749588845
Output voltage of DG-1 (q axis)	Overshoot (%)	211	0
	Oscillation frequency (Hz)	46139.9805662976	50408.3033958121
Output voltage of DG-2 (q axis)	Overshoot (%)	1.09e−8	0
	Oscillation frequency (Hz)	0.358162462584608	0

14.4.3 Statistical tests

A statistical analysis was performed using SPSS software to validate the performance analysis of the algorithms on the basis of obtained values. Independent-sample t-test was performed to determine how the algorithms are unique from each other. The test

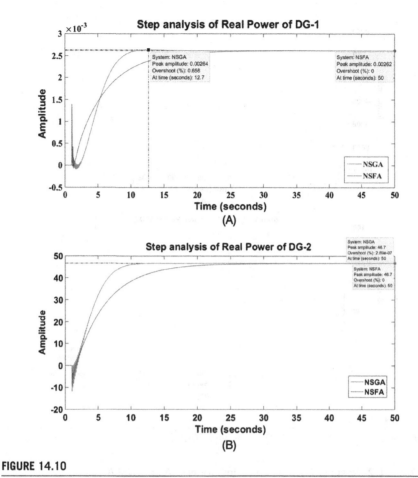

FIGURE 14.10

Step analysis of real power of DG-1 (A) and DG-2 (B).

was performed on the basis of total number of iterations required by each algorithm for convergence, total time for convergence, and total summation of eigenvalues. The F-test was also obtained along with the t-test. The F-test basically finds out the variances of the sample datasets of NSFA and NSGA and determines whether equal variances prevail between them or not. If the $Sigma(p)$ value from the F-test is greater than 0.05, then the group variances of the datasets are said to have equal variances, or else group variances of the datasets are not taken as equal. Similarly, in the t-test if the $Sigma(2\text{-}tailed)(p)$ value is greater than 0.05, then the null hypothesis (H_o) cannot be rejected, i.e., it is considered that mean values of the group data are equal, but if is less than 0.05, then the null hypothesis might be rejected and the alternative hypothesis H_1 can be considered, where mean values of group data are not equal. From Table 14.2 it can be seen that except for total summation of eigenvalues in the t-test, all other cases have $Sigma(p)$ values less than 0.05. So, it can be said

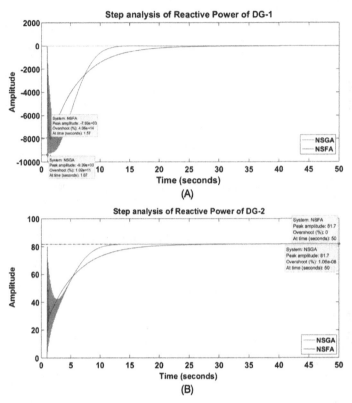

FIGURE 14.11

Step analysis of reactive power of DG-1 (A) and DG-2 (B).

Table 14.2 Results of F-test and t-test for NSGA and NSFA.

Parameters for test	Levene's test for equality of variances (F-test)		Equality means test (t-test)			
	F	Sig.	t	df	Sig. (two-tailed)	Mean difference
Total number of iterations	62.115019	0.000	5.383	30.164	0.000	6.50000
Total time required	45.776540	0.000	4.354	29.841	0.000	7.99043
Total summation of eigenvalues	6.342609	0.015	0.923	58	0.360	21204.25933

that there are significant differences between the NSGA and NSFA algorithms, and both are unique in their own way.

Table 14.3 shows the mean, standard deviation, and standard mean error in total number of iterations, total time, and total summation of eigenvalues for NSGA and

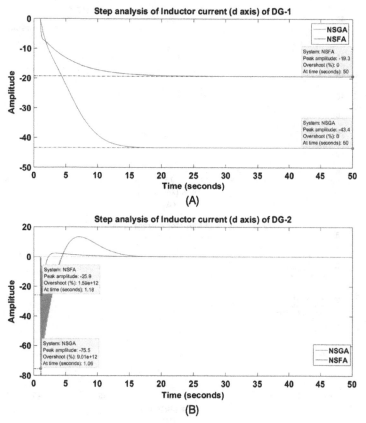

FIGURE 14.12

Step analysis of inductor current (d axis) of DG-1 (A) and DG-2 (B).

Table 14.3 Group statistical data.

Parameters for test	Algorithm	Mean	Standard deviation	Standard mean error
Total number of iterations	NSGA	8.8667	6.54814	1.19552
	NSFA	2.3667	0.92786	0.16940
Total time required	NSGA	12.5364	9.97964	1.82203
	NSFA	4.5459	1.20166	0.21939
Total summation of eigenvalues	NSGA	−12683862.1353	66179.19802	12082.61320
	NSFA	−12705066.3947	107004.22337	19536.20896

NSFA. It is seen that NSFA requires a lower number of iterations and less time to reach convergence and also has more negative eigenvalues compared with NSGA. The above data include 30 independent runs from which the best data are being tabu-

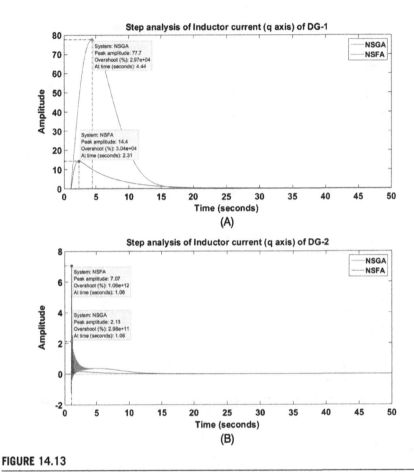

FIGURE 14.13

Step analysis of inductor current (q axis) of DG-1 (A) and DG-2 (B).

lated. The above performance analysis and statistical analysis suggests that the NSFA algorithm performs significantly better compared with NSGA.

14.5 Conclusion

This chapter mainly addresses a new nature-inspired swarm intelligence technique, i.e., NSFA. In the abovementioned sections a typical microgrid system model including both static and dynamic load (induction motor) was developed, optimization of the control parameters was carried out for the proposed system using NSFA and NSGA techniques, and finally performance of the system on the basis of both optimization techniques was analyzed. The analysis done in the Results and discussion section indicates that NSFA is able to stabilize a system with improved damping

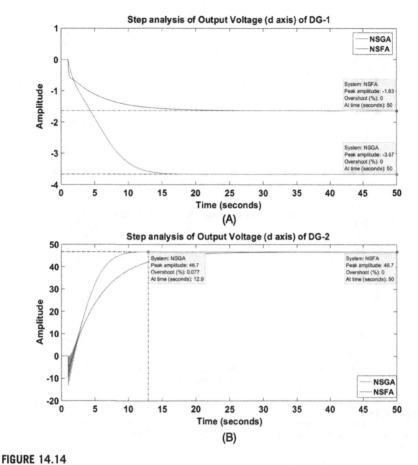

FIGURE 14.14

Step analysis of output voltage (*d* axis) of DG-1 (A) and DG-2 (B).

characteristics, requiring much less time than NSGA. The results shows that overshoot and oscillation frequency response are much better with the NSFA algorithm, and it also requires a significantly lower number of iterations and less time to reach convergence to obtain the best value. Moreover, from the *F*-test and the *t*-test it is evident that NSFA and NSGA algorithms are different from each other and both are unique.

FA itself evolved as an efficient optimization technique, and combining it with a nondominated sorting method to optimize multiobjective functions opened a new opportunity for the researchers in the field of optimization. NSGA was earlier considered as one of the prominent multiobjective optimization techniques, but from the above discussion and analysis it can be said that NSFA has the ability to perform better for multiobjective problems compared with NSGA.

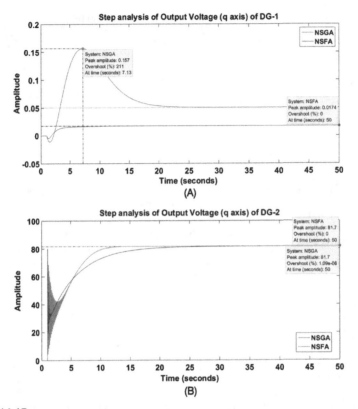

FIGURE 14.15

Step analysis of output voltage (q axis) of DG-1 (A) and DG-2 (B).

References

Abdullah, A., Deris, S., Mohamad, M.S., Hashim, S.Z.M., 2012. A new hybrid firefly algorithm for complex and nonlinear problem. In: Distributed Computing and Artificial Intelligence. Springer, pp. 673–680.

Ahmed, M., Vahidnia, A., Meegahapola, L., Datta, M., 2017. Small signal stability analysis of a hybrid AC/DC microgrid with static and dynamic loads. In: 2017 Australasian Universities Power Engineering Conference (AUPEC), pp. 1–6.

Apostolopoulos, T., Vlachos, A., 2010. Application of the firefly algorithm for solving the economic emissions load dispatch problem. International Journal of Combinatorics 2011.

Basu, B., Mahanti, G.K., 2011. Fire fly and artificial bees colony algorithm for synthesis of scanned and broadside linear array antenna. Progress in Electromagnetics Research 32, 169–190.

Chatterjee, A., Mahanti, G.K., Chatterjee, A., 2012. Design of a fully digital controlled reconfigurable switched beam concentric ring array antenna using firefly and particle swarm optimization algorithm. Progress in Electromagnetics Research 36, 113–131.

Chen, H.C., Chen, P.H., Chang, L.Y., Bai, W.X., 2013. Stand-alone hybrid generation system based on renewable energy. International Journal of Environmental Science and Development 4 (5), 514.

Chung, I.Y., Liu, W., Cartes, D.A., Schoder, K., 2008. Control parameter optimization for a microgrid system using particle swarm optimization. In: 2008 IEEE International Conference on Sustainable Energy Technologies, pp. 837–842.

Chung, I.Y., Liu, W., Cartes, D.A., Moon, S.I., 2011. Control parameter optimization for multiple distributed generators in a microgrid using particle swarm optimization. European Transactions on Electrical Power 21 (2), 1200–1216.

Dorigo, M., Birattari, M., 2010. Ant Colony Optimization. Springer.

dos Santos Coelho, L., de Andrade Bernert, D.L., Mariani, V.C., 2011. A chaotic firefly algorithm applied to reliability-redundancy optimization. In: 2011 IEEE Congress of Evolutionary Computation (CEC), pp. 517–521.

Du, W.B., Gao, Y., Liu, C., Zheng, Z., Wang, Z., 2015. Adequate is better: particle swarm optimization with limited-information. Applied Mathematics and Computation 268, 832–838.

Farahani, S.M., Abshouri, A., Nasiri, B., Meybodi, M., 2011. A Gaussian firefly algorithm. International Journal of Machine Learning and Computing 1 (5), 448.

Fogel, L.J., Owens, A.J., Walsh, M.J., 1966. Artificial Intelligence through Simulated Evolution. Wiley, New York.

Gao, Y., Du, W., Yan, G., 2015. Selectively-informed particle swarm optimization. Scientific Reports 5 (1), 9295.

Goldberg, D.E., Holland, J.H., 1988. Genetic algorithms and machine learning. Machine Learning 3 (2), 95–99.

Hassan, M., Abido, M., 2010. Optimal design of microgrids in autonomous and grid-connected modes using particle swarm optimization. IEEE Transactions on Power Electronics 26 (3), 755–769.

Horng, M.H., 2012. Vector quantization using the firefly algorithm for image compression. Expert Systems with Applications 39 (1), 1078–1091.

Horng, M.H., Lee, Y.X., Lee, M.C., Liou, R.J., 2012. Firefly metaheuristic algorithm for training the radial basis function network for data classification and disease diagnosis. Theory and New Applications of Swarm Intelligence 4 (7), 115–132.

Jain, T., et al., 2018. Impact of load dynamics and load sharing among distributed generations on stability and dynamic performance of islanded ac microgrids. Electric Power Systems Research 157, 200–210.

Kahrobaeian, A., Mohamed, Y.A.R.I., 2013. Analysis and mitigation of low-frequency instabilities in autonomous medium-voltage converter-based microgrids with dynamic loads. IEEE Transactions on Industrial Electronics 61 (4), 1643–1658.

Karaboga, D., Basturk, B., 2007. A powerful and efficient algorithm for numerical function optimization: artificial bee colony (ABC) algorithm. Journal of Global Optimization 39 (3), 459–471.

Katiraei, F., Iravani, R., Hatziargyriou, N., Dimeas, A., 2008. Microgrids management. IEEE Power & Energy Magazine 6 (3), 54–65.

Kennedy, J., 2010. Particle swarm optimization. Encyclopedia of Machine Learning, 760–766.

Kennedy, J., Eberhart, R., 1995. Particle swarm optimization (PSO). In: Proc. IEEE International Conference on Neural Networks. Perth, Australia, pp. 1942–1948.

Li, X., Hui, D., Lai, X., 2013. Battery energy storage station (BESS)-based smoothing control of photovoltaic (PV) and wind power generation fluctuations. IEEE Transactions on Sustainable Energy 4 (2), 464–473.

Marwali, M.N., Jung, J.W., Keyhani, A., 2004. Control of distributed generation systems - Part II: load sharing control. IEEE Transactions on Power Electronics 19 (6), 1551–1561.

Mohamed, Y.A.R.I., El-Saadany, E.F., 2008. Adaptive decentralized droop controller to preserve power sharing stability of paralleled inverters in distributed generation microgrids. IEEE Transactions on Power Electronics 23 (6), 2806–2816.

Palit, S., Sinha, S.N., Molla, M.A., Khanra, A., Kule, M., 2011. A cryptanalytic attack on the knapsack cryptosystem using binary firefly algorithm. In: 2011 2nd International Conference on Computer and Communication Technology (ICCCT-2011), pp. 428–432.

Pogaku, N., Prodanovic, M., Green, T.C., 2007. Modeling, analysis and testing of autonomous operation of an inverter-based microgrid. IEEE Transactions on Power Electronics 22 (2), 613–625.

Prodanovic, M., 2004. Power quality and control aspects of parallel connected inverters in distributed generation. Ph.D. thesis. Imperial College London (University of London).

Radwan, A.A.A., Mohamed, Y.A.R.I., 2013. Stabilization of medium-frequency modes in isolated microgrids supplying direct online induction motor loads. IEEE Transactions on Smart Grid 5 (1), 358–370.

Sayadi, M., Ramezanian, R., Ghaffari-Nasab, N., 2010. A discrete firefly meta-heuristic with local search for makespan minimization in permutation flow shop scheduling problems. International Journal of Industrial Engineering Computations 1 (1), 1–10.

Srinivas, N., Deb, K., 1994. Muiltiobjective optimization using nondominated sorting in genetic algorithms. Evolutionary Computation 2 (3), 221–248.

Tilahun, S., Ong, H.C., 2012. Modified firefly algorithm. Journal of Applied Mathematics 2012, 467631. https://doi.org/10.1155/2012/467631.

Uudrill, J.M., 1968. Dynamic stability calculations for an arbitrary number of interconnected synchronous machines. IEEE Transactions on Power Apparatus and Systems 3, 835–844.

Wang, R., Wu, S., Wang, C., An, S., Sun, Z., Li, W., Xu, W., Mu, S., Fu, M., 2018. Optimized operation and control of microgrid based on multi-objective genetic algorithm. In: 2018 International Conference on Power System Technology (POWERCON), pp. 1539–1544.

Yang, X.S., 2008. Nature-Inspired Metaheuristic Algorithms. Luniver Press, Beckington, UK, pp. 242–246.

Yang, X.S., 2009. Firefly algorithms for multimodal optimization. In: International Symposium on Stochastic Algorithms, pp. 169–178.

Yu, K., Ai, Q., Wang, S., Ni, J., Lv, T., 2015. Analysis and optimization of droop controller for microgrid system based on small-signal dynamic model. IEEE Transactions on Smart Grid 7 (2), 695–705.

Zaman, M.A., Matin, A., et al., 2012. Nonuniformly spaced linear antenna array design using firefly algorithm. International Journal of Microwave Science and Technology 2012.

Swarm robotics – a case study: bat robotics

15

Andrés Iglesias[a,b], **Akemi Gálvez**[a,b], **Patricia Suárez**[b]

[a]*Toho University, Department of Information Sciences, Faculty of Sciences, Funabashi, Japan*
[b]*University of Cantabria, Department of Applied Mathematics and Computational Sciences, Santander, Spain*

CONTENTS

15.1 Swarm intelligence in nature

For centuries, the concept of *intelligence* has been invariably linked to human beings, neglecting the notion that other natural creatures could develop sophisticated strategies other than merely instinctive behaviors. However, a simple observation of nature provides multiple cases that contradict such assertion. A paradigmatic example is given by the collective behavior of colonies of several social insects (ants, termites, bees, wasps, fireflies). These groups exhibit a highly sophisticated social behavior that goes well beyond the simple aggregation of the behavioral routines of

Nature-Inspired Computation and Swarm Intelligence. https://doi.org/10.1016/B978-0-12-819714-1.00026-9

their individual members. The mysterious dance of honeybees to communicate the location of promising food sources, the construction of large arboreal nests by wasps, the creation of impressive cathedral mounds by termites, and the trail followed by ants during food foraging are all good examples of that complex collective behavior of the swarm, unknown to its individual members. A recent study has shown that fire ants are capable of building strong structures such as bridges to crossover. Fire ants have also drawn attention from the public during the hurricane Harvey in Texas in 2017, when social media were filled with images of clumps of fire ants, known as rafts, clumped together on the surface of the water. These clumps can host as many as 100,000 individual ants, which use their waxy, water-resistant bodies to link together around their queen. In this way, they form a temporary structure as they travel in search of a new place to create the tunnels and chambers that make up their nests. Other examples of swarms in nature include fish schooling, bird flocking, animal herding, hawk hunting, and bacterial growth.

Swarms in nature are extremely varied in terms of size: they can vary from a few individuals living in a small area to large colonies spread on a broad territory and consisting of millions of individuals. They are also *decentralized*, meaning that there is no leader in the swarm guiding the other members to carry out the intended tasks. In general, the swarm is comprised by a collection of identical (or quasihomogeneous) individuals evolving in a nonsynchronized way. Such individuals show poor abilities when compared to those of the whole swarm: they have a very limited intelligence, and cannot accomplish the swarm goals without the rest of the group. Surprisingly, it has been shown that the individuals do not need any representation or global understanding of the swarm to produce complex collective behaviors. Members of the swarm do not have information about the general status of the swarm. Instead, the interaction between individuals is performed on a purely local basis. An illustrative example is fish schooling: when schooling, individual fish use the eyes on the sides of their heads and some marks on the bodies of other fishes to keep track of its close neighbors exclusively, ignoring the general evolution of the swarm. Yet, the school can change its shape and direction of movement at impressive speeds, and without collision among its members. Another example is bird flocking: birds in the flock are able to follow a common tendency in their movement, and migrate thousands of kilometers to a given target location, even though each bird exclusively pays attention to its local neighbors.

It is hard to imagine how these complex behaviors can arise from a swarm of simple individuals with limited cognitive abilities. Amazingly, this sophisticated collective behavior emerges from a relatively small set of rather simple behavioral rules exploiting only low-level local interactions between individuals and with the environment, using decentralized control and self-organization. The *self-organization* is given by the interplay among four basic rules: positive feedback, negative feedback, randomness, and multiple interactions (see Bonabeau et al. (1999) for details). A key factor of this process is the implicit communication between individuals through changes in the environment, a process known as *stigmergy*. Stigmergy is a form of self-organization able to produce intelligent behavioral patterns in the absence of any planning, control, or even direct communication between the agents. It supports efficient collaboration by means of indirect coordination between extremely simple agents, who usually lack any memory, intelligence, or even individual awareness of each other. For instance, ants communicate with each other using pheromones, which are released in the environment. At the beginning, the ants move randomly in search of food. However, when an ant finds a potential food source, it returns to the colony, leaving pheromone trails on the way back. If other ants perceive the pheromones, they also follow the trail until the food source and return to the colony, releasing themselves new pheromones, thus reinforcing that particular route. These pheromone trails will evaporate over time, reducing its attractive strength. Obviously, shorter paths are less affected in short term by this evaporation process, so they eventually are visited more frequently than the longer ones. In this way, nature provides a solution to the problem of finding the shortest path between two points: the colony and the food source, in this case. This mechanism is the inspiration for the popular ant colony optimization method (Dorigo, 1992). Another illustrative example of stigmergy arises in nest building by termites, where the structure of the nest will determine the behavioral patterns of the workers in the colony. Each insect collects a mud ball or similar material from its surrounding environment. Then, it releases some pheromones on the ball, which is left on the ground, initially at random locations. These pheromones stimulate other termites to drop their own mud balls (also with their own pheromones) on top of the previous ones. As the pile of mud increases, it becomes more attractive, and hence more new mud balls will be added, a kind of positive feedback reinforcement. Over time, this will lead to the construction of complex nests with many structural elements (pillars, arches, tunnels, chambers) by following a simple decentralized rule set.

Another primary (but not necessarily required) factor for swarm intelligence in nature is the *eusociality*, commonly considered as the highest level of organization in social groups. Eusociality is defined by a set of distinctive characteristics or features for the swarm: cooperative brood care (the group members also raise offspring from other individuals), the existence of overlapping generations within the colony, and an efficient division of labor (often into reproductive and nonreproductive groups, although other specialized groups can be found as well). In many cases, this division of labor is evidenced by the existence of subgroups of individuals that have lost a behavioral ability that is characteristic of other members of the swarm. This behavioral specialization makes the swarms not only more efficient but also more sophisticated, leading to the emergence of different subgroups exhibiting different behavioral roles, usually labeled as castes. Examples of eusocial groups include several species of bees and wasps, and almost all species of ants and termites.

15.2 Computational swarm intelligence

Swarm intelligence is a subfield of artificial intelligence based on the collective behavior of decentralized and self-organized systems comprised of relatively simple agents interacting locally with one another and with the environment, much like the natural swarms actually do (Blum and Merkle, 2008; Hassanien, 2016; Eberhart et al., 2001). In fact, the inspiration for this field typically comes from nature, where different biological systems show very similar features (Yang, 2014, 2016). The concept of swarm intelligence was firstly proposed in the 1980s. Since then, it has attracted increasing attention from the scientific community in a variety of fields, including engineering, economics, computer science, artificial intelligence, and many others. Simultaneously, a myriad of swarm intelligence methods have been developed and widely applied to solve complex problems such as optimization problems. For instance, all swarm behaviors mentioned in the previous section and many others observed in nature have been wisely used as an inspiration to develop different swarm intelligence methods. Classical examples include the ant colony optimization (Colorni et al., 1991; Dorigo, 1992), particle swarm optimization (Kennedy and Eberhart, 1995), and differential evolution (Storn and Price, 1997), all proposed in the 1990s. In contrast to many classical approaches in artificial intelligence focused on the structure of the agents, swarm intelligence methods are instead designed to promote and exploit the local interactions between agents of the swarm. The agents, typically driven by a small set of rules, are not highly intelligent but are still able to complete difficult tasks through strong cooperation, division of labor, and local interaction, leading to the emergence of sophisticated behaviors never observed in a single agent. The interested reader is referred to the books by Bonabeau et al. (1999), Engelbrecht (2005), Eberhart et al. (2001), and Yang (2010a) for a general overview on swarm intelligence and its main methodologies and techniques, along with several interesting features and applications.

15.3 Swarm robotics: definition and main features

From the very beginning, swarm intelligence attracted the attention from the scientific community because of its potential applications in several fields. One of the most relevant applications is the coordination and collective behavior of groups of very simple self-organized robots, which can potentially accomplish complex tasks and therefore replace sophisticated and expensive robots by simple inexpensive drones (Arvin et al., 2014; Faigl et al., 2013; Zahugi et al., 2012). This exciting research field is commonly referred to as swarm robotics. In short, *swarm robotics* is the field that studies how to manage and coordinate large groups (swarms) of relatively simple physical robots through the use of local rules. This includes the design and construction of their physical body and components (sensors, motors, actuators) and their controlling behaviors (Sauter et al., 2007; Saska et al., 2014). The reader is kindly referred to Wagner and Bruckstein (2001) for some illustrative early examples and applications of swarm robotics; see also Navarro and Matía (2013) and Tan and Zheng (2013) for two papers discussing interesting features and developments in the field.

According to the principles of swarm intelligence, an important issue is that the swarm of robots must be designed such that a desired collective behavior emerges from the local interactions among agents themselves, and also between the agents and the environment.

However, this simple definition is not enough, as it can be confused with other approaches of multirobot systems. In general, the following criteria are also imposed on swarm robotics:

1. The robots of the swarm are relatively small, and preferably low-cost, so they can be manufactured and deployed in large amounts.
2. The robots must be autonomous, with the ability to sense and operate in a real environment by themselves.
3. Ideally, the individual robots should all be identical. But even if not, the robotic swarm should be homogeneous. Different subgroups of robots are still allowed, but the number of such subgroups should be low.
4. The robots are simple and cannot solve the intended problems individually, or they exhibit a poor performance in doing so. In other words, they necessarily have to cooperate in order to either solve the problem or do it more efficiently.
5. The robots of the swarm have only local communication and sensing capabilities. The communication among the robots operates normally at a local level, ensuring that the swarm is scalable and robust enough to withstand failures of any individual member.
6. The rules controlling the members of the swarm are usually simple and are performed at the individual level, similarly to the cooperative behavior commonly found in natural swarms, and able to produce a large set of complex collective behaviors.
7. The swarm is decentralized, self-organized, and distributed. As a result, it shows high efficiency, parallelism, scalability, and robustness.

15.4 Advantages and limitations of swarm robotics

As remarked by several authors (Arkin, 1998; Bonabeau et al., 1999), swarm robotic systems offer several interesting advantages, most of which are similar to those commonly found in biological swarms in nature. In this section, they are briefly discussed and compared with those of a single sophisticated robot and of multirobot systems containing several robotic units.

15.4.1 Advantages over a single sophisticated robot

Accomplishing a difficult task with a robot typically requires that the robot has a sophisticated structure and configuration, involving a large number of different components (mechanical, electronic, optical, etc.) and several control modules for motors, sensors, and actuators. This results in high costs for the design, construction, testing, operation, enhancement, and maintenance of the robot and its components, leading to budgets that exceed those typically available for many groups and institutions. In addition, this expensive robot also becomes very vulnerable and prone to errors, as even the smallest component may affect the general performance of the whole robotic structure. On the contrary, a swarm of simple, inexpensive robots can perform similar tasks through strong intergroup cooperative work. The robots can be deployed in large numbers on different areas, increasing the exploratory ability of a single robot, taking advantage of high parallelism. Also, they become less sensitive to errors or accidents, as the loss of an individual robot (or even some of them) does not affect, or affects very little, the general performance of the swarm.

An illustrative example is given by the well-known environmental problem that occurred in 2011, after a strong earthquake, and the subsequent tsunami, reached the shores of the Fukushima prefecture and the nuclear plant in the area, leading to one of the most severe nuclear disasters in recent history. This disaster showed the limits of robot technologies. The hostile radioactive environment proved to be too hard for many robotic units deployed at the nuclear plant. High gamma radiation levels scrambled the electrons within the semiconductors serving as the robots' computing units. All of a sudden, radiation inactivated sophisticated machines that were considered breakthrough technology of that time. Complex and expensive autonomous robots disconnected or got shut down suddenly. Others got snared by fallen structures and deformed obstacles in unexpected places. About 5.5 years after the disaster, in December 2016, a new, powerful and very sophisticated autonomous 24-inch-long robot called Scorpion, developed by Toshiba (investing 2.5 years and an undisclosed high sum), was sent to the plant. The robot was equipped with cameras and sensors to gauge radiation levels and temperatures, including a tail camera mounted for better viewing angles. Although initially planned to operate for 10 hours, the robot became stuck by blocks of melted metal in its way, just 2 hours after the robot entered Unit 2 of the plant. After this failed attempt, Toshiba kept improving its robots, and went back to the plant several times with new, smaller robots working cooperatively. Eventually, they were successful, not only to survive the strong radiation from the plant, but also to provide very useful information about the plant conditions, such as radiation and temperature readings, without disturbing the surrounding environment. The robots were even able to grip small objects with their hand-like attachment to free some indoor paths.

As shown in this hostile scenario, the advantages of robotic swarms over a single sophisticated robot are the following:

- *Improved performance by parallelization*: swarm intelligence systems are very well suited for parallelization, because the swarm members operate on an individual basis, according to their own individual rules, and can perform different actions at different locations simultaneously. This feature makes the swarm more flexible and efficient for complex tasks, as individual robots (or groups of them) can solve different parts of a complex task independently.
- *Task enablement*: groups of robots can do certain tasks that are impossible or very difficult for a single robot (e.g., collective transport of too heavy items, dynamic target tracking, cooperative environment monitoring, autonomous surveillance of large areas).

- *Scalability*: inclusion of new robots into a swarm does not require reprogramming the whole swarm. Furthermore, because interactions between robots involve only neighboring individuals, the total number of interactions within the system does not increase dramatically by adding new units.
- *Distributed sensing and action*: a swarm of simple interconnected mobile robots deployed throughout a large search space possesses greater exploratory capacity and a wider range of sensing than a sophisticated robot. This makes the swarm much more effective in several tasks: exploration and navigation (e.g., in disaster rescue missions), nanorobotics-based manufacturing, microrobotics for human body diagnosis, and many others.
- *Stability and fault tolerance*: due to the decentralized and self-organized nature of the swarm, the failure of a single unit does not affect the completion of the given task. If one or several individuals fail or quit the task, the swarm can adapt to the population size change through implicit task reallocation without the need of any external operation.
- *Economy*: the cost of building the individual robots of a swarm is very low compared with single, sophisticated robots. In general, swarm robotic units are designed to be very economical, even when produced in large amounts. Maintenance costs are also very low: since the robots are created similar to each other, their components are usually cheap and highly interchangeable. Also, having simple homogeneous robots simplifies their maintenance as it minimizes the expertise required to fix any failed robotic unit.
- *Energy efficiency*: since the swarm robots tend to be small, they do not require large power sources, but a small battery or similar. These small batteries increase the life time of the swarm as a whole.

15.4.2 Advantages over a multirobot system

There are many areas where groups of robots are used simultaneously to carry out a common task (Dudek et al., 1996; Iocchi et al., 2001). Notable examples include multirobot systems, sensor networks, and multiagent systems. However, they are not considered as robotic swarms, as they do not generally follow the rules and principles of swarm robotics. Table 15.1, borrowed from Tan and Zheng (2013), summarizes the main features and differences among these four systems of robotic units.

As remarked in Tan and Zheng (2013), the main differences between swarm robotics and the other systems are the population size, control, homogeneity, flexibility, and scalability. Population size in swarm robotics can vary broadly, whereas it is kept small for multirobot and multiagent systems, and is fixed for sensor networks. Robotics swarms are also decentralized and autonomous, while systems from the other paradigms are mostly centralized, remote, or hierarchical. On the other hand, multirobot and multiagent systems generally involve heterogeneous robots, usually associated with specialized roles. As a result, they may achieve high performance for specialized tasks, but, at the same time, they miss the inherent adaptability, flexibil-

Table 15.1 Comparison of swarm robotics and other systems of robots.

	Swarm robotics	Multirobot system	Sensor network	Multiagent system
Population size	Variation in great range	Small	Fixed	In a small range
Control	Decentralized and autonomous	Centralized or remote	Centralized or remote	Centralized or hierarchical or network
Homogeneity	Homogeneous	Usually heterogeneous	Homogeneous	Homogeneous or heterogeneous
Flexibility	High	Low	Low	Medium
Scalability	High	Low	Medium	Medium
Environment	Unknown	Known or unknown	Known	Known
Motion	Yes	Yes	No	Rare
Typical applications	Postdisaster relief	Transportation	Surveillance	Net resources management
	Military applications	Sensing	Medical care	Distributed control
	Dangerous applications	Robot football	Environmental protection	

ity, and scalability of swarm robotic setups, making them less suited to be adapted to other problems. This distinction is important, and often misleading. For example, the robotic teams competing at popular events such as *Robocup* do not qualify as swarm robotics as long as different robots in the team are assigned very specialized roles. The flexibility allows the swarm to cope with different tasks using the same hardware and minimal changes in software, in a similar way to how the natural swarms deal with different tasks in real life. In swarm robotics, the robots exhibit flexibility in changing their problem solving strategies to adapt to changes in the environment. Such changes do not require reprogramming the whole swarm, just incremental changes are performed, based on machine learning, in order to improve the current strategy. Finally, the scalability is a consequence of the local sensing and communication. Robots in a swarm have a limited range for sensing, and the communication among robots is performed locally. Other robotic systems follow a global communication approach, which is a critical limiting factor to the growth of the system, since the communication cost when adding new individuals to the group increases exponentially.

All these advantages have motivated a great interest in swarm robotics during the 2000s and 2010s, as evidenced by the large number of large-scale research projects in this area. In Section 15.5, we will discuss some of the most relevant projects carried out in the field during this period.

15.4.3 Limitations of swarm robotics

Of course, the robotic swarms have also their own limitations and drawbacks. The most important ones are the following:

- *Interference and collisions*: since the robots in a swarm share a similar scenario and do not communicate to each other on a global scale, they can occasionally interfere with each other, leading to unexpected collisions, occlusions, and other issues.
- *Uncertainty*: the degree of efficiency of a swarm in completing a task is related to the collective behavior of its members, which requires coordination at some extent for better assignment of tasks, based on location, availability, and other factors. But coordination means to know where other robots are located and what are they doing. Owing to the local communication among members of the swarm, each individual robot keeps always some uncertainty about the status of the other robots. This can potentially lead to conflicts such as robots of the swarm competing against each other instead of cooperating.
- *Lack of specialization*: the homogeneous nature of the swarm can be disadvantageous for highly specialized tasks, where an alternative sophisticated, goal-oriented robot might outperform the swarm.
- *Lack of understanding of the behavioral pattern of the swarm*: since the collective behavior of the swarm emerges from the local interactions among its members and with the environment through implicit rules, sometimes it is difficult to understand the global behavioral patterns of the group and to determine the best strategies leading to a successful completion of a task.

15.5 Swarm robotic projects

Several swarm robotics projects have been developed during the 2000s and 2010s. Ideally, robotic swarms are expected to be developed at a hardware level, creating a physical robotic prototype comprised of different electronic and mechanical components, such as sensors, motors, and actuators. Such physical prototypes will be globally referred to in this chapter as *robotic platforms*. However, creating these robotic platforms is not always practical or affordable, so some projects rely on computer simulations instead. This approach is also common for the feasibility analysis and assessment typically carried out at early stages of large-scale projects in order to determine the best configuration before mass production. These software-based approaches will be called *robotic simulators and frameworks* in this chapter. Following this simple classification, in the following sections we will describe some of the most relevant swarm robotic projects, grouped as robotic platforms and robotic simulators and frameworks, respectively. We remark however that this list is by no means exclusive, as many other swarm robotic projects have been developed during the last few years. Its description in detail is out of the scope of this chapter, and an in-depth explanation would probably require a separate whole chapter. Here we will restrict our discussion to summarize some of the most popular approaches in the field.

15.5.1 Swarm robotic platforms

Swarm robotics has been a focal point of interest for the academic and the industrial sectors during the last decades. Early projects in the 1980s and 1990s set the principles and foundations of the field. Pioneering research projects in this regard were, for instance, CEBOT, SWARMS, or ACTRESS, which showed the potential of this field rather than creating impressive developments on their own. This research effort was followed up by more ambitious, large-scale projects, such as the European-funded projects Swarm-bots[1] (2001–2005), i-Swarm[2] (2004–2008), and Swarmanoid[3] (2006–2010). Several initiatives were also carried out in USA universities such as Harvard, MIT, Stanford, UPenn, ASU, Texas A&M, and many others, leading to a golden era in swarm robotics.

One of the early approaches in the field was the robot *Khepera* (Mondada et al., 1999), developed at the EPFL (Lausanne, Switzerland) in the mid-1990s. It was quite popular, having been sold to a thousand research labs worldwide and serving the swarm robotic research community for more than a decade. Subsequent versions (such as *Khepera III* (Pugh et al., 2009)) were released for the following decade along with some simulation platforms (see Section 15.5.2 for further details). The current version, *Khepera IV*[4] integrates the Linux core running on an 800 MHz ARM Cortex-A8 Processor with 512 MB of RAM and with 802.11 b/g Wifi, Bluetooth 2.0 EDR, and several sensors (three-axis accelerometer, three-axis gyroscope), including an array of eight infrared sensors for obstacle detection, four more for tasks such as fall avoidance and line following, and five ultrasonic sensors for long-range object detection. The model also comes with a color camera (752×480 pixels, 30 FPS), an embedded microphone, three programmable RGB LEDs on top of the robot, and an internal battery providing a running time of about 7 hours. Different extensions are also available for nearly unlimited configurations.

Another evolution of the robot *Khepera* is given by *Alice* (Caprari et al., 2000; Caprari and Siegwart, 2005), a small, autonomous "sugarcube" robot developed by Gilles Caprari at the Autonomous Systems Lab at EPFL. *Alice* became quite popular, as it was rather small for that time (under 1 cubic inch), and relatively inexpensive, making it affordable to construct and operate a large crowd of robots simultaneously. Arrays of up to 90 robots working cooperatively were built up, a remarkable achievement at that time. These microrobots were even for sale, not only for researchers but also for hobbyists. Equipped with a microcontroller PIC16LF877 with 8K Flash

---

[1] http://www.swarm-bots.org/

[2] http://www.i-swarm.org/

[3] http://www.swarmanoid.org/

[4] http://www.k-team.com

program memory and four infrared proximity sensors, they were also highly configurable, with a 24 pin connector for extension and numerous extension modules available for different research purposes. Some illustrative examples of applications of *Alice* include navigation and map building (Caprari et al., 1998), the soccer kit (consisting of two teams of three *Alice* robots playing soccer cooperatively on an A4 page), and the embodiment of cockroach aggregation (Garnier et al., 2008). Other interesting applications can also be found in Caprari and Siegwart (2005).

A larger (CD size) robot was *Kobot*, developed at the Middle East Technical University, Turkey (Turgut et al., 2008a,b). *Kobot* is a mobile robot with several sensors (distance, bearing, vision, compass) designed for several swarm robotic research tasks, like coordinated motion. Some remarkable features, such as the infrared short-range sensors and the ability of sensing the relative headings of neighboring robots, were applied to analyze a self-organized flocking scenario of a swarm of robots, moving as a coherent group and avoiding obstacles as if they were a single "superorganism."

E-puck was a popular and successful research project for the development of a miniature mobile robot for educational purposes. The project released a small wheeled, cylindrical-shaped microrobot designed to be robust and affordable enough to allow intensive classroom use (Mondada et al., 2009). The robots have a simple mechanical structure easy to understand, operate, and maintain. The robot was also very flexible, with many options for further enhancement and updating, such as several sensors, processing units, and other extensions. It was quite popular for academic purposes, particularly in small amounts, with several tens of research projects based on this robot in fields such as mobile robotics, signal processing, image and sound feature extraction, real-time programming, human–computer interaction, embedded systems, collective behavioral patterns, and so on. However, its size and price were limiting factors for its massive deployment, as required largely in swarm robotics.

Another popular microrobot was *Jasmine*,[5] a public open hardware robot developed by the University of Stuttgart to create a simple and cost-effective microrobotics platform (Schmickl et al., 2008). This microrobot has been widely used in swarm robotics studies, such as playing the role of a honeybee in aggregation scenarios (Kernbach et al., 2009). A differential feature of *Jasmine* with respect to other swarms of robots is that *Jasmine* only supports local communication, while long-distance communication is neither intended nor implemented, making it a good test-bed for many problems. In particular, *Jasmine* robots communicate through infrared LEDs, so that robots have to be in direct vision, and direct communication is not possible with robots not in sight. The potential advantage of this approach is that the robots do not perceive information that does not come from their direct environment, and hence, this design encapsulates the concept of stigmergy by design. Also, its infrared equipment provides support for directional communication. In this way, not only the message but also the spatial context (the direction from which the message is received, the intensity of the signal, the neighborhood topology, and so on) can be

---

[5] http://www.swarmrobot.org/

received and processed. However, the communication can be potentially disturbed by sun light and other sources of light.

Another interesting approach in swarm robotics is given by the creation of self-assembly robots, as addressed in the *Sambots* project (Wei et al., 2011). Multiple Sambots can be linked to build new structures through self-assembly and self-configuration. To this aim, a docking mechanism was designed so that the docking interface can be rotated around the robot body. In this way, several robots can be freely connected to create varied complex structures and configurations, ranging from linear structures (e.g., snakes, worms) to looped structures (e.g., triangles, rings), and so on. Other microrobots well suited for swarm robotics are *S-bot* (Mondada et al., 2005), a very versatile robot equipped with several actuators and developed by the swarmbots project; *i-Swarm robot* (Seyfried et al., 2005; Valdastri et al., 2006), developed for the *i-Swarm* project, which combines microrobotics, adaptive systems, and self-organizing swarm systems by creating small microrobots that can be used in large numbers, of even more than 100 individuals, working cooperatively; *Swarm-Bot* (McLurkin, 2004), a successful robotic platform where the robots are able to locate and move to dock charging stations attached to the walls where they can be charged; *AMiR* (Autonomous Miniature Robot) (Arvin et al., 2009), a wheeled affordable, low-cost swarm robotic platform developed under an open source and open hardware architecture for studies about honeybee aggregation and fuzzy-based models; and *Colias* (Arvin et al., 2014) and its evolution, *Colias-Φ* (Arvin et al., 2015), both developed at the University of Lincoln. They are small low-cost (about 25 British pounds each) open source microrobots for swarm robotic applications and bio-inspired vision systems (e.g., moving according to one or several gradient light sources), respectively. With a diameter of 4 cm and moving at up to 35 cm/second, they can sense each other and communicate through infrared sensors. A set of long-range infrared proximity sensors allows the robot to communicate with its neighbors at a distance range of 0.5 cm to 2 m, while a set of three short-range sensors and an independent processor are used for obstacle detection and avoidance. For communication purposes, the researchers also developed a low-bitrate messaging system requiring only 10 bits per message and 200 bits/second speed.

The *Swarmanoid* project (2006–2010) extended the work carried out in the previous project *Swarm-bots* to three-dimensional scenarios, and introduced three types of insect robots: *Eye-bot* (with the ability to fly in order to explore large areas quickly and even attach to the ceiling to provide a general top view of the environment); *Hand-bot* (designed to manipulate objects and with the capacity to climb walls and vertical surfaces of objects, but in need of other robots to walk around); and *Foot-bot* (a wheeled robot able to move through irregular terrains and equipped with a gripper so that it can carry objects, or other robots, and form physical connections with other foot-bots or hand-bots). The combination of the three types of robots, along with some communication protocols and a simulation framework, provides the capability to work in the whole three-dimensional space.

A recent trend in swarm robotics concerns the scalability of the robotic swarms. Projects such as *iRobot* (McLurkin et al., 2006), developed at MIT, worked on the development of distributed algorithms for robotic swarms consisting of hundreds of robots. Crowds of up to 100 robots (and even more) have been analyzed in other projects as well, such as the *I-swarm* (Seyfried et al., 2005), *SwarmBot*, and other microrobotic models (McLurkin and Smith, 2007; McLurkin et al., 2013; Şahin, 2005). An interesting step in this issue was the small *kilobot*,[6] by Michael Rubenstein, at the Self Organizing Systems Research Group at Harvard University (Rubenstein et al., 2012). The robot is 3.3 cm tall and, instead of using wheels, it stands on three rigid legs. The movement is driven through two vibrators; if one is activated, the robot turns at about 45 degrees per second; when both are activated, the robot moves forward at about 1 cm/s. A three-color (red, green, and blue) LED is used to display information to the user. Communication between robots is performed using infrared sensors, limiting the working area to flat surfaces and short distances. The robot can only perform three simple tasks: respond to light, measure a distance, and sense the presence of other robots. However, when combined, they can organize themselves into shapes, such as grouping into clusters based on their own color light (or that of their neighbors) or dispersing to fill a space. Kilobots are programmed all at once, as a group, using infrared light. Each kilobot gets the same set of instructions as the next. With just a few lines of programming, they can execute together complex natural processes. For instance, they can synchronize their flashing lights like a swarm of fireflies, or move together towards a light source similar to the way bacteria search for food (Rubenstein et al., 2013, 2014a). Recently, a swarm of 1000 kilobots has been widely reported in the literature (Rubenstein et al., 2014b). With this huge number, individual units are not really important; it does not even matter if one or a few robots break down, as the collective behavior of the swarm still prevails. In other words, a large robotic swarm provides a lot of flexibility and robustness, as the collective tendency of the swarm is highly immune to individual failures or breakages (Gauci et al., 2017). It also gives room to analyze some interesting configurations, for instance, gradient formation, where a source robot generates a gradient value that is incremented as it propagates through the swarm, giving each robot a metric of the distance value from the source. Another interesting application is for localization tasks, where robots determine their position in the coordinate system by communicating with already localized robots (Rubenstein et al., 2014a; Gauci et al., 2017).

A new robot with amazing motion features called *Salto* was presented in May 2019 at the *International Conference on Robotics and Automation-iCRA* in Montreal (Yim et al., 2019). Developed by researchers at the University of California, Berkeley, a previous version of the robot, with high-flying capabilities, was presented in 2016. Since then, the robot has been upgraded with a large array of new skills, such as the ability to bounce in place like a pogo stick and to jump through obstacle courses like a skilled jumping dog. The small robot, with a size of less than a foot in height, can vault over three times its height in a single bound. The inspiration for this robot comes from the galago, or Senegalese bush baby, a primate whose muscles and tendons store

---

[6] https://ssr.seas.harvard.edu/kilobots

energy in a way that gives it the ability to string together multiple jumps in a matter of seconds. Mimicking this action, *Salto* can navigate through complex environments impossible to cross without jumping or flying. In addition, *Salto* can now "feel" its own body (e.g., determine what angle it is pointing or the bend of its leg), so that it can be used in outdoor environments with a radio controller used to tell it where to go.

An even more recent new addition (published in July 2019) to the family of swarm robotic platforms is given by the exciting *Tribots*, three-legged T-shaped origami robots developed at EPFL (Zhakypov et al., 2019). They are autonomous multilocomotion insect-scale robots inspired by trap-jaw ants, designed to reproduce the complex strategies followed by ants to complete sophisticated tasks and evade larger predators. The inspiration was given by the movement of the *Odontomachus* ants, which are capable of hurling themselves through the air using their jaws. In that movement, used primarily to escape a predator, they snap their powerful jaws together to jump backwards from leaf to leaf. This movement pattern has been replicated and then extended with the *Tribots*. Although each robot has a minimal physical intelligence and weighs only 10 grams, it supports five distinct gaits: vertical jumping for height, horizontal jumping for distance, somersault jumping to clear obstacles, walking on textured terrain, and crawling on flat surfaces. *Tribots* can be assembled, with minimal components and few assembly steps, in only a matter of minutes, by folding a quasi-two-dimensional metamaterial sandwich constituted by easily integrated mechanical, material, and electronic layers, making them very well suited for mass production. They are equipped with infrared and proximity sensors for detection and communication purposes. Additional sensors might be added for different tasks. In spite of its apparent simplicity, they are able to communicate and act collectively. For instance, they can cooperate to overcome obstacles and move objects much larger and heavier than themselves. They arc also suitable for sophisticated eusocial behaviors, where different robots are assigned different specialized tasks, such as explorers, workers, leaders, and so on.

15.5.2 Swarm robotic simulators and frameworks

Swarm robotics is a very complex research field, as it involves many different disciplines and technologies that should be almost perfectly intertwined for good performance. They include electronics, mechanics, physics, and computer hardware and software. However, it is very difficult to make all these technologies work perfectly together in real-world experiments. And even if that happens, it requires a budget and resources that are not commonly available for research groups and institutions. A feasible solution in this regard is to rely on realistic simulations and fast prototyping by software to reduce the burden in developing a swarm of physical robotic units as well as make improvements on preliminary designs at early stages. It also helps to speed up the experiments, as researchers can focus on the most critical aspects of the project without wasting time and resources on physical components, technological issues, or environmental conditions.

Owing to these reasons, several robotic simulators and even programming frameworks for robots and robotic swarms have been developed, with different user interfaces and technical features. Some have been created exclusively for experimentation on particular robotic platforms and are not available for public use. Others are publicly available and very popular. In this section, we describe some of the most popular robot simulators and programming frameworks suitable for robotic swarms.

One of the first popular open source simulators for multirobot simulation (including swarm robotics) is *Player/Stage/Gazebo* (Gerkey et al., 2003), a freeware open source programming framework comprised of three main components:

- *Player:* a language-independent and platform-independent network server for robot control providing a clean and simple interface to the robot's sensors and actuators over the IP network.
- *Stage*: a multirobot simulator of a population of mobile robots, sensors, and actuators in a two-dimensional bitmapped environment. It is often used as a *Player* plugin module, providing populations of virtual devices for *Player*. As such, it has been successfully applied to two-dimensional simulations of robots (see, for instance, Vaughan (2008)), including populations up to 1000 mobile robots.
- *Gazebo*: a multirobot simulator for outdoor environments in three dimensions. Originally designed as a three-dimensional alternative to *Stage*, *Gazebo* has evolved into a powerful robot simulator supporting three-dimensional indoor and outdoor environments, and providing access to several physics and graphical engines. It provides a realistic simulation of sensors and applies the ODE physics engine, instead of that in *Stage*. It also includes modules and support for a variety of sensors and robot models.

Two more specialized robotic simulators are *UberSim* and *USARSim*. Developed at Carnegie Mellon, *UberSim* (Browning and Tryzelaar, 2003) was originally designed for fast validation of soccer robots. The programs created for the soccer robots and sensors can be written in C and uploaded to the robotic units via TCP/IP. The simulator relies on the ODE physics engine for realistic simulations of motions and interactions among the physical elements. Built on the popular commercial game engine *Unreal Engine 2.0*, *USARSim* (Carpin et al., 2007) has been developed for search and rescue research activities of the annual international robotics competition *Robocup*. It uses a reliable physics engine to simulate the physics, geometric models, and even noise of the robotic units, and has become one of the most complete simulators for swarm robotics.

Several general purpose robotic simulators supporting different robotic models have also been developed. *Teambots* is a Java-based simulator supporting simulation and execution of multirobot systems and compatible with models of some popular robotic companies. Another popular multirobot simulator is *Webots* (Michel, 2004), a cross-platform commercial product to simulate real robots through realistic models of many of the most popular commercial robots. In *Webots* the objects in the scene can be customized by the user. It also provides functionalities for a remote controller to test the real robots. *SwarmBot3D* is a simulation tool for the S-bot robot of the

Swarmbot project (McLurkin et al., 2006; Mondada et al., 2005). It was developed on top of *Vortex*, a commercial physics simulation engine, and describes both robots and worlds as XML text files. It also includes all hardware functionalities (sensors and mechanics) of a real S-bot as well as support for handling a group of robots either as independent units or in a swarm configuration. Another popular robot simulation program is *Microsoft Robotics Developer Studio* (RDS), a Windows-based framework in C# for control and simulation of robotic units (Jackson, 2007). *ARGoS*[7] is a freely available open source multiphysics engine for real-time simulation of heterogeneous robotic swarms (Pinciroli et al., 2012). It has a pluggable architecture that supports migration among multiple physics engines and even parallel simulation. For instance, it is possible to subdivide the simulated space into several subsets, managed by different physics engines running in parallel. Also, its multithreading architecture optimizes the use of multicore CPUs. The system has been widely used to simulate thousands of wheeled robots. The system is easily customizable by adding new plugins. It was the official simulator of the *Swarmanoid* project, as well as the main robot simulation tool in the European projects *ASCENS, H2SWARM, E-SWARM*, and *Swarmix*.

Some of the most popular robotic platforms described in Section 15.5.1 are also equipped with powerful simulation frameworks. For instance, several robot simulator programs have been developed for Khepera III and subsequent versions of this popular robot. Those programs include the *Khepera-Lisp Interface* (KHLI), *Khepera Simulator, YAKS, KiKS* (a *Khepera* simulator for MATLAB®), and the *Khepera III Toolbox* (a software toolbox for the *Khepera III* robot). Another example is *Jasmine*, with a simulation system based on *Breve* (a simulation package of large distributed artificial life systems evolving in a continuous three-dimensional world), an object-oriented programming language called *Steve* and sensor models for *Jasmine III* robot, and two robot simulators developed at EPFL for the popular *e-puck* robotic platform: the *ENKI* system, an open source two-dimensional physics-based robot simulator in C++, and *V-REP*, a three-dimensional robot simulator based on a distributed control architecture, where scripts can be attached to the objects in the scene, providing modules for sensor simulation, forward and inverse kinematics, physics engines, path planning, computational geometry calculations, and so on. Other notable examples include *Mona*, an open source open hardware platform developed at the University of Manchester; the *R-one* robotic initiative from Rice University providing an operating system and accompanying software to program the *R-one* robotic platform; a widely used robotic kit for educational and research purposes, *Kilobot*, from Harvard University; and the recently released software code of the University of Colorado Boulder crowdfunding *Droplets* project including a publicly accessible robot simulator. *Kilogrid* is an open source virtualization environment for the *Kilobot* robot. *Kilogrid* is a recent modular system composed of a grid of computing nodes providing a two-way communication channel between the robots and a remote workstation (Allwright et

---

[7] https://www.argos-sim.info/

al., 2018). The system is designed to extend the sensory–motor abilities of the *Kilobot*, to simplify data collection, and also for parameter fine-tuning and proper setup during experiments. The performance of the system has been successfully tested and validated with swarms of up to 100 *Kilobots* (Valentini et al., 2018).

15.6 A case study: bat robotics

Most swarm intelligence methods are computational metaphors based on the dynamics of natural groups. A classical example is the *ant colony optimization* method, based on the behavior of colonies of ants, which are able to carry out difficult tasks, impossible for individual ants (Dorigo, 1992; Dorigo and Gambardella, 1997). Another example is the behavior of a flock of birds when moving all together following a common tendency in their displacements, an inspiration to the *particle swarm optimization* method (Eberhart, 2001; Kennedy and Eberhart, 1995). Other examples include artificial bee colony, firefly algorithm, cuckoo search, and many others.

Among them, the *bat algorithm* is an increasingly popular swarm intelligence algorithm. It was proposed originally by Prof. Xin-She Yang in 2010 (Yang, 2010b; Yang and Gandomi, 2012). The algorithm is based on the echolocation behavior of bats. The author focused particularly on microbats, as they use a type of sonar called *echolocation*, with varying pulse rates of emission and loudness, to detect prey, avoid obstacles, and locate their roosting crevices in the dark. Similarly, in this chapter, the term bats should be understood to refer to microbats. The interested reader is referred to the general paper by Yang and He (2013) for a comprehensive review of the bat algorithm, its variants, and applications. This algorithm is very efficient to solve complex optimization problems, such as those described, for instance, by Iglesias et al. (2015a,b), Iglesias et al. (2017), Iglesias et al. (2018), and Yang (2011).

An important observation is that the basic principle of the bat algorithm (the echolocation through ultrasound) and many of its most important features and parameters can be easily reproduced with current hardware. This makes the bat algorithm an excellent method for potential applications in swarm robotics. We call this approach *bat robotics*. We remark that some recent papers have reported real implementations of robotic platforms to study the flight of bats (Ramezani et al., 2017) and successful applications of the bat algorithm to robotics (Fister et al., 2016a,b). However, they are mostly focused on specialized problems, and have no direct relationship with the area of swarm robotics. Very recently, we developed the first real prototype for bat robotics (Suárez et al., 2019). The following sections will summarize some of the basic ideas behind this bat robotics development.

15.6.1 The bat algorithm

The idealization of the echolocation of microbats can be summarized as follows (see Yang (2010b) for details):

1. Bats use echolocation to sense distance and distinguish between food, prey, and background barriers.
2. Each virtual bat flies randomly with a velocity v_i at position (solution) x_i with a fixed frequency f_{min} and varying wavelength λ and loudness A_0 to search for prey. As it searches and finds its prey, it changes wavelength (or frequency) of the emitted pulses and adjusts the rate of pulse emission r, depending on the proximity of the target.
3. It is assumed that the loudness will vary from an (initially large and positive) value A_0 to a minimum constant value A_{min}.

In order to apply the bat algorithm for optimization problems more efficiently, some additional assumptions are strongly advisable. In general, we assume that the frequency f evolves on a bounded interval $[f_{min}, f_{max}]$. This means that the wavelength λ is also bounded, because f and λ are related to each other by the fact that the product $\lambda \times f$ is constant. For practical reasons, it is also convenient that the largest wavelength is chosen such that it is comparable to the size of the domain of interest (the search space, for optimization problems). For simplicity, we can assume that $f_{min} = 0$, so $f \in [0, f_{max}]$. The rate of pulse can simply be in the range $r \in [0, 1]$, where 0 means no pulses at all and 1 means the maximum rate of pulse emission. With these idealized rules, the basic pseudocode of the bat algorithm is shown in Algorithm 1.

Basically, the algorithm considers an initial population of $\mathcal{P}$ individuals (bats). Each bat, representing a potential solution of the optimization problem, has a location x_i and velocity v_i. The algorithm initializes these variables with random values within the search space. Then, the pulse frequency, pulse rate, and loudness are computed for each individual bat. Then, the swarm evolves in a discrete way over generations, like time instances until the maximum number of generations, $\mathcal{G}_{max}$, is reached. For each generation g and each bat, new frequency, location and velocity are computed according to the following evolution equations:

$$f_i^g = f_{min}^g + \beta(f_{max}^g - f_{min}^g), \tag{15.1}$$

$$v_i^g = v_i^{g-1} + [x_i^{g-1} - x^*] f_i^g, \tag{15.2}$$

$$x_i^g = x_i^{g-1} + v_i^g, \tag{15.3}$$

where $\beta \in [0, 1]$ follows the random uniform distribution and x^* represents the current global best location (solution), which is obtained through evaluation of the objective function at all bats and ranking of their fitness values. The superscript $(.)^g$ is used to denote the current generation g.

The best current solution and a local solution around it are probabilistically selected according to some given criteria. Then, search is intensified by a local random walk. For this local search, once a solution is selected among the current best solutions, it is perturbed locally through a random walk of the form

$$x_{new} = x_{old} + \epsilon \mathcal{A}^g, \tag{15.4}$$

Algorithm 1 Bat algorithm pseudocode.

REQUIRE: (Initial parameters);
Population size: $\mathcal{P}$;
Maximum number of generations: $\mathcal{G}_{max}$;
Loudness: $\mathcal{A}$;
Pulse rate: r;
Maximum frequency: f_{max};
Dimension of the problem: d;
Objective function: $\phi(\mathbf{x})$, with $\mathbf{x} = (x_1, \dots, x_d)^T$;
Random number: $\theta \in U(0, 1)$;
$g \leftarrow 0$;
Initialize the bat population $\mathbf{x}_i$ and $\mathbf{v}_i$, $(i = 1, \dots, n)$;
Define pulse frequency f_i at $\mathbf{x}_i$;
Initialize pulse rates r_i and loudness $\mathcal{A}_i$;
while $g < \mathcal{G}_{max}$ **do**
    for $i = 1$ **to** $\mathcal{P}$ **do**
        Generate new solutions by adjusting frequency, and updating velocities and locations //Eqs. (15.1)–(15.3);
        if $\theta > r_i$ **then**
            $\mathbf{s}^{best} \leftarrow \mathbf{s}^g$ //select the best current solution;
            $\mathbf{ls}^{best} \leftarrow \mathbf{ls}^g$ //generate a local solution around $\mathbf{s}^{best}$;
        Generate a new solution by local random walk;
        if $\theta < \mathcal{A}_i$ *and* $\phi(\mathbf{x}_i) < \phi(\mathbf{x}^*)$ **then**
            Accept new solutions;
            Increase r_i and decrease $\mathcal{A}_i$;
    $g \leftarrow g + 1$
Rank the bats and find current best $\mathbf{x}^*$;
Return x*

---

where ϵ is a random number with uniform distribution on the interval $[-1, 1]$ and $\mathcal{A}^g = <\mathcal{A}_i^g>$ is the average loudness of all the bats at generation g.

If the new solution achieved is better than the previous best one, it is probabilistically accepted depending on the value of the loudness. In that case, the algorithm increases the pulse rate and decreases the loudness. This process is repeated for the given number of generations. In general, the loudness decreases once a bat finds its prey (in our analogy, once a new best solution is found), while the rate of pulse emission decreases. For simplicity, the following values are commonly used: $\mathcal{A}_0 = 1$ and $\mathcal{A}_{min} = 0$, assuming that the latter value means that a bat has found the prey and temporarily stops emitting any sound. The evolution rules for loudness and pulse rate are as follows:

$$\mathcal{A}_i^{g+1} = \alpha \mathcal{A}_i^g, \tag{15.5}$$

$$r_i^{g+1} = r_i^0[1 - exp(-\gamma g)], \tag{15.6}$$

FIGURE 15.1

Physical prototypes for bat robotics.

where α and γ are constants. Note that for any $0 < \alpha < 1$ and any $\gamma > 0$ we have

$$\mathcal{A}_i^g \to 0, \quad r_i^g \to r_i^0, \quad \text{as } g \to \infty. \tag{15.7}$$

In general, each bat should have different values for loudness and pulse emission rate, which can be computationally achieved by randomization. To this aim, we can take an initial loudness $\mathcal{A}_i^0 \in (0, 2)$ while the initial emission rate r_i^0 can be any value in the interval $[0, 1]$. Loudness and emission rates will be updated only if the new solutions are improved, an indication that the bats are moving towards the optimal solution. As a result, the bat algorithm applies a parameter tuning technique to control the dynamic behavior of a swarm of bats. Similarly, the balance between exploration and exploitation can be controlled by tuning algorithm-dependent parameters.

15.6.2 Robotic platform for bat robotics

This section summarizes some of the most important ideas regarding the practical realization of bat robotics through a swarm of simple robotic units. This bat robotic approach is implemented at the physical and logical levels, as described in the next paragraphs. The interested reader is referred to (Suárez et al., 2019) for further details about this implementation and some additional experiments.

Our robotic swarm for bat algorithm consists of a set of identical wheeled robots, as those shown in Fig. 15.1. Each robot comes with a kit of simple yet powerful hardware components replicating the most relevant features of the bat algorithm (the others being replicated by software). The hardware and software implementation of the robotic platform are described in Sections 15.6.2.1 and 15.6.2.2, respectively.

15.6.2.1 Hardware implementation

The robots are assembled on a rigid chassis, hosting the battery and the electronics of the robotic unit. The chassis and all other mechanical parts (e.g., wheels, holders) are generated by 3D printing using polylactic acid filament. The robot is equipped with *Freaduino UNO*, an *Arduino*-compatible single-board microcontroller operating at both 3.3V and 5V. This feature allows us to connect several 3.3V modules (e.g., Xbee, Bluetooth transmitters, LCD screens, accelerometers, gyroscopes). This microcontroller is responsible for all programming tasks and the connectivity among the different components of the robot. We also add a protoshield and a mini breadboard for further connectivity of electronic components. This combination has proved to be a low-cost solution well suited to meet our needs.

Regarding the sensors, in our implementation we use the ultrasound sensor *HC-SR04*, operating at 5V DC that uses sonar to compute the distance to an object for collision avoidance purposes, much like bats or dolphins actually do. We also use the triple-axis magnetometer board *HMC-5883L* for global spatial orientation of the robotic units of the swarm and the *HC-05 Bluetooth card* for wireless communication and data exchange over short distances (about a range of 10 m) among the robots and with a central server for tracking purposes.

15.6.2.2 Software implementation

This robotic platform was programmed on a personal computer through the open source *Arduino IDE* and then compiled with *AVR-GCC* using the library *AVR Libc*. Other external libraries and files have been used for controlling the servomotors, communication via Bluetooth and with I2C devices, and other tasks. The resulting binary code was transferred to the microcontroller through an RS232 serial port to TTL converter for final loading and execution.

15.6.3 Robotic simulation framework for bat robotics

The bat robotics physical implementation described above can also be simulated computationally. To this aim, the authors developed a simulation framework using *Unity 5*, a popular multiplatform game engine with a high portability. This game engine comes with all features and tools required for high-quality rendering and accurate simulation of three-dimensional scenes, along with a powerful physics engine for all motion routines. It also provides a nice graphical editor and support for the programming languages *C#* and *JavaScript*. The robotic simulation framework we developed for bat robotics has been created in *JavaScript* using the *Visual Studio* integrated programming environment. This source code can be further compiled to generate standalone applications for different hardware configurations and operating systems.

15.6.4 Real bats, bat algorithm, and bat robotics: analogies

As discussed before, the most relevant features of the real bats and the bat algorithm can be replicated by either hardware or software. Here, we discuss the analogies and differences found in the process. Our discussion is briefly summarized in Table 15.2, where we compare the real microbats, the bat algorithm, and our bat robotics implementation (arranged in columns) for different features (arranged in rows). For each feature, the symbol ✓ indicates that it is supported; otherwise, the symbol ✗ is used.

Table 15.2 Analogies and differences among real microbats, the bat algorithm, and our implementation for bat robotics.

	Real microbats	Bat algorithm	Our bat robotics approach
No centralized behavior	✓	✓	✓
No vision abilities	✓	Not applicable	✓
Ultrasound based	✓	✓	✓
Operating frequency	25–150 kHz	✓	40 kHz
# cycles burst	10–20	Not applicable	8
Operating time	in ms	Not applicable	~60 ms
Accuracy range	mm–cm	Not applicable	~3 mm
Traveling range of pulses	m	Not applicable	0.02~5 m
Loudness tuning	✓	✓	✓
Pulse rate tuning	✓	✓	✓
Flying abilities	✓	Not applicable	✕

The following features are replicated by hardware:

1. *Decentralized behavior:* Microbats in nature are decentralized, as they neither have a leader nor take centralized decisions. Instead, they synchronize their individual decisions for many important communal issues (Kerth, 2008). This is also a central ingredient of the bat algorithm. The same happens in our implementation. All robotic units evolve on an individual basis: the bat algorithm is encoded in the microcontroller board and executed locally. There is a central server, only used for tracking and monitoring purposes, but it does not take any decision about the swarm or its members.

2. *No vision abilities:* Many microbats are blind and so are our robotic units for bat robotics. Accordingly, we do not include any vision device such as optical cameras, thermal camera, or infrared sensors.

3. *Ultrasound based:* Both real microbats and the bat algorithm rely on echolocation. This important feature is also a key component in our bat robotics implementation, through the ultrasound sensors indicated above.

4. *Operating frequency:* Typical frequencies for microbats vary in the range of 25–150 kHz. Our robotic units for bat robotics operate at a constant frequency of 40 kHz, well within the range of real bats.

5. *# of cycles of burst:* The ultrasound sensors of our robots send an eight-cycle burst of ultrasound pulses, very similar to the normal value of real bats, which typically range about 10–20 cycles of burst.

6. *Operating time:* The signals of our robots are captured in the order of milliseconds, similar to what the real bats actually do.

7. *Accuracy range:* The accuracy range of the sensor is about 3 mm, very similar to the typical ranges of real microbats.

8. *Traveling range of pulses:* In our robots this value is 2–500 cm, matching well the range of a few meters of several bats.

Some other interesting bat algorithm features have been replicated by software:

1. *Loudness tuning:* The loudness can readily be modulated in the source programming code (as expected, since our source code is actually an implementation of the bat algorithm).
2. *Pulse rate tuning:* The pulse rate of our ultrasound signals can be modified by software to adapt to different conditions, as it typically happens with bats in the real world and in good agreement with the bat algorithm.

Finally, it is worthwhile to mention that, in contrast to real microbats, our robots lack the ability to fly. Therefore, our bat robotics implementation is constrained to move on a purely two-dimensional space. Research in this area is out of the scope of this chapter and hence is not discussed here.

15.6.5 Some applications of bat robotics

Although the bat robotics approach described here has been introduced very recently, it has already some interesting potential and actual applications. Obviously, most of them are similar to those of other swarm robotics schemes and will not be discussed here. However, others are more specialized and better suited for the particular features of bat robotics.

In Suárez and Iglesias (2017), the bat robotics swarm was applied to the problem of coordinated exploration in an unknown physical three-dimensional environment. In this problem, the goal is to reach a target point without any explicit knowledge about its location or the possible paths leading to the target. The experiment was conducted at both computational and real-world levels, with 100 computer executions and 10 real-world executions due to battery constraints, respectively. In all cases, the robots are placed at random initial locations. The results showed that the bat robotics approach is very well suited for this particular task, actually much better than we anticipated.

We found surprisingly intelligent behaviors, where some units in a robotic formation leave the group to explore other alternative ways, in cases of traffic jam owing to overcrowding, challenging scenarios (for instance, narrow passages and corridors where only a single robot could advance at once) or difficult configurations, such as U and V shapes, where the robotic units would be forced to come back. Sometimes, we observed that the robots turned around without any apparent advantage for doing so; however, the reason became clear after some iterations, when the unit found a way to reach a better location than its robotic mates in the formation, even although it initially implies to move further away from better locations. Finally, we did not find any configuration so far unsolvable for the robots, provided that at least one two-dimensional solution exists. These results were later extended to more exhaustive experiments in Suárez et al. (2019) and in Suárez et al. (2017) for the case of indoor scenarios. In particular, we considered a fully furnished house with eleven

FIGURE 15.2

Two bat robotic teams used to solve a target point search in a noncooperative mode.

rooms (with different geometric shapes, configurations, and spatial distribution of objects and obstacles within) and a garage. The environment was configured to be highly dynamic, with many changes for different executions, making it particularly well suited to check the performance of our approach.

Suárez et al. (2018b) extended the previous examples to analyze the interplay of two robotic swarms applied to solve a target point search in a noncooperative mode. In particular, we consider the case of two identical robotic swarms (shown in Fig. 15.2) deployed within the same environment to perform dynamic exploration seeking for two different unknown target points. The movement of the robots is driven by the bat algorithm but also affected by three factors: the complex and irregular geometry of the scene, the collisions with robots of the other swarm and their own, and the fact that the map is unknown to the robots. Some illustrative videos showed the ability of the robots to avoid the collisions and move forward towards the target. We also observed many interesting behavioral patterns for the robotic swarms. They include moving in a formation, aggregation patterns for intensive exploration near the optima, bifurcation moving patterns for simultaneous exploration and obstacle avoidance, cooperation among robots to force a robot of the other team to move away, and the ability to escape from U and V configurations such as dead ends, among others. This work was then generalized by Suárez et al. (2018a) to deal with the *self-centered* mode, in which each swarm tries to solve its own goals with little (or no) consideration of any other factor external to the swarm. From our experiments, we found that the bat robotics swarm allows the robotic units to find their targets in a reasonable time. Furthermore, some configurations were found to be impossible to overcome for individual robots, but solvable for the robotic swarms, supporting the notion that the cooperation among the robotic units is a key factor for solving the (otherwise unsolvable) problem.

15.7 Conclusions

This chapter summarizes some interesting ideas regarding swarm robotics, understood as the application of the swarm intelligence principles and methodology to robotics. In particular, swarm robotics is concerned with the interplay and coordination of a swarm of simple homogeneous robotic units cooperating together to accomplish a task through local communication in a distributed and decentralized way. The source of inspiration for this field comes from social groups in nature, such as social insects, for which sophisticated behaviors arise from individuals with a very limited intelligence through local interaction among individuals and with the environment. The chapter also discusses some advantages and limitations of swarm robotics with regard to single sophisticated robots and other multirobot systems. Then, it summarizes some of the most relevant swarm robotics projects developed during the 2000s and 2010s, either as physical prototype platforms or as computational simulators and frameworks.

A common approach in swarm robotics is to design robotic units that are adapted to the intended tasks, with little (if any) consideration to the particular swarm intelligence technique used to drive them. In this chapter, we follow a different reasoning, proposing the development of robotic swarms fully specialized for the swarm intelligence of interest. An illustrative example is described for a popular swarm intelligence technique called bat algorithm, where the robotic units are carefully designed to replicate the most important features of the real microbats and the computational bat algorithm as accurately as possible. We call this approach bat robotics. An actual implementation of bat robotics is presented at both the physical and the logical level. Some recent applications of such implementation are also described.

Finally, we remark that, in spite of the impressive advances in the field, swarm robotics is still far from practical applications in many areas. Most current implementations are carried out as research projects at academic institutions, but there is still a gap between the theoretical and academic developments and their applications to real-world problems. Once the basic grounds of the discipline have been established, we anticipate a renewed interest in applying this exciting technology to many problems in a variety of fields in the coming years. Undoubtedly, we are just scratching the surface when it comes to the potential applications of swarm robotics in several areas. Multivehicle robotics is currently used in a range of applications ranging from search and rescue in the ocean to agricultural monitoring and farming, from aerospace and defense to infrastructure and logistics, and even teaming up with humans for help and assistance in dirty and dangerous tasks. Recent developments for cyborg insects have shown that it is possible to pair robotic units with living organisms by using optical electrodes to inject steering commands directly into the nervous systems, such as the *DragonflEye* project (DraperLab, 2019), or relying on neuromuscular stimulation of muscles for flight and walking control (Doan et al., 2018). The next step is to take advantage of this technology to make them behave as a cyborg swarm. Through these ideas, perhaps our current conception of a cyborg, half-human half-robot, will be reshaped to include swarms of robotic units in the mixture.

Acknowledgments

This work is supported by the Computer Science National Program of the Spanish Research Agency (Agencia Estatal de Investigación) and European Funds, Project

#TIN2017-89275-R (AEI/FEDER, UE), and the project PDE-GIR of the European Union's Horizon 2020 research and innovation program under the Marie Sklodowska-Curie Actions grant agreement #778035. We owe special thanks to our colleague and friend Almudena Campuzano for her careful reading and revision of a preliminary draft version of this chapter and for her many constructive criticisms and suggestions that helped us to improve the writing of the chapter significantly.

References

Allwright, M., Bhalla, N., Pinciroli, C., Dorigo, M., 2018. Simulating multi-robot construction in argos. In: Dorigo, M., Birattari, M., Blum, C., Christensen, A.L., Reina, A., Trianni, V. (Eds.), Swarm Intelligence. Springer International Publishing, Cham, pp. 188–200.

Arkin, R.C., 1998. An Behavior-Based Robotics, first ed. MIT Press, Cambridge, MA, USA. ISBN 0262011654.

Arvin, F., Samsudin, K., Ramli, A., 2009. Development of a miniature robot for swarm robotic application. International Journal of Computer and Electrical Engineering (ISSN 1793-8163) 1 (4), 436–442. https://doi.org/10.7763/IJCEE.2009.V1.67.

Arvin, F., Murray, J.C., Shi, L., Zhang, C., Yue, S., 2014. Development of an autonomous micro robot for swarm robotics. In: 2014 IEEE International Conference on Mechatronics and Automation, pp. 635–640.

Arvin, F., Yue, S., Xiong, C., 2015. Colias-phi: an autonomous micro robot for artificial pheromone communication. International Journal of Mechanical Engineering and Robotics Research. https://doi.org/10.18178/ijmerr.4.4.349-353.

Blum, C., Merkle, D. (Eds.), 2008. Swarm Intelligence. Springer Berlin Heidelberg.

Bonabeau, E., Dorigo, M., Theraulaz, G., 1999. Swarm Intelligence: From Natural to Artificial Systems. Oxford University Press, Inc., New York, NY, USA. ISBN 0-19-513159-2.

Browning, B., Tryzelaar, E., 2003. Übersim: a multi-robot simulator for robot soccer. In: Proceedings of the Second International Joint Conference on Autonomous Agents and Multiagent Systems. AAMAS '03. ACM, New York, NY, USA, pp. 948–949.

Caprari, G., Siegwart, R., 2005. Mobile micro-robots ready to use. Alice. In: 2005 IEEE/RSJ International Conference on Intelligent Robots and Systems. IEEE, Piscataway, N.J., pp. 3295–3300.

Caprari, G., Balmer, P., Piguet, R., Siegwart, R., 1998. The autonomous micro robot "alice". A platform for scientific and commercial applications. In: Proceedings of The 1998 International Symposium on Micromechatronics and Human Science (MHS). IEEE, Piscataway, NJ, pp. 231–235.

Caprari, G., Estier, T., Siegwart, R., 2000. Fascination of down scaling – Alice the sugar cube robot. In: Proceedings of The IEEE International Conference on Robotics and Automation (ICRA), Workshop on Mobile Micro-Robots. ETH-Zurich, Zurich.

Carpin, S., Lewis, M., Wang, J., Balakirsky, S., Scrapper, C., 2007. USARSim: a robot simulator for research and education. Proceedings 2007 IEEE International Conference on Robotics and Automation. IEEE.

Colorni, A., Dorigo, M., Maniezzo, V., 1991. Distributed optimization by ant colonies. In: Actes de la Première Conférence Européenne sur la Vie Artificielle. Paris, France. Elsevier Publishing, pp. 134–142.

Doan, T.T.V., Tan, M.Y., Bui, X.H., Sato, H., 2018. An ultralightweight and living legged robot. Soft Robotics 5 (1), 17–23. https://doi.org/10.1089/soro.2017.0038.

Dorigo, M., 1992. Optimization, Learning and Natural Algorithms. Ph.D. thesis Dipartimento di Elettronica, Politecnico di Milano, Italy. 140 pp.

Dorigo, M., Gambardella, L., 1997. Ant colony system: a cooperative learning approach to the traveling salesman problem. IEEE Transactions on Evolutionary Computation 1 (1), 53–66. https://doi.org/10.1109/4235.585892.

DraperLab, 2019. Draper news. https://www.draper.com/news-releases/equipping-insects-special-service. (Accessed 10 September 2019). Online.

Dudek, G., Jenkin, M.R., Milios, E., Wilkes, D., 1996. A taxonomy for multi-agent robotics. Autonomous Robots 3 (4). https://doi.org/10.1007/bf00240651.

Eberhart, Shi Y., 2001. Particle swarm optimization: developments, applications and resources. In: Proceedings of the 2001 Congress on Evolutionary Computation. IEEE.

Eberhart, R., Shi, Y., Kennedy, J., 2001. Swarm Intelligence, first ed. Series in Artificial Intelligence. Morgan Kaufmann. ISBN 9781558605954.

Engelbrecht, A., 2005. Fundamentals of Computational Swarm Intelligence. John Wiley and Sons, Chichester, England. ISBN 978-0-470-09191-3.

Faigl, J., Krajnik, T., Chudoba, J., Preucil, L., Saska, M., 2013. Low-cost embedded system for relative localization in robotic swarms. In: 2013 IEEE International Conference on Robotics and Automation. IEEE.

Fister, D., Safaric, R., Fister, I., Fister jr, I., 2016a. Parameter tuning of PI-controller with bat algorithm. Informatica 40, 109–116.

Fister, D., Fister, I., Fister, I., Šafarič, R., 2016b. Parameter tuning of PID controller with reactive nature-inspired algorithms. Robotics and Autonomous Systems 84, 64–75. https://doi.org/10.1016/j.robot.2016.07.005.

Garnier, S., Jost, C., Gautrais, J., Asadpour, M., Caprari, G., Jeanson, R., Grimal, A., Theraulaz, G., 2008. The embodiment of cockroach aggregation behavior in a group of micro-robots. Artificial Life 14 (4), 387–408. https://doi.org/10.1162/artl.2008.14.4.14400.

Gauci, M., Ortiz, M.E., Rubenstein, M., Nagpal, R., 2017. Error cascades in collective behavior: a case study of the gradient algorithm on 1000 physical agents. In: Proceedings of the 16th Conference on Autonomous Agents and MultiAgent Systems. AAMAS '17. International Foundation for Autonomous Agents and Multiagent Systems, Richland, SC, pp. 1404–1412.

Gerkey, B.P., Vaughan, R.T., Howard, A., 2003. The player/stage project: tools for multi-robot and distributed sensor systems. In: Proceedings of the 11th International Conference on Advanced Robotics, pp. 317–323.

Hassanien, A.E., 2016. Swarm Intelligence. CRC Press.

Iglesias, A., Galvez, A., Collantes, M., 2015a. Bat algorithm for curve parameterization in data fitting with polynomial Bézier curves. In: 2015 International Conference on Cyberworlds (CW). IEEE.

Iglesias, A., Gálvez, A., Collantes, M., 2015b. Global-support rational curve method for data approximation with bat algorithm. In: Artificial Intelligence Applications and Innovations. Springer, pp. 191–205.

Iglesias, A., Gálvez, A., Collantes, M., 2017. Multilayer embedded bat algorithm for b-spline curve reconstruction. Integrated Computer-Aided Engineering (ISSN 1069-2509) 24 (4), 385–399. https://doi.org/10.3233/ICA-170550.

Iglesias, A., Gálvez, A., Collantes, M., 2018. Iterative sequential bat algorithm for free-form rational Bézier surface reconstruction. International Journal of Bio-Inspired Computation 11 (1), 1. https://doi.org/10.1504/ijbic.2018.090093.

Iocchi, L., Nardi, D., Salerno, M., 2001. Reactivity and deliberation: a survey on multi-robot systems. In: Balancing Reactivity and Social Deliberation in Multi-Agent Systems. Springer Berlin Heidelberg, pp. 9–32.

Jackson, J., 2007. Microsoft robotics studio: a technical introduction. IEEE Robotics & Automation Magazine 14 (4), 82–87. https://doi.org/10.1109/m-ra.2007.905745.

Kennedy, J., Eberhart, R., 1995. Particle swarm optimization. In: Proceedings of ICNN 95 – International Conference on Neural Networks. Institute of Electrical & Electronics Engineers (IEEE).

Kernbach, S., Thenius, R., Kernbach, O., Schmickl, T., 2009. Re-embodiment of honeybee aggregation behavior in an artificial micro-robotic system. Adaptive Behavior 17 (3), 237–259. https://doi.org/10.1177/1059712309104966.

Kerth, G., 2008. Causes and consequences of sociality in bats. BioScience 58 (8), 737–746. https://doi.org/10.1641/b580810.

McLurkin, J.D., 2004. Stupid robot tricks: a behavior-based distributed algorithm library for programming swarms of robots. M.S. Thesis. Massachusetts Institute of Technology.

McLurkin, J., Smith, J., 2007. Distributed algorithms for dispersion in indoor environments using a swarm of autonomous mobile robots. In: Distributed Autonomous Robotic Systems 6. Springer, Japan, pp. 399–408.

McLurkin, J., Smith, J., Frankel, J., Sotkowitz, D., Blau, D., Schmidt, B., 2006. Speaking swarmish: human-robot interface design for large swarms of autonomous mobile robots. In: AAAI Spring Symposium: to Boldly Go Where No Human-Robot Team Has Gone Before. AAAI, pp. 72–75.

McLurkin, J., Lynch, A.J., Rixner, S., Barr, T.W., Chou, A., Foster, K., Bilstein, S., 2013. A low-cost multi-robot system for research, teaching, and outreach. In: Springer Tracts in Advanced Robotics. Springer Berlin Heidelberg, pp. 597–609.

Michel, O., 2004. Cyberbotics ltd. webots™: professional mobile robot simulation. International Journal of Advanced Robotic Systems 1 (1), 5. https://doi.org/10.5772/5618.

Mondada, F., Franzi, E., Guignard, A., 1999. The development of Khepera. In: Proceedings of the 1st International Khepera Workshop, pp. 7–13.

Mondada, F., Gambardella, L., Floreano, D., Nolfi, S., Deneubourg, J., Dorigo, M., 2005. The cooperation of swarm-bots – physical interactions in collective robotics. IEEE Robotics & Automation Magazine 12 (2), 21–28. https://doi.org/10.1109/mra.2005.1458313.

Mondada, F., Bonani, M., Raemy, X., Pugh, J., Cianci, C., Klaptocz, A., Magnenat, S., christophe Zufferey, J., Floreano, D., Martinoli, A., 2009. The e-puck, a robot designed for education in engineering. In: Proceedings of the 9th Conference on Autonomous Robot Systems and Competitions, pp. 59–65.

Navarro, I., Matía, F., 2013. An introduction to swarm robotics. ISRN Robotics 2013, 1–10. https://doi.org/10.5402/2013/608164.

Pinciroli, C., Trianni, V., O'Grady, R., Pini, G., Brutschy, A., Brambilla, M., Mathews, N., Ferrante, E., Caro, G.D., Ducatelle, F., Birattari, M., Gambardella, L.M., Dorigo, M., 2012. ARGoS: a modular, parallel, multi-engine simulator for multi-robot systems. Swarm Intelligence 6 (4), 271–295. https://doi.org/10.1007/s11721-012-0072-5.

Pugh, J., Raemy, X., Favre, C., Falconi, R., Martinoli, A., 2009. A fast onboard relative positioning module for multirobot systems. IEEE/ASME Transactions on Mechatronics 14 (2), 151–162. https://doi.org/10.1109/tmech.2008.2011810.

Ramezani, A., Chung, S.J., Hutchinson, S., 2017. A biomimetic robotic platform to study flight specializations of bats. Science Robotics 2 (3), eaal2505. https://doi.org/10.1126/scirobotics.aal2505.

Rubenstein, M., Ahler, C., Nagpal, R., 2012. Kilobot: a low cost scalable robot system for collective behaviors. 2012 IEEE International Conference on Robotics and Automation. IEEE.

Rubenstein, M., Cabrera, A., Werfel, J., Habibi, G., McLurkin, J., Nagpal, R., 2013. Collective transport of complex objects by simple robots: theory and experiments. In: Proceedings of the 2013 International Conference on Autonomous Agents and Multi-Agent Systems. AAMAS '13. International Foundation for Autonomous Agents and Multiagent Systems, Richland, SC, pp. 47–54.

Rubenstein, M., Ahler, C., Hoff, N., Cabrera, A., Nagpal, R., 2014a. Kilobot: a low cost robot with scalable operations designed for collective behaviors. Robotics and Autonomous Systems 62 (7), 966–975. https://doi.org/10.1016/j.robot.2013.08.006.

Rubenstein, M., Cornejo, A., Nagpal, R., 2014b. Programmable self-assembly in a thousand-robot swarm. Science 345 (6198), 795–799. https://doi.org/10.1126/science.1254295.

Şahin, E., 2005. Swarm robotics: from sources of inspiration to domains of application. In: Swarm Robotics. Springer Berlin Heidelberg, pp. 10–20.

Saska, M., Vonásek, V., Krajník, T., Přeučil, L., 2014. Coordination and navigation of heterogeneous MAV–UGV formations localized by a 'hawk-eye'-like approach under a model predictive control scheme. The International Journal of Robotics Research 33 (10), 1393–1412. https://doi.org/10.1177/0278364914530482.

Sauter, J.A., Matthews, R., Parunak, H.V.D., Brueckner, S.A., 2007. Effectiveness of digital pheromones controlling swarming vehicles in military scenarios. Journal of Aerospace Computing, Information, and Communication 4 (5), 753–769. https://doi.org/10.2514/1.27114.

Schmickl, T., Thenius, R., Moeslinger, C., Radspieler, G., Kernbach, S., Szymanski, M., Crailsheim, K., 2008. Get in touch: cooperative decision making based on robot-to-robot collisions. Autonomous Agents and Multi-Agent Systems 18 (1), 133–155. https://doi.org/10.1007/s10458-008-9058-5.

Seyfried, J., Szymanski, M., Bender, N., Estaña, R., Thiel, M., Wörn, H., 2005. The i-SWARM project: intelligent small world autonomous robots for micro-manipulation. In: Swarm Robotics. Springer Berlin Heidelberg, pp. 70–83.

Storn, R., Price, K., 1997. Differential evolution – a simple and efficient heuristic for global optimization over continuous spaces. Journal of Global Optimization (ISSN 1573-2916) 11 (4), 341–359. https://doi.org/10.1023/A:1008202821328.

Suárez, P., Iglesias, A., 2017. Bat algorithm for coordinated exploration in swarm robotics. In: Advances in Intelligent Systems and Computing. Springer, Singapore, pp. 134–144.

Suárez, P., Gálvez, A., Iglesias, A., 2017. Autonomous coordinated navigation of virtual swarm bots in dynamic indoor environments by bat algorithm. In: Lecture Notes in Computer Science. Springer International Publishing, pp. 176–184.

Suárez, P., Gálvez, A., Fister, I., Fister, I., Osaba, E., Ser, J.D., Iglesias, A., 2018a. Bat algorithm swarm robotics approach for dual non-cooperative search with self-centered mode. In: Intelligent Data Engineering and Automated Learning – IDEAL 2018. Springer International Publishing, pp. 201–209.

Suárez, P., Gálvez, A., Fister, I., Fister, I., Osaba, E., Ser, J.D., Iglesias, A., 2018b. Interplay of two bat algorithm robotic swarms in non-cooperative target point search. In: Highlights of Practical Applications of Agents, Multi-Agent Systems, and Complexity: the PAAMS Collection. Springer International Publishing, pp. 543–550.

Suárez, P., Iglesias, A., Gálvez, A., 2019. Make robots be bats: specializing robotic swarms to the bat algorithm. Swarm and Evolutionary Computation 44, 113–129. https://doi.org/10.1016/j.swevo.2018.01.005.

Tan, Y., Zheng, Z.Y., 2013. Research advance in swarm robotics. Defence Technology 9 (1), 18–39. https://doi.org/10.1016/j.dt.2013.03.001.

Turgut, A., Çelikkanat, H., Gökçe, F., Şahin, E., 2008a. Self-organized flocking with a mobile robot swarm. Swarm Intelligence 1, 39–46.

Turgut, A.E., Çelikkanat, H., Gökçe, F., Şahin, E., 2008b. Self-organized flocking in mobile robot swarms. Swarm Intelligence 2 (2–4), 97–120. https://doi.org/10.1007/s11721-008-0016-2.

Valdastri, P., Corradi, P., Menciassi, A., Schmickl, T., Crailsheim, K., Seyfried, J., Dario, P., 2006. Micromanipulation, communication and swarm intelligence issues in a swarm microrobotic platform. Robotics and Autonomous Systems 54 (10), 789–804. https://doi.org/10.1016/j.robot.2006.05.001.

Valentini, G., Antoun, A., Trabattoni, M., Wiandt, B., Tamura, Y., Hocquard, E., Trianni, V., Dorigo, M., 2018. Kilogrid: a novel experimental environment for the kilobot robot. Swarm Intelligence 12 (3), 245–266. https://doi.org/10.1007/s11721-018-0155-z.

Vaughan, R., 2008. Massively multi-robot simulation in stage. Swarm Intelligence 2 (2–4), 189–208. https://doi.org/10.1007/s11721-008-0014-4.

Wagner, I.A., Bruckstein, A.M., 2001. Annals of Mathematics and Artificial Intelligence 31 (1/4), 1–5. https://doi.org/10.1023/a:1016666118983.

Wei, H., Chen, Y., Tan, J., Wang, T., 2011. Sambot: a self-assembly modular robot system. IEEE/ASME Transactions on Mechatronics 16 (4), 745–757. https://doi.org/10.1109/tmech.2010.2085009.

Yang, X.S., 2010a. Nature-Inspired Metaheuristic Algorithms, second ed. Luniver Press. ISBN 1905986289.

Yang, X.S., 2010b. A new metaheuristic bat-inspired algorithm. In: Nature Inspired Cooperative Strategies for Optimization (NICSO 2010). Springer Berlin Heidelberg, pp. 65–74.

Yang, X.S., 2011. Bat algorithm for multi-objective optimisation. International Journal of Bio-Inspired Computation 3 (5), 267. https://doi.org/10.1504/ijbic.2011.042259.

Yang, X.S. (Ed.), 2014. Nature-Inspired Optimization Algorithms. Elsevier.

Yang, X.S. (Ed.), 2016. Nature-Inspired Computation in Engineering. Springer International Publishing.

Yang, X.S., Gandomi, A.H., 2012. Bat algorithm: a novel approach for global engineering optimization. Engineering Computations 29 (5), 464–483. https://doi.org/10.1108/02644401211235834.

Yang, X.S., He, X., 2013. Bat algorithm: literature review and applications. International Journal of Bio-Inspired Computation 5 (3), 141. https://doi.org/10.1504/ijbic.2013.055093.

Yim, J.K., Wang, E.K., Fearing, R.S., 2019. Drift-free roll and pitch estimation for high-acceleration hopping. In: 2019 International Conference on Robotics and Automation (ICRA). IEEE.

Zahugi, E.M.H., Shabani, A.M., Prasad, T.V., 2012. Libot: design of a low cost mobile robot for outdoor swarm robotics. In: 2012 IEEE International Conference on Cyber Technology in Automation, Control, and Intelligent Systems (CYBER). IEEE.

Zhakypov, Z., Mori, K., Hosoda, K., Paik, J., 2019. Designing minimal and scalable insect-inspired multi-locomotion millirobots. Nature 571 (7765), 381–386. https://doi.org/10.1038/s41586-019-1388-8.

Electrical harmonics estimation in power systems using bat algorithm

16

Serdar Kockanat

Sivas Cumhuriyet University, Department of Electrical and Electronics Engineering, Sivas, Turkey

CONTENTS

16.1 Introduction

In recent years, the demands of electrical energy in many place, such as home, office, and industry, have rapidly increased. This leads to the enlargement and importance of electrical power systems. The most important criterion that increases the efficiency of electrical power systems is power quality. If an electrical power system has a high power quality, this system is considered as an efficient and lossless system. However, the usage of solid state switching elements and devices depending on developing semiconductor technology in households and industrial areas has been increasing in recent years. This causes significant pollution in the harmonic levels of the electrical signals in power systems and grids. The harmonic pollution leads to negative situations such as high energy losses, overheating of the systems, reduced life time of sensitive biomedical and communication devices, data losses, and operational problems of circuit breaker equipment using in electrical grids. In order to keep under control this negative effect occured in modern electric networks and power systems, various harmonic distortion calculation, elimination, and tracking applications are determined by certain rules (Wiczynski, 2008). Since many electrical networks and grids use a fundamental operating frequency of 50 Hz or 60 Hz, harmonics occur on integer multiples of these frequencies and the abovementioned applications accept these frequencies as reference. Also, harmonics are unwanted disturbing waveforms

Nature-Inspired Computation and Swarm Intelligence. https://doi.org/10.1016/B978-0-12-819714-1.00027-0

and reduce the power quality in the electrical grids and power systems in which it occurs (Chen and Chen, 2014).

In order to reduce the abovementioned negative effects of harmonics in electrical grids and power systems, different methods have been proposed by researchers and practitioners. In these applications, studies were carried out for modeling, monitoring, estimating, and eliminating electrical harmonics. However, in all reported approaches, estimation of amplitude and phase values is considered as an important criterion especially in both voltage and current signals used in electrical grids and power systems. Discrete Fourier transform (DFT) and Fast Fourier transform (FFT) algorithms are the primary applied techniques in harmonic estimation studies (Chu, 2008; Lin, 2012). However, these techniques succeeded in the harmonic estimation of stationary signals, but did not achieve the desired success in estimation of nonstationary signals. Furthermore, they have several disadvantages such as insufficient memory and picket fence effect problems in practical applications (Wang and Sun, 2006). In order to overcome these disadvantages, new approaches such as Kalman filtering (Zadeh et al., 2010), adaptive notch filtering (Yazdani et al., 2009), and wavelet transformation (Saleh et al., 2011) have been proposed. These reported approaches have been successful in electrical harmonic estimation according to FFT and DFT methods, but they have deficiencies such as wavelet waveform selection problem and need for prior knowledge.

In recent decades, researchers and practitioners have successfully applied optimization algorithms to propose alternative approaches in the solutions of many linear and nonlinear engineering problems. Mentioned alternatives have also been employed to the electrical harmonic estimation. In addition, new hybrid approaches have been proposed to improve the solution performance since the electrical harmonic estimation problem has a nonlinear characteristic. In hybrid applications, to improve estimation accuracy, least square (LS) and recursive LS (RLS) algorithms are especially used for amplitude estimation and evolutionary and swarm algorithms are employed for phase estimation. The genetic algorithm (GA) (Bettayeb and Qidwai, 2003), particle swarm optimization (PSO) (Lu et al., 2008; Narayan et al., 2009), bacterial foraging optimization (BFO) (Ray and Subudhi, 2012), the artificial bee colony algorithm (Biswas and Amitava Chatterjee, 2013; Kabalci et al., 2018), biogeography-based optimization (Singh et al., 2016a), the firefly algorithm (Singh et al., 2016b), and the gravity search algorithm (Singh et al., 2017) have been reported for estimating harmonics occurring in electrical grids and power systems. For aforementioned methods based on evolutionary and swarm algorithms, sinusoidal electrical test signals including fundamental, sub-, and inter-harmonics were performed in the literature. Also, varying noise conditions such as 10 dB and 20 dB have been taken into account in the estimation operations. The achieved harmonics estimation results have been compared with each other with the aim to develop the most accurate and effective harmonic estimation approaches for electrical grids and power systems.

To the best of our knowledge, no approach based on the bat algorithm (BA) have been reported for estimating electrical harmonics in electrical grids and power systems. In this chapter, an estimation approach based on BA (Yang, 2010) is suggested for the harmonic estimation problem in electrical grids and power systems. BA is a heuristic and swarm-based algorithm that has been presented by Yang in 2010 and which models the echolocation behaviors of natural living bats. In addition, BA has been applied to solve different engineering problems. For performance and accuracy analyses of the proposed method, the aforementioned literature problems reported in hybrid applications are considered. In addition, the effects of noise on the harmonic estimator performance are investigated. The results are compared with previous studies in the literature, and the strengths and weaknesses of the proposed harmonic estimator are revealed.

16.2 Bat algorithm

Bats are animals in nature that usually live in enclosed areas such as caves and can be seen especially at night or in the dark. Bats are the only mammals that can fly and have wings. Although there are many different species of bats, wing sizes and weights vary in an interesting way. There are bats of different weights, ranging from 2 g to 1 kg. In the dark, bats have the ability to determine the direction and distance in which an object or a living organism is located by making use of a kind of sonar called echolocation. They use echolocation behavior successfully at every stage of their lives, from communication to hunting. Inspired by these behaviors of bats, an optimization algorithm that can be used to solve many mathematical problems has been developed. A swarm-based optimization algorithm called BA, which models the behavior of bats, has been presented to the literature by Yang (2010).

BA is based on the following three basic rules when modeling the echolocation behavior of bats.

Rule 1: All bats obtain their position information by echolocation and thus determine the location of the prey.

Rule 2: Each bat flies randomly at position x_i with velocity v_i, and with frequency f_{min}. They change their wavelength (λ) and loudness (A_0) and search for their prey.

Rule 3: The loudness value can be adjusted in many different ways. However, when performing modeling, we can assume that the loudness value changes between the minimum and maximum values.

In the modeled echolocation behavior, the position, velocity, and frequency updates are performed with the following equations:

$$f_i = f_{min} + (f_{max} - f_{min})\beta, \tag{16.1}$$

$$v_i^t = v_i^{t-1} + (x_i^t - x_*)f_i, \tag{16.2}$$

$$x_i^t = x_i^{t-1} + v_i^t. \tag{16.3}$$

Here f_i is the frequency value that a bat can produce in the frequency range. It is a randomly distributed vector in the range [0,1]. In addition, v_i^t and x_i^t are the current velocity and position values, and x_* is the current global best location. In Fig. 16.1, the basic flowchart of BA is presented.

16.3 Mathematical model of electrical harmonic in power systems

In electrical grids and power systems, electrical signals are modeled as sine and cosine functions. Accordingly, the general waveform of an electrical harmonic current or voltage can be defined as

$$s(t) = \sum_{n=1}^{N} A_n \sin(\omega_n t + \theta_n) + A_{dc} \exp(-\alpha_{dc} t) + \mu(t), \tag{16.4}$$

where N shows the harmonic order, ω_n is the angular frequency of the component n, f_0 is the fundamental frequency, $\mu(t)$ presents additive white Gaussian noise, and $A_{dc} \exp(-\alpha_{dc} t)$ indicates the expected decaying term. The unknown magnitude and phase values are given as A_n and θ_n. The general waveform is sampled by the sampling period T_s and obtained as a discrete-time electrical harmonic signal. Then, applying the Taylor series expansion for the expected decaying term and using the sine and cosine function, the newly obtained equation is given as

$$s(k) = \sum_{n=1}^{N} [A_n \sin(\omega_n k T_s) \cos(\theta_n) + A_n \cos(\omega_n k T_s) \sin(\theta_n)]$$
$$+ A_{dc} - A_{dc} \alpha_{dc} k T_s + \mu(k). \tag{16.5}$$

Then, the waveform is suggested in the parametric form as

$$s(k) = y(k) H(k), \tag{16.6}$$

where

$$y(k) = [\sin(\omega_1 k T_s) \cos(\omega_1 k T_s) \ldots \sin(\omega_n k T_s) \cos(\omega_n k T_s) 1 - k T_s]^T, \tag{16.7}$$

$$H = [A_1 \cos(\theta_1) A_1 \sin(\theta_1) \ldots A_n \cos(\theta_n) A_n \sin(\theta_n) A_{dc} A_{dc} \alpha_{dc}]^T. \tag{16.8}$$

At the optimization stages, all unknown amplitude and phase values are updated with the following equation:

$$H(k) = [H_1(k) H_2(k) \ldots H_{(2n-1)}(k) H_{(2n)}(k) H_{(2n+2)}(k)]^T. \tag{16.9}$$

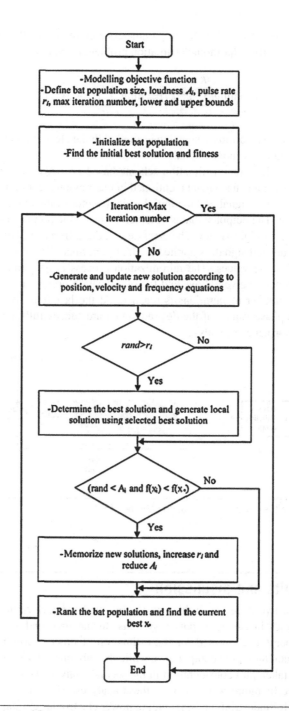

FIGURE 16.1

The basic flowchart of BA.

In order to optimize the unknown parameters weight vector, the objective function (J) of the electrical harmonic estimation problem is expressed as

$$J = \min\left(\sum_{k=1}^{K} e^2(k)\right) = \min\left(\sum_{k=1}^{K} (s_k - s_{k_{est}})^2\right), \qquad (16.10)$$

where s_k is the actual harmonic signal measured with the electrical power system and $s_{k_{est}}$ is the estimated harmonic signal.

The electrical harmonic estimation scheme based on BA is shown in Fig. 16.2. The literature harmonic test signal mentioned in the previous section is applied to the system as the desired signal and the initial weight vector of the amplitude and phase values is formed. The output signal is obtained by using the proposed sinusoidal harmonic model and weight vector. The error signal, the difference between the output signal and the desired signal, is applied to BA to optimize the weight vectors of the amplitude and phase values. The minimization process is performed according to the objective function given above and the optimization process is completed when the maximum number of iterations is reached. At the last step, the best estimated amplitude and phase values of the desired signal are successfully obtained and the estimation performance is analysed.

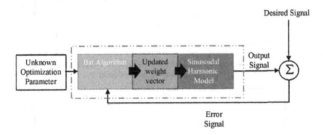

FIGURE 16.2

Block diagram of the harmonic estimation scheme of BA.

16.4 Results and discussion

In this section, two different problems are presented for performing harmonic estimation analyses in electrical power systems. In the first analysis, the harmonic estimation process is examined for the estimation of power harmonics containing only fundamental harmonic components, while the sub- and inter-harmonics components are also taken into consideration in the second analysis. Harmonic test signals accepted in the literature were used for these analyses. These test signals are generally defined for industrial loads, including power electronics circuits and nonlinear systems. In addition, in both experiments, performance evaluations were analysed for cases with a signal-to-noise ratio (SNR) of 10 dB and 20 dB. When performing Analysis 1 and Analysis 2, the estimation performance of the suggested approach based on BA was analysed and the results were compared with those of the GA, PSO, and BFO algorithms previously reported in the literature (Ray and Subudhi, 2012). Since BA is not used as hybrid for electrical harmonic estimation, a fair comparison was made with nonhybrid algorithms. In addition, population size was selected as 100 for GA-, PSO-, BFO-, and BA-based approaches in all analyses. The proposed algorithm is performed 30 times with different random initial values.

Analysis 1: (Harmonic estimation including fundamental harmonics)

In the first analysis, for electrical harmonic estimation, the literature test signal containing random noise and a decaying DC component is given as follows:

$$s(t) = 1.5\sin(2\pi f_1 t + 80^o) + 0.5\sin(2\pi f_3 t + 60^o) + 0.2\sin(2\pi f_5 t + 45^o)$$
$$+ 0.15\sin(2\pi f_7 t + 36^o) + 0.1\sin(2\pi f_{11} t + 30^o) + 0.5\exp(-5t) + \mu(t).$$
$$(16.11)$$

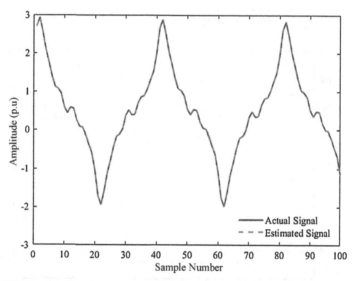

FIGURE 16.3

Comparison of actual signal and estimated signal by BA (0.01 random noise).

Fig. 16.3 shows the actual signal and estimated signal by BA for 0.01 random noise and decaying DC component. As can be seen from Fig. 16.3, for 100 samples, the alignment between the actual signal and estimated signal is very high and the difference errors between them are very low. Fig. 16.4 and Fig. 16.5 show the variation of the amplitude and phase values of each harmonic depending on the cycle number. After the 80th cycle, the amplitude and phase values of all harmonics are estimated.

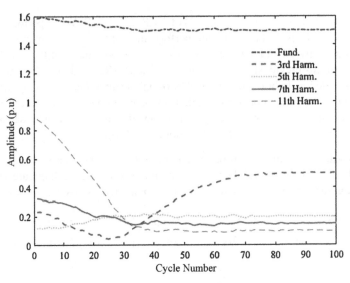

FIGURE 16.4

Amplitude estimation of harmonics using BA (0.01 random noise).

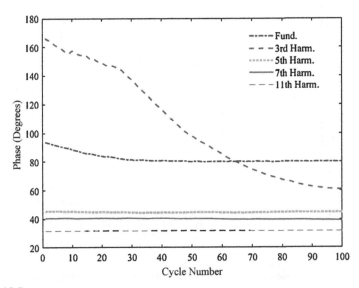

FIGURE 16.5

Phase estimation of harmonics using BA (0.01 random noise).

In addition, Table 16.1 shows the obtained results of harmonic estimation performed with BA. The abbreviations F, A, P, and E in Table 16.1 and Table 16.2 refer to frequency, amplitude, phase, and error respectively. As can be seen from Table 16.1, the amplitude values of the fundamental, third, fifth, seventh, and eleventh harmonics are more accurate than those of the GA, PSO, and BFO algorithms. The estimated phase values of the fundamental, third, and fifth harmonics were more accurate than the others. This proves that the estimation approach based on BA shows more accurate and effective estimation performance than other algorithms.

Table 16.1 Comparison of the results of harmonic estimation algorithms for Analysis 1 (0.01 noise).

Algorithm	Parameters	Fund	3rd	5th	7th	11th
Actual	F (Hz)	50	150	250	350	550
	A (V)	1.5	0.5	0.2	0.15	0.1
	P (o)	80	60	45	36	30
GA	A (V)	1.48	0.485	0.18	0.158	0.0937
	E (%)	1.33	3.0	10.0	5.33	6.3
	P (o)	80.61	62.4	47.03	34.354	26.7
	E (%)	0.61	2.4	2.03	1.646	3.3
PSO	A (V)	1.482	0.488	0.182	0.1561	0.0948
	E (%)	1.2	2.4	9.0	4.06	5.2
	P (o)	80.54	62.2	46.6	34.621	27.31
	E (%)	0.54	2.2	1.6	1.379	2.69
BFO	A (V)	1.4878	0.5108	0.1945	0.1556	0.1034
	E (%)	0.8147	2.1631	2.7267	3.7389	3.4202
	P (o)	80.4732	57.9005	45.8235	34.5606	29.1270
	E (%)	0.4732	2.0995	0.8235	1.4394	0.873
Proposed BA	A (V)	1.4990	0.5001	0.1997	0.1493	0.1004
	E (%)	**0.0648**	**0.0229**	**0.1625**	**0.4787**	**0.3625**
	P (o)	80.0019	59.9685	45.0392	39.1413	31.4496
	E (%)	**0.0024**	**0.0526**	**0.0870**	8.7258	4.8321

After examining the success of the proposed approach based on BA by using the test harmonic signal in the literature, two different noises, 10 dB and 20 dB, were applied to the test harmonic signal. Fig. 16.6 and Fig. 16.9 show actual signals and estimated signal by BA for 20 dB and 10 dB noise ratios. It can be observed that the increasing noise level negatively affects the performance of the harmonic estimation approach, but even under these conditions, the predicted signals are quite similar to the noiseless literature test harmonic signal. This shows the resistance of the approach to noise. Fig. 16.7, Fig. 16.8, Fig. 16.10, and Fig. 16.11 show the variation of amplitude and phase values depending on the cycle number.

Analysis 2: (Harmonic estimation including sub- and inter-harmonics)

In the second analysis, the literature test signal containing the proposed sub- and inter-harmonics for electrical harmonic estimation is expressed as follows:

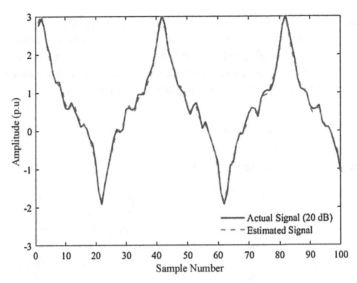

FIGURE 16.6

Comparison of actual signal and estimated signal by BA (20 dB noise).

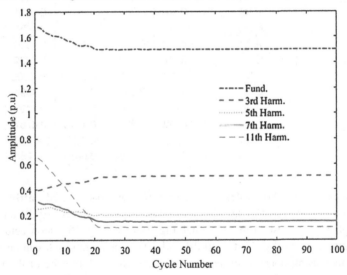

FIGURE 16.7

Amplitude estimation of harmonics using BA (20 dB noise).

$$s(t) = 0.505 \sin(2\pi f_{sub}t + 75^o) + 1.5 \sin(2\pi f_1 t + 80^o) + 0.5 \sin(2\pi f_3 t + 60^o)$$
$$+ 0.25 \sin(2\pi f_{int1}t + 65^o) + 0.35 \sin(2\pi f_{int2}t + 20^o) + 0.2 \sin(2\pi f_5 t + 45^o)$$
$$+ 0.15 \sin(2\pi f_7 t + 36^o) + 0.1 \sin(2\pi f_{11}t + 30^o) + 0.5 \exp(-5t) + \mu(t).$$
$$(16.12)$$

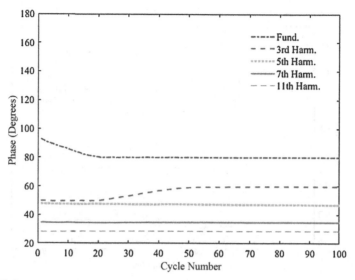

FIGURE 16.8

Phase estimation of harmonics using BA (20 dB noise).

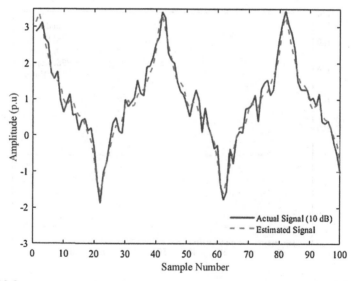

FIGURE 16.9

Comparison of actual signal and estimated signal by BA (10 dB noise).

Fig. 16.12 shows the actual signal and estimated signal by BA for the literature test harmonic signal including 0.01 random noise, a DC component, and sub- and inter-harmonics. This figure is drawn for 100 samples. As can be seen from this fig-

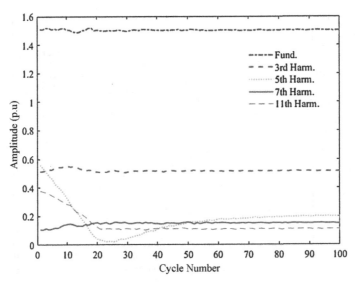

FIGURE 16.10

Amplitude estimation of harmonics using BA (10 dB noise).

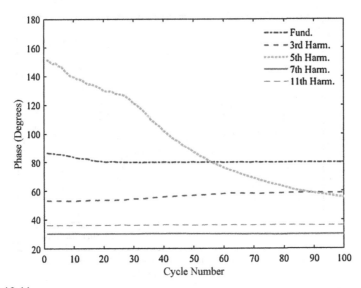

FIGURE 16.11

Phase estimation of harmonics using BA (10 dB noise).

ure, the alignment between real and estimated signals is quite high and there are no fluctuations between them. Fig. 16.13 and Fig. 16.14 show the variation of the amplitude and phase values of each harmonic depending on the cycle number. Estimation

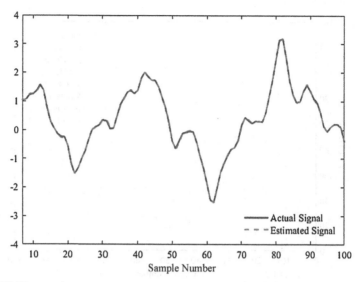

FIGURE 16.12

Comparison of actual signal and estimated signal by BA (0.01 random noise).

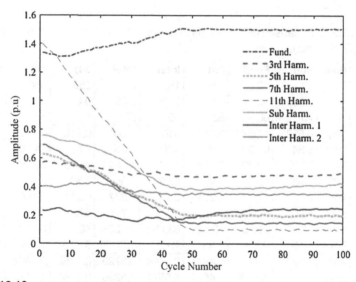

FIGURE 16.13

Amplitude estimation of harmonics using BA (0.01 random noise).

of amplitude and phase values of all harmonics continued until the 100th cycle. In addition, Table 16.2 gives the obtained result of harmonic estimation performed with BA. As can be seen from this table, the amplitude values estimated using BA, in-

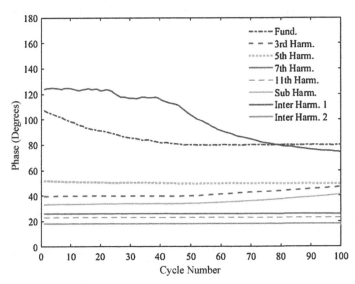

FIGURE 16.14

Phase estimation of harmonics using BA (0.01 random noise).

Table 16.2 Comparison of the results of harmonic estimation algorithms for Analysis 2 (0.01 noise).

Alg.	Para.	Sub	Fund	3rd	Inter1	Inter2	5th	7th	11th
	F(Hz)	20	50	150	180	230	250	350	550
Actual	A(V)	0.505	1.5	0.5	0.25	0.35	0.2	0.15	0.1
	P(o)	75	80	60	65	20	45	36	30
GA	A(V)	0.532	1.5083	0.472	0.238	0.381	0.215	0.172	0.117
	E(%)	5.34	0.553	5.6	4.8	8.85	7.5	14.66	17
	P(o)	73.02	79.23	57.55	62.41	17.64	48.33	38.78	32.56
	E(%)	1.98	0.77	2.45	3.59	2.36	3.33	2.78	2.56
PSO	A(V)	0.53	1.5049	0.281	0.24	0.377	0.211	0.165	0.11
	E(%)	4.95	0.326	3.8	4	7.7	5.5	10	11.0
	P(o)	73.51	79.45	58.12	63.28	18.23	48.1	37.109	31.87
	E(%)	1.49	0.55	1.88	1.72	1.77	3.1	1.109	1.87
BFO	A(V)	0.525	1.4788	0.4877	0.2664	0.3729	0.2052	0.1464	0.1016
	E(%)	3.995	1.4103	2.4575	6.5574	6.5295	2.5764	2.4170	1.5531
	P(o)	74.48	79.8361	61.2316	63.9910	19.6887	47.698	36.737	29.393
	E(%)	0.514	0.1639	1.2316	1.0090	0.3113	2.6983	0.7462	0.6072
Proposed BA	A(V)	0.5045	1.5009	0.4995	0.2498	0.3509	0.1985	0.1478	0.0995
	E(%)	**0.0993**	**0.0632**	**0.1049**	**0.0675**	**0.2704**	**0.7524**	**1.4763**	**0.5420**
	P(o)	74.945	79.997	59.753	66.023	18.254	48.184	26.105	22.898
	E(%)	**0.0736**	**0.0032**	**0.4120**	1.5736	8.7313	7.0764	27.4875	23.6732

cluding the sub- and inter-harmonics, are more accurate than those of the GA, PSO, and BFO algorithms, because the lowest relative error values were obtained. The estimated phase values of the sub-, fundamental, and third harmonics are more accurate than the others. As can be seen from these results, the estimation approach based on BA is more accurate and effective than other algorithms.

After analysing the success of the estimation approach based on BA by using the test harmonic signal containing sub- and inter-harmonics, two different noises, 10 dB and 20 dB, were applied to the same used signal. Fig. 16.15 and Fig. 16.18 show actual signal and estimated signal using BA for 20 dB and 10 dB noise ratios. When these figures are examined, it is seen that the increasing noise level negatively affects the performance of the harmonic estimation approach, but even under these conditions, the estimated signals are quite similar to the noiseless literature test harmonic signal. This shows the resistance of the approach to noise. Fig. 16.16, Fig. 16.17, Fig. 16.19, and Fig. 16.20 show the variation of amplitude and phase values depending on the cycle number. In these figures, it is seen that the amplitude and phase estimation values are completed in approximately the 70th cycle but some harmonics continue until the 100th cycle.

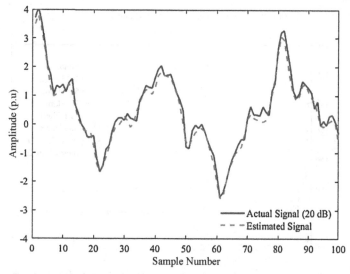

FIGURE 16.15

Comparison of actual signal and estimated signal by BA (20 dB noise).

16.5 **Conclusions**

In order to effectively design the active filters used to eliminate the negative effects of harmonics in electrical networks and power systems, it is necessary to provide

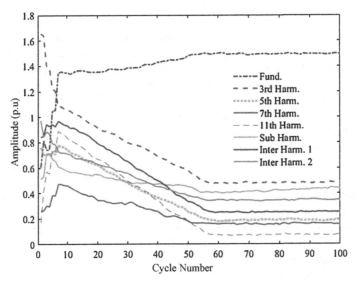

FIGURE 16.16

Amplitude estimation of harmonics using BA (20 dB noise).

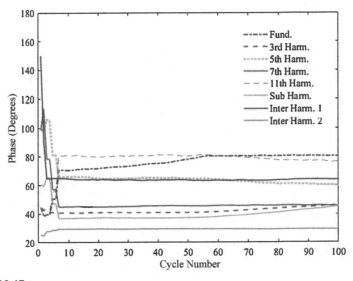

FIGURE 16.17

Phase estimation of harmonics using BA (20 dB noise).

estimation of electrical harmonics that is as accurate as possible. In this chapter, an approach based on BA is presented for estimating electrical harmonics. Estimation performance analyses were evaluated for two different harmonic test problems pro-

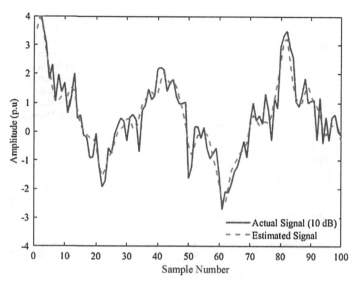

FIGURE 16.18

Comparison of actual signal and estimated signal by BA (10 dB noise).

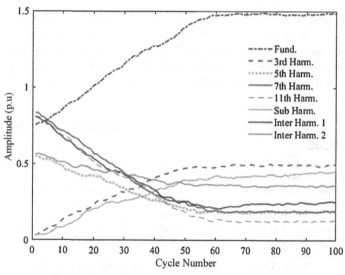

FIGURE 16.19

Amplitude estimation of harmonics using BA (10 dB noise).

posed in the literature. In addition, the analyses at two different noise levels, 10 dB and 20 dB, were reassessed. The results were compared with those of well-known algorithms such as GA, PSO, and BFO algorithm. To sum up, when all analyses and

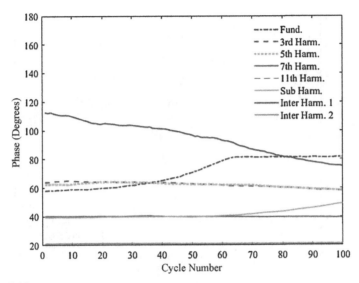

FIGURE 16.20

Phase estimation of harmonics using BA (10 dB noise).

results are taken into account, it is concluded that the harmonic estimation approach based on BA shows very good performance. In addition, the resistance of the proposed approach to noise is verified. In future studies, it is planned to combine BA with LS and RLS algorithms to propose BA-LS and BA-RLS hybrid approaches and to increase phase prediction performances in particular.

References

Bettayeb, M., Qidwai, U., 2003. A hybrid least squares-GA-based algorithm for harmonic estimation. IEEE Transactions on Power Delivery 18 (2), 377–382.

Biswas, S., Amitava Chatterjee, Goswami S.K., 2013. An artificial bee colony-least square algorithm for solving harmonic estimation problems. Applied Soft Computing 13 (5), 2343–2355.

Chen, C.I., Chen, Y.C., 2014. Comparative study of harmonic and interharmonic estimation methods for stationary and time-varying signals. IEEE Transactions on Industrial Electronics 61 (1), 397–404.

Chu, E., 2008. Discrete and Continuous Fourier Transforms Analysis: Applications and Fast Algorithms. Taylor and Francis, New York.

Kabalci, Y., Kockanat, S., Kabalci, E., 2018. A modified ABC algorithm approach for power system harmonic estimation problems. Electric Power Systems Research 154, 160–173.

Lin, H.C., 2012. Power harmonics and interharmonics measurement using recursive group-harmonic power minimizing algorithm. IEEE Transactions on Industrial Electronics 59 (2), 1184–1193.

Lu, Z., Ji, T.Y., Tang, W.H., Wu, Q.H., 2008. Optimal harmonic estimation using a particle swarm optimizer. IEEE Transactions on Power Delivery 23 (2), 1166–1174.

Narayan, R., Chatterjeeb, R.D., Goswami, S.K., 2009. An application of PSO technique for harmonic elimination in a PWM inverter. Applied Soft Computing 9 (4), 1315–1320.

Ray, P.K., Subudhi, B., 2012. BFO optimized RLS algorithm for power system harmonics estimation. Applied Soft Computing 12 (8), 1965–1977.

Saleh, S.A., Moloney, C.R., Rahmani, M.A., 2011. Analysis and development of wavelet modulation for three-phase voltage-source inverters. IEEE Transactions on Industrial Electronics 58 (8), 3330–3348.

Singh, S.K., Sinha, N., Kumar, A., Sinha, G.N., 2016a. Power system harmonic estimation using biogeography hybridized recursive least square algorithm. International Journal of Electrical Power and Energy Systems 83, 219–228.

Singh, S.K., Sinha, N., Kumar, A., Sinha, G.N., 2016b. Robust estimation of power system harmonics using a hybrid firefly based recursive least square algorithm. International Journal of Electrical Power and Energy Systems 80, 287–296.

Singh, S.K., Kumari, D., Sinha, N., Goswami, A.K., Sinha, N., 2017. Gravity search algorithm hybridized recursive least square method for power system harmonic estimation. Engineering Science and Technology, an International Journal 20 (3), 874–884.

Wang, M., Sun, Y., 2006. A practical method to improve phasor and power measurement accuracy of DFT algorithm. IEEE Transactions on Power Delivery 21 (3), 1054–1062.

Wiczynski, G., 2008. Analysis of voltage fluctuations in power networks. IEEE Transactions on Instrumentation and Measurement 57 (11), 2655–2664.

Yang, X.S., 2010. A new metaheuristic bat-inspired algorithm. In: Cruz, C., González, J.R., Pelta, D.A., Terrazas, G. (Eds.), Nature Inspired Cooperative Strategies for Optimization (NISCO 2010). In: Studies in Computational Intelligence, vol. 284. Springer, Berlin, pp. 65–74.

Yazdani, D., Bakhshai, A., Joos, G., Mojiri, M., 2009. A real time three phase selective harmonic extraction approach for grid connected converters. IEEE Transactions on Industrial Electronics 56 (10), 4097–4106.

Zadeh, R.A., Ghosh, A., Ledwich, G., 2010. Combination of Kalman filter and least-error square techniques in power system. IEEE Transactions on Power Delivery 25 (4), 2868–2880.

CSBIIST: cuckoo search-based intelligent image segmentation technique

17

Kirti, Anshu Singla

Chitkara University Institute of Engineering and Technology, Chitkara University, Punjab, India

CONTENTS

17.1 Introduction

Image segmentation is the exercise of dissociating the backdrop objects from the forefront objects. It can be categorized according to certain attributes: (i) based on texture, (ii) based on thresholds, (iii) based on clusters, and (iv) based on a given region (Pal and Pal, 1993; Sezgin and Sankur, 2004). Thresholding is to ascertain the distinct zones from the background (Sahoo et al., 1988; Hertz and Schafer, 1988; Kohler, 1981). Thresholding techniques must incorporate: (i) contextual knowledge of an image, (ii) the most favorable threshold, and (iii) the count of distinct zones. Histogram-based thresholding techniques (Glasbey, 1993; Tobias and Seara, 2002; Otsu, 1979; Akay, 2013) do not incorporate contextual knowledge of the image. Abutaleb (1989) and Hammouche et al. (2008) proposed a technique to consider the contextual knowledge of an image. But the aforementioned techniques are computationally demanding.

Nature-Inspired Computation and Swarm Intelligence. https://doi.org/10.1016/B978-0-12-819714-1.00028-2

In order to overcome the computational overhead occurring due to multiple thresholds, bio-inspired algorithms were employed (Tuba, 2014; Bhandari et al., 2016; Cuevas et al., 2013; Alihodzic and Tuba, 2014). Innumerable nature-inspired algorithms from the literature, such as the genetic algorithm (GA), particle swarm optimization (PSO), bumble bee mating optimization, artificial bee colony (ABC), ant colony optimization, the bat algorithm, elephant herding optimization, the firefly algorithm, the cuckoo search algorithm, gray wolf optimization, honey bee optimization, the lion optimization algorithm, and many more, can be employed to procure the most favorable solutions in less time (Ye et al., 2015; Sathya and Kayalvizhi, 2010; Tao et al., 2003; Maitra and Chatterjee, 2008; Horng, 2011; Raja et al., 2014; Oliva et al., 2019; Dhal et al., 2019). The former techniques in literature overcame the computational overhead but are not intelligent enough to reckon the number of thresholds. Singla and Patra (2017a) proposed a technique that incorporates the contextual knowledge as well as the count of objects in an image. Singla and Patra (2017b) and Singla et al. (2019) deployed the PSO and ABC algorithms, respectively, to automatically segment the image.

In this chapter, an intelligent technique named cuckoo search-based intelligent image segmentation is proposed. The proposed technique employs a cuckoo search algorithm based on the breeding etiquette of the cuckoo species. The remainder of the chapter is structured as follows. A brief and epigrammatic outline of cuckoo search and histograms is bestowed in Sections 17.2 and 17.3, respectively. In Section 17.4, the proposed intelligent technique is delineated. Section 17.5 describes the entire setup for quantitative and qualitative analysis, and assessment of acquired results is presented in Section 17.6. At last, in Section 17.7 addenda for the proposed technique and the future scope are discussed.

17.2 Cuckoo search

In 2009, Yang and Deb (2009) proposed an algorithm named cuckoo search. The root of the cuckoo search algorithm is the breeding etiquette of cuckoo birds. To understand the cuckoo search in a preferable way, brooding etiquette and Lévy flights must be studied first.

17.2.1 Breeding etiquette of cuckoo species

The species of cuckoos are enthralling due to their hostile approach of reproduction. There are diverse species of cuckoos, such as roadrunners, anis, koels, guira, malkohas, *Tapera*, etc. The aforementioned species of cuckoos stow their eggs in the general nest. The idea behind stowing in a general nest is to strengthen the incubating possibility of their own eggs (Payne and Sorensen, 2005). The birds who reside in the communal nest are known as scroungers. The breeding scroungers are classified into three categories: (i) nest takeover, (ii) intraspecific, and (iii) cooperative. In the case the host bird learns about the egg of the other species, they either throw the eggs out or relinquish the nest. *Tapera* species are experts in imitating color and eggs of the host birds. This characteristic of cuckoos decreases the possibility of abandonment. To increase the incubation possibility, cuckoos stow their eggs before the host bird stows its own eggs in the nest.

17.2.2 Lévy flights

Animals such as birds and insects hunt for their chow randomly or in an apparently random manner (Brown et al., 2007; Pavlyukevich, 2007; Reynolds and Frye, 2007). Lévy flight is random walk to estimate the distribution of the step size using a power law also known as Lévy distribution. Light is also relevant to Lévy flights. The behavior of hunter foragers is similar to the behavior of Lévy flights. The power law (Viswanathan et al., 1999) can be written as follows:

$$P(s) = s^{-\mu}, 1 < \mu \le 3, \tag{17.1}$$

where s represents the length of the step size. Lévy flights provide recommended results for optimization problems.

There are certain rules for the cuckoo search algorithm that need to be followed:

1. Each cuckoo can stow only one egg at once in a randomly opted nest.
2. Only the fittest eggs will uproot to the next level.
3. The count of host nests is fixed.

17.2.3 Cuckoo search algorithm

1. Initialize the count of the host nest x_i, $i = 1, 2....n$, and utmost iterations Max_I and fitness functions $f\_f(x)$, $x = \{x_1, x_2, ..., x_d\}^T$.

2. Repeat until stop criterion encountered,

    While $(t \le Max_I)$

        Acquire cuckoo by employing Lévy flights.

            Reckon the fitness fit of cuckoos.

        Opt the best nest, say, k, from the host n in a quasirandom manner.

        If $(Fit_i \ge Fit_k)$

            Substitute k with new solution.

    End

        Renounce the inadequate nests.

        Carry the best nest to the next level.

        Order the nests according to their caliber.

    End

3. Procure the best outcome.

17.2.4 **Variants**

The first cuckoo search was tried utilizing numerical capacity advancement bench-marks. As a rule, this sort of issues speaks to a proving ground for new created calculations. See Fig. 17.1.

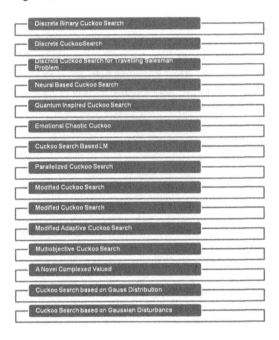

Discrete Binary Cuckoo Search

Discrete CuckooSearch

Discrete Cuckoo Search for Travelling Salesman Problem

Neural Based Cuckoo Search

Quantum Inspired Cuckoo Search

Emotional Chaotic Cuckoo

Cuckoo Search Based LM

Parallelized Cuckoo Search

Modified Cuckoo Search

Modified Cuckoo Search

Modified Adaptive Cuckoo Search

Multiobjective Cuckoo Search

A Novel Complexed Valued

Cuckoo Search based on Gauss Distribution

Cuckoo Search based on Gaussian Disturbance

FIGURE 17.1

Variants of cuckoo search.

17.3 **Histogram-based thresholding**

Karl Pearson is the pioneer of the histogram (Kapur et al., 1985; Pizer et al., 1987; Borenstein and Koren, 1991; Kim, 1997). A histogram is the distribution of frequency of gray level of pixel position of an image. The gorge point acts as threshold to segment an image. Histogram can be built by distributing the range of pixels into uniform class. Then the frequency for each class is enumerated. The x plane and y plane will represent the evenly distributed classes and frequency of pixel position, respectively.

Assume the histogram $h(m, n)$ relative to an image incorporates a dark object in a light background. Let us opt the threshold to be T, in such a way to dissociate the backdrop objects from forefront objects. Any pixel position $h(m, n)$ for $h(m, n) \geq T$ is extracted as forefront objects and for $h(m, n) \leq T$ is extracted as backdrop objects. By choosing a proper threshold intensity value on the x axis, all pixels with

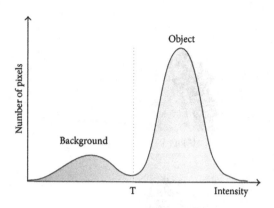

FIGURE 17.2

Histogram representation with single threshold given by Despotović et al. (2015).

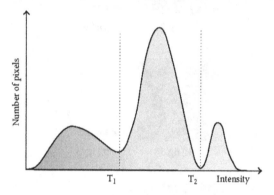

FIGURE 17.3

Histogram representation with multiple thresholds given by Despotović et al. (2015).

intensity below the threshold can be considered as forefront area, and those above are considered as backdrop. This procedure is known as thresholding. There are numerous approaches to threshold an image subject to the errand close by. Histogram-based thresholding can be performed in two ways: a single threshold or multiple thresholds. Fig. 17.2 and Fig. 17.3 show the histogram representations with single and multiple thresholds. Fig. 17.4 demonstrates the histogram for some images of the dataset.

17.4 Proposed intelligent segmentation technique

In this chapter, we focus on a cuckoo search-based intelligent image segmentation technique. The goal of the technique is to dissociate the backdrop objects from the forefront objects using the population-based metaheuristic cuckoo search algorithm. The steps involved in the aforementioned intelligent technique are shown in Fig. 17.5.

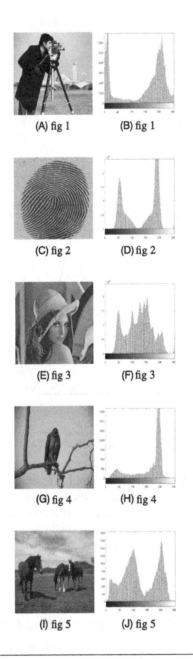

(A) fig 1   (B) fig 1

(C) fig 2   (D) fig 2

(E) fig 3   (F) fig 3

(G) fig 4   (H) fig 4

(I) fig 5   (J) fig 5

FIGURE 17.4

Original images and their respective histograms.

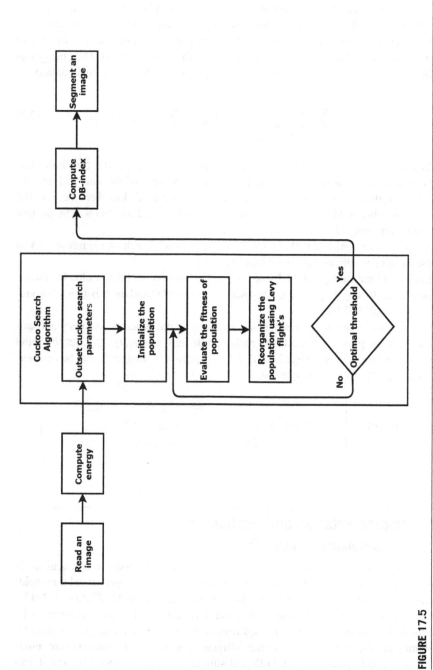

FIGURE 17.5

Proposed CSBIIST.

1. Read an image I of extent $R \times C$, where $I = G_{mn}$, $1 \leq m \leq R$, $1 \leq n \leq C$, and G_{mn} represents the intensity level, or gray level, of each pixel position of image I. Pixel position can be represented as (m, n) and its value can be in the range of 0 to N, where N represents the upper bound of the intensity that a pixel can have.

2. For each intensity level, enumerate the energy of the input image I by implementing energy function developed by Singla and Patra (2017a). The energy can be enumerated for each intensity level ($0 \leq j \leq N$) by the following equation:

$$E_j = \sum_{m-1}^{R} \sum_{n-1}^{C} \sum_{pq \in s_{mn}^2} b_{mn} \times b_{pq} + \sum_{pq \in s_{mn}^2} C_{mn} \times C_{pq}. \qquad (17.2)$$

3. The reckoned energy incorporates vertex and gorge. A pretender threshold may present in between two seriate vertices. The pivot point of the line unifying the seriate vertices is intended as initial pretender threshold. Let there be k initial pretender thresholds $t_1, t_2.....t_k$. Then $k + 1$ experimental regions will be present in an input image I.

4. Supply the pretender threshold set T_k to the cuckoo search algorithm to reckon the most favorable threshold for input image I.

5. Enumerate the entropy and DB-index t, and reckon optimal thresholds for the input image I using Kapur's entropy (Kapur et al., 1985) and the DB-index (Davies and Bouldin, 1979).

6. Segment an input image to equate the acquired output based on accuracy and caliber. Let the pretender thresholds reckoned in step 3 be $t_1, t_2....t_k$. Let the input image and segmented image be I and S_I, respectively. Image S_I can be enumerated using the following equation:

$$S_I(m, n) = \begin{cases} 0, \ 0 \leq I(m, n) \leq t_1, \\ (j - 1) \times (\frac{256}{k} - 1), \ t_{j-1} \leq I(m, n) \leq t_j \forall 2 \leq j \leq k - 1, \\ 255, \ t_k \leq I(m, n) \leq 255. \end{cases}$$
$$(17.3)$$

17.5 Implementation and evaluation

17.5.1 Experimental explanation

The proposed techniques such as cuckoo search-based intelligent image segmentation, as well as the state-of-the-art techniques such as cuckoo search-based thresholding and histogram-based thresholding, have been deployed in MATLAB® R2017a. For the proposed technique, there are certain parameters that are required to be fine-tuned to acquire the recommended results. We performed analytical testing by implementing the techniques to set the following parameters: population size, number of nests, and discovery rate. Different simulations were carried out, and it was

Table 17.1 Parameters required for the proposed CSBIIST.

Method	Parameter	Value
CSBIIST	Population	100
	Number of nests	25
	Discovery rate	0.25

(A) fig 1   (B) fig 2   (C) fig 3

(D) fig 4   (E) fig 5

FIGURE 17.6

Original gray images dataset.

observed that results obtained are efficient and have negligible variations. The parameters selected for simulation are described in Table 17.1.

The standard dataset of gray images as demonstrated in Fig. 17.6 has been gathered from http://imageprocessingplace.com/root_files_V3/image_databases.htm. To assess the caliber of the proposed technique, we have contemplated the images that incorporate multiple thresholds for experimental implementation.

17.6 Results and discussion

To endorse the accuracy and consistency of acquired results, entropy and DB-index have been computed for automatically encountered threshold. It can be discerned that with increasing threshold count, entropy proportionally increases. The increase in threshold count strongly reduces the quality of an image. That is why the proposed technique computes the DB-index, which makes the technique intelligent by computing the automatic thresholds.

Table 17.2 Optimal thresholds, entropy acquired using CS-BIIST for different images.

Image	CSBIIST		
	Optimal threshold	*DB-index*	*Entropy*
Cameraman	90	0.0417	0.0817
Fingerprint	120	0.0290	0.0792
Lena	80,115,179	0.1492	0.0804
Crow	118	0.0876	0.0753
Horses	147	0.1085	0.0754

Table 17.3 Optimal thresholds, entropy acquired using CSBT for different images.

Image	Number of thresholds	CSBIIST		
		Thresholds	*DB-index*	*Entropy*
Cameraman	1	90	0.0417	0.0817
	2	94,190	0.2084	0.0815
	3	83,142,186	0.1656	0.0820
	4	45,101,136,189	0.1692	0.0842
	5	27,54,107,144,199	0.2000	0.0859
Fingerprint	1	125	0.0290	0.0886
	2	101,156	0.1611	0.0889
	3	87,129,172	0.1247	0.0898
	4	64,109,137,170	0.1668	0.0939
	5	81,109,153,170,202	0.2670	0.0904
Lena	1	146	0.2656	0.0779
	2	91,155	0.2054	0.0788
	3	91,140,180	0.1676	0.0788
	4	80,111,148,185	0.1612	0.0800
	5	61,108,144,172,232	0.1721	0.0838
Crow	1	128	0.0800	0.0819
	2	82,156	0.1298	0.0845
	3	92,145,188	0.2091	0.0836
	4	72,102,134,186	0.1956	0.0855
	5	43,83,116,175,229	0.1771	0.0914
Horses	1	139	0.1095	0.0758
	2	62,133	0.1271	0.0771
	3	60,135,201	0.2408	0.0773
	4	70,114,155,221	0.2500	0.0764
	5	19,41,89,145,232	0.2281	0.0904

Table 17.2 depicts the optimal thresholds, the DB-index, and their respective entropy for each image reckoned using the proposed technique. Both the DB-index

Table 17.4 Optimal thresholds, entropy acquired using HBISTUCS for different images.

Image	Number of thresholds	HBISTUCS		
		Thresholds	*DB-index*	*Entropy*
Cameraman	1	133	0.0555	0.0804
	2	121,193	0.1922	0.0808
	3	45,108,194	0.1995	0.0842
	4	56,97,145,196	0.1754	0.0834
	5	39,99, 137,189,210	0.2101	0.0847
Fingerprint	1	159	0.0713	0.0884
	2	78,162	0.2808	0.0908
	3	68,107,141	0.1495	0.0927
	4	77,118,128,179	0.1381	0.0910
	5	60,84,101,164,174	0.3819	0.0958
Lena	1	126	0.2242	0.0775
	2	99,165	0.1811	0.0782
	3	81,131,174	0.1630	0.0799
	4	84,151,124,199	0.1896	0.0795
	5	68,108,155,182,198	0.1899	0.0821
Crow	1	130	0.0831	0.0819
	2	53,138	0.1848	0.0887
	3	53,117,176	0.2049	0.0887
	4	64,109,169,230	0.1371	0.0866
	5	35, 63, 85, 146,189	0.2448	0.0942
Horses	1	116	0.1492	0.0754
	2	65,148	0.1357	0.0768
	3	78,145,185	0.2043	0.0760
	4	36,75,123,165	0.1623	0.0817
	5	31,82,110, 135,178	0.1932	0.0833

and entropy have been acquired by deploying cuckoo search-based intelligent image segmentation, cuckoo search-based thresholding, and histogram-based image segmentation using cuckoo search.

It can be discerned from Table 17.2 that entropy reckoned by deploying CSBIIST is lower compared with entropy reckoned using CSBT and HBISTUCS, as shown in Table 17.3 and Table 17.4, respectively.

To estimate the potency of the proposed technique, acquired results were quantitatively inspected. The caliber and quality of the proposed technique are evaluated in Fig. 17.7 and Fig. 17.8, which depict the segmented images acquired using CSBIIST and HBISTUCS, respectively.

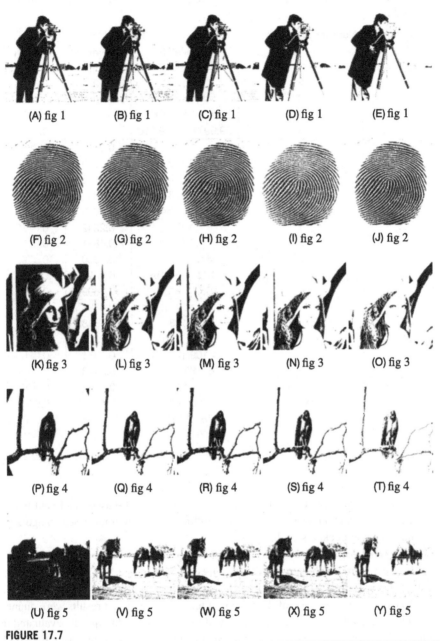

(A) fig 1   (B) fig 1   (C) fig 1   (D) fig 1   (E) fig 1

(F) fig 2   (G) fig 2   (H) fig 2   (I) fig 2   (J) fig 2

(K) fig 3   (L) fig 3   (M) fig 3   (N) fig 3   (O) fig 3

(P) fig 4   (Q) fig 4   (R) fig 4   (S) fig 4   (T) fig 4

(U) fig 5   (V) fig 5   (W) fig 5   (X) fig 5   (Y) fig 5

FIGURE 17.7

Segmented images using CSBIIST.

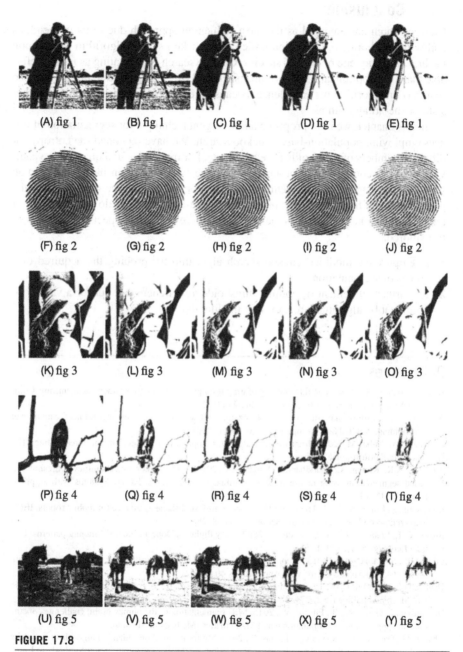

(A) fig 1 (B) fig 1 (C) fig 1 (D) fig 1 (E) fig 1

(F) fig 2 (G) fig 2 (H) fig 2 (I) fig 2 (J) fig 2

(K) fig 3 (L) fig 3 (M) fig 3 (N) fig 3 (O) fig 3

(P) fig 4 (Q) fig 4 (R) fig 4 (S) fig 4 (T) fig 4

(U) fig 5 (V) fig 5 (W) fig 5 (X) fig 5 (Y) fig 5

FIGURE 17.8

Segmented images using HBISTUCS.

17.7 Conclusion

Cuckoo search has emerged as the most influential approach due to its tremendous ability to ameliorate optimization issues. The cuckoo search algorithm relies upon the breeding etiquette of cuckoo species. Cuckoo search is emanating as more preferable than other optimization algorithms, such as ABC, PSO, GA, and others. Cuckoo search is an algorithm par excellence, because it entails fewer parameters than other state-of-the-art algorithms.

In this chapter, we have proposed an intelligent technique for segmentation of images employing population-based cuckoo search. We have reckoned the entropy for different numbers of thresholds for each image. For qualitative and quantitative analysis, we have compared the proposed technique with histogram-based and cuckoo search-based state-of-the-art techniques. The proposed technique contemplates the pixel position of the image while other state-of-the-art techniques do not bother about the pixel position. Nevertheless, besides these there are some aspects that are yet to be explored:

1. to employ the modified cuckoo search algorithm for probing the acquired outcomes of segmentation,
2. for further refinement of optimal thresholds, researchers can opt for a variety of optimization algorithms that exist in the literature.

References

Abutaleb, A.S., 1989. Automatic thresholding of gray-level pictures using two-dimensional entropy. Computer Vision, Graphics, and Image Processing 47 (1), 22–32.

Akay, B., 2013. A study on particle swarm optimization and artificial bee colony algorithms for multilevel thresholding. Applied Soft Computing 13 (6), 3066–3091.

Alihodzic, A., Tuba, M., 2014. Improved bat algorithm applied to multilevel image thresholding. The Scientific World Journal 2014.

Bhandari, A.K., Kumar, A., Chaudhary, S., Singh, G.K., 2016. A novel color image multilevel thresholding based segmentation using nature inspired optimization algorithms. Expert Systems with Applications 63, 112–133.

Borenstein, J., Koren, Y., 1991. The vector field histogram-fast obstacle avoidance for mobile robots. IEEE Transactions on Robotics and Automation 7 (3), 278–288.

Brown, C.T., Liebovitch, L.S., Glendon, R., 2007. Lévy flights in Dobe Ju/'hoansi foraging patterns. Human Ecology 35 (1), 129–138.

Cuevas, E., Sossa, H., et al., 2013. A comparison of nature inspired algorithms for multi-threshold image segmentation. Expert Systems with Applications 40 (4), 1213–1219.

Davies, D.L., Bouldin, D.W., 1979. A cluster separation measure. IEEE Transactions on Pattern Analysis and Machine Intelligence 2, 224–227.

Despotović, I., Goossens, B., Philips, W., 2015. MRI segmentation of the human brain: challenges, methods, and applications. Computational and Mathematical Methods in Medicine.

Dhal, K.G., Das, A., Ray, S., Gálvez, J., Das, S., 2019. Nature-inspired optimization algorithms and their application in multi-thresholding image segmentation. Archives of Computational Methods in Engineering, 1–34.

Glasbey, C.A., 1993. An analysis of histogram-based thresholding algorithms. CVGIP: Graphical Models and Image Processing 55 (6), 532–537.

Hammouche, K., Diaf, M., Siarry, P., 2008. A multilevel automatic thresholding method based on a genetic algorithm for a fast image segmentation. Computer Vision and Image Understanding 109 (2), 163–175.

Hertz, L., Schafer, R.W., 1988. Multilevel thresholding using edge matching. Computer Vision, Graphics, and Image Processing 44 (3), 279–295.

Horng, M.H., 2011. Multilevel thresholding selection based on the artificial bee colony algorithm for image segmentation. Expert Systems with Applications 38 (11), 13785–13791.

Kapur, J.N., Sahoo, P.K., Wong, A.K., 1985. A new method for gray-level picture thresholding using the entropy of the histogram. Computer Vision, Graphics, and Image Processing 29 (3), 273–285.

Kim, Y.T., 1997. Contrast enhancement using brightness preserving bi-histogram equalization. IEEE Transactions on Consumer Electronics 43 (1), 1–8.

Kohler, R., 1981. A segmentation system based on thresholding. Computer Graphics and Image Processing 15 (4), 319–338.

Maitra, M., Chatterjee, A., 2008. A hybrid cooperative–comprehensive learning based PSO algorithm for image segmentation using multilevel thresholding. Expert Systems with Applications 34 (2), 1341–1350.

Oliva, D., Elaziz, M.A., Hinojosa, S., 2019. Multilevel thresholding for image segmentation based on metaheuristic algorithms. In: Metaheuristic Algorithms for Image Segmentation: Theory and Applications. Springer, pp. 59–69.

Otsu, N., 1979. A threshold selection method from gray-level histograms. IEEE Transactions on Systems, Man and Cybernetics 9 (1), 62–66.

Pal, N.R., Pal, S.K., 1993. A review on image segmentation techniques. Pattern Recognition 26 (9), 1277–1294.

Pavlyukevich, I., 2007. Lévy flights, non-local search and simulated annealing. Journal of Computational Physics 226 (2), 1830–1844.

Payne, R.B., Sorensen, M.D., 2005. The Cuckoos, vol. 15. Oxford University Press.

Pizer, S.M., Amburn, E.P., Austin, J.D., Cromartie, R., Geselowitz, A., Greer, T., ter Haar Romeny, B., Zimmerman, J.B., Zuiderveld, K., 1987. Adaptive histogram equalization and its variations. Computer Vision, Graphics, and Image Processing 39 (3), 355–368.

Raja, N., Rajinikanth, V., Latha, K., 2014. Otsu based optimal multilevel image thresholding using firefly algorithm. Modelling and Simulation in Engineering 2014, 37.

Reynolds, A.M., Frye, M.A., 2007. Free-flight odor tracking in drosophila is consistent with an optimal intermittent scale-free search. PLoS ONE 2 (4), e354.

Sahoo, P., Soltani, S., Wong, A., Chen, Y., 1988. A survey of thresholding techniques. Computer Vision, Graphics, and Image Processing.

Sathya, P., Kayalvizhi, R., 2010. PSO-based Tsallis thresholding selection procedure for image segmentation. International Journal of Computer Applications 5 (4), 39–46.

Sezgin, M., Sankur, B., 2004. Survey over image thresholding techniques and quantitative performance evaluation. Journal of Electronic Imaging 13 (1), 146–166.

Singla, A., Patra, S., 2017a. A fast automatic optimal threshold selection technique for image segmentation. Signal, Image and Video Processing 11 (2), 243–250.

Singla, A., Patra, S., 2017b. PSO based context sensitive thresholding technique for automatic image segmentation. In: Proceedings of Sixth International Conference on Soft Computing for Problem Solving, pp. 151–162.

Singla, A., et al., 2019. Context-sensitive thresholding technique using ABC for aerial images. In: Soft Computing and Signal Processing. Springer, pp. 85–93.

Tao, W.B., Tian, J.W., Liu, J., 2003. Image segmentation by three-level thresholding based on maximum fuzzy entropy and genetic algorithm. Pattern Recognition Letters 24 (16), 3069–3078.

Tobias, O.J., Seara, R., 2002. Image segmentation by histogram thresholding using fuzzy sets. IEEE Transactions on Image Processing 11 (12), 1457–1465.

Tuba, M., 2014. Multilevel image thresholding by nature-inspired algorithms: a short review. Computer Science Journal of Moldova 22 (3).

Viswanathan, G.M., Buldyrev, S.V., Havlin, S., Da Luz, M., Raposo, E., Stanley, H.E., 1999. Optimizing the success of random searches. Nature 401 (6756), 911.

Yang, X.S., Deb, S., 2009. Cuckoo search via Lévy flights. In: 2009 World Congress on Nature & Biologically Inspired Computing (NaBIC), pp. 210–214.

Ye, Z.W., Wang, M.W., Liu, W., Chen, S.B., 2015. Fuzzy entropy based optimal thresholding using bat algorithm. Applied Soft Computing 31, 381–395.

Improving genetic algorithm solution performance for optimal order allocation in an e-market with the Pareto-optimal set

18

Jacob Hunte[a], Mechelle Gittens[b], Curtis Gittens[b], Hanan Lutfiyya[a], Thomas Edward[b]

[a]*Western University, Computer Science Department, Middlesex College, London, Ontario, Canada*
[b]*The University of the West Indies Cave Hill Campus, Department of Computer Science, Mathematics and Physics, Bridgetown, Barbados*

CONTENTS

Nature-Inspired Computation and Swarm Intelligence. https://doi.org/10.1016/B978-0-12-819714-1.00029-4

18.1 Introduction

This chapter presents a genetic algorithm (GA) and application designed to help local farmers in small island states sell their goods by assisting them with produce distribution. There are many real-world applications of GAs but there is still a need for more documented applications in real-world scenarios. Real-world applications include work by Altiparmak et al. (2006) which is of particular interest to our work because it introduces GA as an approach to handling supply chain networks (SCNs). Other real-world applications of interest include the work by Costa et al. (2013), which recognizes the need to combine and customize methods to suit the domain. The authors created a hybrid approach using simulated annealing and GA. This is done to minimize the variance of overall energy loss from illegal use of energy generated by energy distribution companies. We see another hybrid approach by Rika et al. (2019), where GA and deep learning are combined to automatically reconstruct Portuguese tile panels. This is a nontrivial variation of the jigsaw puzzle problem that significantly impacts Portuguese heritage. Our work is also intended to impact a social problem for small island developing states and specifically the Caribbean.

This chapter looks at the use of GA within an SCN; specifically, it uses a Pareto Optimal Set (POSet) approach to solving a multiobjective SCN problem. It also investigates the impact of their solution procedure on performance and compares the quality of the Pareto-optimal solutions generated by this procedure with those generated by simulated annealing. Such case studies would allow certain aspects of GA to be addressed in their implementation in existing systems. For example, GA is known to require a large amount of resources and the requirement for resources increases with the size of the problem being solved. In a real-world application we understand that there are limits to the resources available because of questions such as:

- How would one implement a GA while considering absolute limits on available resources?
- What is the impact of such limits on the GA's performance?
- When is a GA feasible?

With the numerous applications of GA for optimization we would like to add to this knowledge in the area of practical applications of GA. Our work therefore examines whether a GA, when integrated into an existing farmers-to-consumer e-market system, can facilitate optimal distribution of fresh produce in the farmer-to-consumer segment of the agricultural supply chain.

We focus on the cost of the GA from a resource point of view. We also look at the impact that the restrictions tied to our specific application have on the GA's performance. We then determine if the GA can perform in our scenario.

The chapter proceeds as follows. Section 18.2 provides a general overview of the problem and the literature associated with each area of the system and its development. Section 18.3 offers a detailed explanation of the distribution manager (DM) and GA. Section 18.4 explains how the decision module and the e-market should interact to provide the final product. We look at the initial results generated by the system in Section 18.5 and draw final conclusions in Section 18.6.

18.2 **The problem overview**

Mentzer et al. (2001) broadly defines a supply chain as a combination of a firm's raw material producers, product assemblers, and transportation companies that are involved with creating produce and delivering it to the final consumer. Supply chain management (SCM) is the balancing of customer service, inventory management, and costs (Stevens, 1989). This definition accurately captures the notion that a producer must address the trade-off between cost reduction and customer satisfaction when performing optimizations within a supply chain.

De Keizer (de Keizer et al., 2012) notes that fresh produce SCNs have more complex relationships that must also manage trade-offs between transportation costs and inventory costs due to perishability. Ahumada and Villalobos (2009) support this in their paper on planning models for agrifood supply chains. This trade-off is also present in the distribution optimization that we are performing and is interesting for our GA and system design.

We present a produce distribution management system that helps local farmers sell their goods to individual customers. The DM module of the system attempts to optimize the distribution process with the goal of maximizing customer satisfaction. Hoffmann et al. (2019) demonstrates a domain-specific language called Athos facilitating the declarative specification of problems such as the vehicle routing problems concerned with time windows. Although this is an application of GA to vehicle routing, we have not seen the balancing of this routing based on the individuals' demand for the product being distributed.

SCM allows us to optimize business resources for distribution. Our focus is on the development of a distribution system for a local e-market. This is an SCN for fresh produce where complex relationships between actors (parameters) can be managed through optimization.

The goal is to minimize delivery costs for both customers and farmers. Therefore, at some point in our optimization we would reach a point where a reduction in cost made to one of the two entities causes the other entity's cost to increase, for example, attempting to reduce both the number of distribution centers and the distance traveled by the customers. The fewer distribution centers a farmer must deliver to, the smaller the distance the farmer must travel. However, as we reduce the total number of distribution centers, we increase the number of customers serviced by a distribution center and the distance that each customer must travel.

To develop the DM, we must optimize with respect to a number of objectives that may require using calculus-based, enumerative or guided random techniques (Bandyopadhyay and Saha, 2013). Maharana et al. (2017) focused on a number of objectives when addressing the efficient generation and acquisition of biofuel to supply many users considering several factors such as the use of various technologies and the procurement of biomass. They also had to factor in the optimal location of physical plants to produce the fuel and the costs for fuel transportation. This is similar to our consideration of a supply network for many individual farmers to many individual customers. They modeled the combinatorial problem to allow the incorporation of computational intelligence-based optimization approaches. They recognized that the delivery module's objectives require a technique that can balance many objectives. We have the same considerations and therefore pay attention to guided random techniques and specifically GA.

We focus on GA because, as discussed in Bandyopadhyay and Saha (2013), they outperform other techniques when it comes to optimizing multiple conflicting objectives. We must note, however, that though the work in Maharana et al. (2017) has similar considerations and seeks similar outcomes, the application is different and the combination of parameters are not as dynamic as those that we must optimize for. That is, it is neither clear how many customers we will have to work with, nor how many farmers, nor how much produce or orders. Hence our solution must be somewhat more generalized. Maharana et al. (2017) present a proposed strategy that works for the case of three feedstock sources, two central processing facilities, four industrial facilities, and eight users. We are able to show combinations that cover quantities and scenarios that may change. It is important to do this in a real-world situation and to understand the complexity of accommodating for dynamic quantities, because it is important to understand whether cases will scale. The work by Azad et al. (2019) is more generalized, since it offers a two-phase GA. In the first phase, the inventory and production decisions are solved. The vehicle routing and transportation concerns are addressed in phase two. The researchers conducted sensitivity analysis to examine the efficiency for large-scale inventory and routing situations. Though the researchers are successful with the two-phase approach, the dynamic nature of our SCN problem, with individual customers and individual farmers, is not suited to this industrial proposal for managing inventory and production.

18.3 **Genetic algorithms**

Fonseca and Fleming (1993) describe the application of GA to multiobjective problems. They specifically look at the scenario where combining the objectives to generate a single criterion which is then optimized by using a utility function is difficult or infeasible. Their work (Fonseca and Fleming, 1993) characterizes the POSet as an alternate approach to handling multiple objectives in optimization. Our delivery module must be optimized to meet several objectives. This chapter highlights the concept of the POSet as an alternative to combining objectives. This alternative solution generates a set of optimal solutions instead of a single solution. The objectives have complex relationships and conflicts that cannot combine into a single objective. The Pareto-optimal set would allow the GA to appropriately represent the objectives and generate multiple solutions.

18.3.1 **The Pareto-optimal set**

It is this conflict that has caused us to look towards Pareto-optimality. The notion of Pareto-optimality is built around the trade-off of one objective over another in terms of performance. The POSet will contain optimal solutions in the sense that no improvement in performance towards any objective can be made without degrading the performance of another (Fonseca and Fleming, 1993). A solution X is said to dominate solution Y if X outperforms Y in every objective. A solution is inserted into the POSet only if it is not dominated by any other solution, including those solutions already in the POSet. In each iteration of the GA, the nondominated solutions are identified in the current population and added to the POSet.

GAs also have potential when it comes to multiobjective optimization because of their ability to maintain a set of possible solutions during execution which allows them to find multiple noninferior solutions in parallel. This is possible because GAs generate a population of potential solutions in each iteration. Each solution in the population can be evaluated against the objectives allowing multiple noninferior solutions to be found in a single iteration (Fonseca and Fleming, 1993).

18.3.2 **The final solution set**

Our system provides a set of solutions to the farmer from which they choose the final solution. There are two reasons for this approach. The first is due to the number and complexity of variables surrounding agrifood supply chains, the second is due to the need for the system to perform an advisory role.

We understand that in our environment there are factors that are unquantifiable that would determine which solution would be best. These factors stem from the large differences between farmers in how each of them runs their business. The environment is highly dynamic and due to the ad hoc way the distribution decisions are made by farmers, it may be infeasible to attempt to incorporate them into the DM module. Each farmer has a unique motivation behind their distribution choices and these rapidly change as they conduct business. As such we do not attempt to account for each farmer's unique business model and attempt to generate a single solution, but rather we generate a set of solutions that would cover a range of distribution models.

This is critical to developing a system that farmers would want to use. Removing control over the distribution of their produce would require a large change in mindset and business practices. As such, we do not require the algorithm to find a final solution from our solution set. However, due to our POSet being unrestrained in size it is possible to have many solutions in the final POSet. As a result, we propose to filter the final solution set to ensure a more manageable set of solutions is provided to the farmer for their final selection. Farmers would essentially be giving up control over a large portion of their business and may not wish to do so. Allowing them to retain control by selecting from a set of solutions gives them a sense of control.

18.4 The agricultural e-market system overview

The e-market is a previously developed online system developed within our research group that uses both mobile and web-based interfaces called FarmersConnect. The original goal of the mobile and web interfaces was to facilitate the purchase and sale of fresh produce. It also allowed customers to give and view feedback about the quality of each farmer's produce and customer satisfaction. The FarmersConnect project consists of two systems: FarmersConnect:Buy and FarmersConnect:Sell. The Sell system is the interface provided to farmers that have registered with the system. It allows them to list their produce, its quantity, and its price for sale. This was primarily done by landline phone using Voxeo Prophecy VoIP solution. We implemented a landline solution since the system interfaces with both technologically savvy and technologically inexperienced users.

The Buy system is a web interface for customers to purchase produce. Once a customer has registered online, they can buy and rate a farmer's produce and leave comments on produce they have purchased. This provides feedback for farmers and offers future customers a guide to the quality of produce and service that farmers provide.

The goal of adding a distribution algorithm to the FarmersConnect system is to ensure ease of access to ordered produce to reduce fulfillment barriers. LocalBuy, like FarmersConnect, outlines the incorporation of distribution management and an e-market for fresh produce (Li et al., 2009). It proposes a web-based virtual farmers market with the main goal of connecting consumers directly with producers, cutting out the middle man, increasing the nutritional value of the food they consume, and supporting the local economy. LocalBuy provides an online (virtual) market for the sale of local fresh produce to consumers and a physical processing location called a hubsite. The farmer receives orders and is notified of the hubsite. The farmer packages the orders and notifies the hubsite that they are ready. The farmer then takes the orders to the hubsite for customer pickup. The LocalBuy system also allows customers to give feedback about product quality and records customer satisfaction and selling history for each farmer.

Although LocalBuy and our proposed system share some functional characteristics, we must note that their operations are dependent on hubsites where all farmers would essentially be dependent on the existence of a single permanent shared facility for distribution. This fact generates two main issues in the target environment for our system. One issue is the need for investments to be made by the farmers, governments, or third parties to fund the construction and maintenance of the hubsites. In our environment, government investment in the region is drastically reducing as interest in the industry falls. As a result of reduction in government investment and the movement of interest away from the industry, obtaining funding for the construction of permanent hubsites may pose a problem.

The other issue is the fact that farmers using LocalBuy must change the way they distribute their produce. Currently farmers employ a number of methods in order to distribute their produce. Farmers go to marketplaces and attempt to generate sales, drive their produce to locations they believe they will find customers, and deliver to their customers in some cases. Using a system like LocalBuy farmers would no longer handle the delivery of produce to their customers. Instead, both the farmers and the customers who pick up their orders would have to now go to a single hubsite. Not only would the farmers have to adjust, but their consumers would have to do so as well.

In work by Lehmann et al. (2012) on the use of technology in agriculture, the authors noted that the reason why the application of technology in a particular environment may fail is not because technology is incapable of addressing the need. They state that it fails because the technology used cannot be integrated without forcing changes in the existing structure and processes of the environment in which it is implemented. Noting that this is the main cause of failure in applying technology in an environment, we focus on minimizing the changes farmers and customers must make in order to use the system.

The DM addresses the issues associated with needing hubsites and the forced change that this will cause. It allows farmers to independently handle their own deliveries without needing the construction and operation of shared hubsites. This stops a farmer from having to make such a large adjustment in order to use the system. It also does not require large investments for implementation.

18.4.1 System details

The goal of this research is to construct a system that allows users, both farmers and their customers, to interact with the GA residing in a DM. The goal of this interaction is to collect transaction information for orders and available resources and advise farmers of the most efficient ways to dynamically service these orders. There are three main components to the total system that contribute towards this goal. These components are the website, the distribution module (GA), and the database. For the explanation of these three components they will be treated as if they exist on their own independent servers. The operating system of all three servers used in this research is Ubuntu server 14.04.3 LTS.

Fig. 18.1 shows the three components and the general flow of data between each component.

- The web server houses and runs the primary interface for the farmers and consumers; this is the e-market.
- The application server houses the GA and the database holds all information used in the system.
- The database server hosts the database which stores all user data from the farmer and customer. In addition to these data, the database stores all optimization results generated by the GA.

The following sections outline the specifics for each server in depth. The outlines provide details of the software components used in the system and the utility that each component offers to the functionality of the entire system. The farmer's resource in-

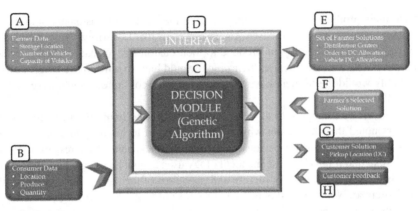

FIGURE 18.1

Components of the GA system.

formation for the algorithm is listed in (**A**) and is collected via the e-market interface (**D**). The order information supplied by customers in (**B**) contains all data relating to the orders made, like the produce requested and the quantity desired by the customer. As shown in (**A**) and (**B**), the interface would not only have to facilitate collection and storage of available produce information and the requests for produce, but it collects supplementary optimization information such as the following:

- the customer's and farmer's location,
- the available delivery vehicles,
- the capacity of each vehicle from the farmer.

The information collected in A and B by the e-market is passed to the distribution module and the GA (**C**). The GA then generates the distribution options. Each distribution option contains:

- the cluster grouping for all of the orders,
- the distribution locations for each cluster grouping,
- the delivery vehicle to be used to service each cluster.

The e-market presents the distribution options generated by the algorithm to the farmer in (**E**). The farmer then selects the desired solution that they wish to use to distribute. This selection is passed back to the e-market as shown in (**F**). The e-market logs this selection and notifies the customer of the location where their produce is available for pick up as shown in (**G**). Once they have received the produce they will update the system with this fact and provide any feedback on their customer experience (**H**). The feedback would be input using the preexisting rating system already made available in the e-market at the start of this project.

18.4.2 **The distribution manager**

To build the DM, we determined the technical requirements and analyzed those requirements to extract the objectives tied to the optimization that the DM must perform. Finally, we design the chromosome that accurately represents the problem in a form that can be manipulated by the genetic operators easily. Generally, the GA must be aware of the client's needs and the farmer's ability to address them. It must also be able to track the impact that the optimization of the resources has on the satisfaction of the consumer, while making practical and realistic decisions about how the farmer's resources are to be used.

The abstract states that the overall problem can be broken down into two distinct optimization tasks, the clustering of farmer's orders into groups and the allocation of vehicles to service these groups. Our GA performs both tasks simultaneously. This is because they both share the same performance metric of distance traveled. The goal of the clustering is to optimize the grouping of orders to minimize the distance traveled by both farmers and customers. The goal of optimizing the vehicle allocation is to reduce the distance traveled by the farmer. This links the two tasks and provides the motivation behind allowing the algorithm to combine both problems. As a result, we would solve these optimization problems together.

18.5 **The objectives and optimization functions**

The two main entities impacted by the optimization performed are the farmer and consumer. As discussed, both entities' costs would be evaluated by the distance each must travel. The distance a customer travels is simply the distance to their allocated distribution center. Each distribution center is generated based on the locations of the customers assigned to it. The distribution center is placed in the center of the customer locations assigned to it. The distance function (the sum of all orders to their locations) would then represent the cost, or performance of a solution towards minimizing customer costs. It would be the sum of all distances traveled by each customer and is shown in Eq. (18.2). We have

$$D_{c,d} = distance(Customer\,Location, Assigned\,DC), \tag{18.1}$$

where $D_{c,d}$ is the distance calculated between the customer and their assigned distribution center, and

$$C_{c,d} = \Sigma_{i=1}^{n} distance(Customer\,Location_i, Assigned\,DC_i), \tag{18.2}$$

where $C_{c,d}$ is the customer distance cost.

The farmer must deliver to each distribution center. However, if the amount of produce needed at a given distribution center exceeds the capacity of the vehicle, multiple trips will be needed. The distance cost is calculated as follows:

$$N_i = weight\ of\ product\ for\ DC_i/max(VehicleCapacity),$$
$$C_f = \Sigma_{i=1}^{n} distance(Farmer\ Location, DC_i) \times N_i, \tag{18.3}$$

where N_i is the number of trips and C_f is the distance cost of the farmer.

In order to solve these three equations, the algorithm requires the following information:

- each order's weight,
- each order's location,
- the farmer's location,
- the number of vehicles available for delivery,
- the capacity of the available vehicles for delivery.

This information is passed to the GA in a JSON variable using the GA's constructor. The order's weight is generated by multiplying the weight-to-size of the produce by its size value. Each item ordered will be matched to a weight-to-size value per kilogram as stored in our database. For example, say the weight-to-size value of a potato is three cubic inches (size) per kilogram (weight). This means that every kilogram of potatoes will take three cubic inches of space in the delivery vehicle. So say an order is for 10 kilograms of potatoes, using Eq. (18.3) and the size value of potatoes (three cubic inches), the order's size would be 30 cubic inches. Taking lettuce as another example, one would have a larger weight to size value, of 16 inches per kilogram, as lettuce would be lighter and require more space per kilogram. These weight-to-size values would allow all orders to be converted from a list of items and weights into the space that they would take in a transport vehicle.

It is also important to note that the distribution centers may be placed in locations that are invalid. Recall that the locations for our distribution centers are generated in the center of the customers assigned to that distribution center. This means that the algorithm does not consider whether the locations are feasible. It would therefore be possible for the algorithm to generate a location for a distribution center that is located for example in a body of water like a lake or a river.

To address this issue, we will use a list of locations that are acceptable and simply map the distribution center locations generated by the algorithm to the nearest acceptable location. The approach of using these locations in the optimization phase was also considered. The implications of using all of these locations in the algorithm supported our decision to exclude these locations from the algorithm. As the algorithm stands now, we limit the number of locations that must be evaluated to a maximum of the number of customers. Including all of the acceptable locations (which may be in the thousands) in the optimization process itself would have an extremely negative impact on the GA's performance.

At present, our combinatorial optimization solution (the GA) has to consider each of the different ways that each customer can be grouped and there is one distribution center generated for each group. Using all the acceptable locations the algorithm now has to test each grouping with each acceptable location to determine which location would be best. Given the large number of acceptable locations that there might be, the computational cost for calculating the cost of one grouping would be drastically increased. With each grouping being evaluated for every acceptable location, our total evaluation cost would be some multiple of the number of acceptable locations. This additional cost would be too expensive. Therefore, not all of the acceptable locations for meeting customers are used during the optimization process.

As mentioned, the number of locations considered by the GA is limited to the number of customers in that problem. Since the worst case is that the farmer has to deliver to each customer. This limit is enforced naturally due to the constraints defined for this optimization problem. These constraints stipulate that a customer can only be assigned to one distribution center which generates a limit equivalent to the number of customers in the given problem. If a customer can only be assigned to one distribution center or group, then the most groups that can exist would be equivalent to the number of customers to be grouped. The limit is the number of customers in the problem because each customer can be assigned to their own group. Using the limit and then mapping to the nearest meeting point ensures that the computational cost is kept as small as possible while still considering all available meeting points.

18.6 Chromosome structure

The first element needed for the GA is the set of chromosomes themselves. These are the structures that will be manipulated by the genetic operators (crossover and mutation) and are the basis of the GA. The chromosome representation has two segments of genes, the vehicle-cluster allocation and the cluster-order allocation. This coincides with the main tasks of grouping orders into clusters and assigning vehicles to service each cluster.

18.6.1 Vehicle-cluster allocation

The vehicle-cluster segment of the chromosome represents the allocation of a delivery vehicle to a cluster or distribution center. Its length is equal to the number of orders in the current optimization problem. This is because it is possible for each order location to require a delivery to it individually in the case that each order's size maximizes the capacity of the delivery vehicles. In this case, each order location would essentially become a distribution center.

The value of an element in the chromosome represents the vehicle number. The location of that element, i.e., slot $1-N$ (N is the length of the segment), determines which cluster a vehicle will service. The maximum value is the number of available vehicles and the minimum is 1.

- The value at any position in the segment must be less that the number of vehicles.
- The value at any position in the segment must be greater than 1.

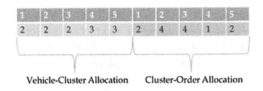

Vehicle-Cluster Allocation   Cluster-Order Allocation

FIGURE 18.2

A chromosome representation in an example of the cluster – order allocation.

18.6.2 Cluster order allocation

The cluster-order segment represents the grouping of orders into clusters. The value represents the cluster that an order is assigned to. The location of the element is the order that is assigned. This segment's length is equivalent to the number of orders in the current problem. The two constraints for this segment of the chromosome are very similar to the first segment and are listed below.

- The value at any position in a segment must be lower than the number of customers.
- The value at any position in a segment must be greater than 1.

Fig. 18.2 shows a chromosome representation of an instance of the problem. In this case a farmer receives five orders and has three vehicles available to deliver them.

The number of locations for each segment is equivalent to the number of orders in the problem instance. The values in segment one will range from 1–3 (the number of vehicles available). Segment two's values will range from 1–5 (the number of orders received) to allow the clustering and separation of orders into groups. In the given example the first segment shows us that clusters 1, 2, and 3 are serviced by vehicle 2 and clusters 4 and 5 are serviced by vehicle 3. The second segment shows that orders 1 and 5 are grouped into cluster 2, orders 2 and 3 are grouped into cluster 4, and order 4 is the only member of cluster 1.

18.6.3 The algorithm

The algorithm requires as input:

- each order's size S_T (generated by the produce types' T weight-to-size value) and their locations (see Eq. (18.4)),
- the farmer's location,
- the number and capacity of the available vehicles for delivery,

$$S_T = Weight * SizeValue_T. \tag{18.4}$$

Each item ordered will be matched to a size value per kilogram. This would allow all orders to be converted from a list of items and weights into the size that they would

```
Genetic algorithm (void)
{
   Declare and initialize population set P;
   Declare empty set POE for elite performers;
   Declare empty set M and X;
   While (termination criterion not met)
   {
      Set M = P + POE;
      Evaluate performance of solutions in M;
      Update POE with new Pareto-optimal solutions;
      Set X = Selected solutions from M for new generation
      Set P = Crossover and mutated X;
   }
   Final Solution Set = Filter (POE);
}
```

FIGURE 18.3

DM GA.

take in a transport vehicle. This size is calculated using Eq. (18.4). The algorithm in Fig. 18.3 shows the general structure of the GA used in the DM.

Since each segment has different values to represent the two problems, we must modify each segment separately even though we will evaluate the chromosomes' performance as one solution. Using the algorithm shown above we would randomly initialize the population set P using two distinct sets of parameters. This would allow us to generate the required values for each segment and produce a valid chromosome.

The Pareto-optimal elite (POE) set would hold all the best performers. This set will be passed back to the farmer after the algorithm has finished. The termination criterion is a maximum iteration value. The fitness evaluation function generates the performance for a chromosome on each objective. The objective performances of each chromosome including the current members of the POE set are then passed to the PO set generator. The POE set is then updated with the chromosomes selected by the PO set generator. This set is unbounded in order to maximize the range of solutions sampled along the Pareto-optimal front.

The rank of each solution is given based on the Euclidean distance the solution has (on an N-dimensional graph) from the origin (N being the number of objectives that the algorithm is optimizing). We first calculate the distance of each solution and assign the ranks by setting the chromosome with the smallest distance to the origin to rank one and continuing.

These ranks are then used to determine the probability of selection for crossover using Eq. (18.5). The chromosomes selected for crossover would be from the POE set and the rejected solutions from the current iteration. This combination is used to avoid premature convergence and adequate exploration of the search space. We have

$$P(selection) = 1/\sqrt{rank_i}. \qquad (18.5)$$

Crossover and mutation are performed separately on each of the two segments. This is to ensure that valid chromosomes are generated as done during population initialization.

Once the algorithm has completed execution the POE set is passed to the filter to reduce the number of solutions that would be passed back to the farmer. The filter simply selects the top 10 ranked solutions.

18.6.3.1 Genetic algorithm performance

A total of eight problem sizes were tested and each problem size was tested 90 times. We tested only eight problem sizes because of the cost of generating the PS_{true}. Even with the reduced computational cost it is too large for a problem size that is greater than 12 for the system available for testing.

As discussed, the size of the solution space, which is the number of solutions that must be tested, follows the sequence of the bell numbers. So for a problem size with 13 customers the 13th bell number would be 27.644437 million. This means that there would be 27.644437 million solutions to test to generate PS_{true}. Now considering that there are three tests in each problem size it would require 82.933311 million solutions to be tested. Additionally the GA would also have to be tested for the problem size. A single execution of problems of this size would not be an issue. However, recall that 90 tests must be performed for each problem size. This would require 2.5 billion evaluations for our initial test and this number would only marginally decrease as different search space percentages are tested. Understanding the implications of this as it relates to the time it will take, 12 customers are selected as the maximum test size for the purpose of the research in this work.

Table 18.1 shows the details of the test plan and the equations used to calculate the results from the data generated by the tests.

Table 18.1 Test plan and equations.

Problem size	Prob inst 1	Prob inst 2	Prob inst 3	Total tests
5	30	30	30	90
6	30	30	30	90
7	30	30	30	90
8	30	30	30	90
9	30	30	30	90
10	30	30	30	90
11	30	30	30	90
12	30	30	30	90
Total	240	240	240	720

Overall 720 tests were done. Each problem size has three different problem instances of that size, of which each is executed in the GA 30 times. The eight different problem sizes and three unique problems for each problem size allow us to cover a large range of problems while the 30 repetitions of each problem test for each problem size provide accurate results that are accurate and consistent with how the GA would perform. Table 18.2 shows the average error percent of the PS_{known} generated for the 720 tests using 50% of each problem's search space as the number of evaluations performed. Each average represents the 30 repetitions on each of the three problems for the eight problem sizes.

Table 18.2 Average error percent of PS_{known}.

Problem size	Test 1	Test 2	Test 3
5	0.000000	0.000000	0.000000
6	0.000000	0.000000	0.017470
7	0.004173	0.003691	0.008769
8	0.209514	0.028279	0.502032
9	0.478971	0.540708	0.077838
10	0.017363	0.275157	0.323019
11	0.043041	0.075364	0.038576
12	0.620054	0.159279	0.061836

It can be seen in Table 18.2 that the percentage error ranges between 0 and 0.6%. This means that the PS_{known} on average is accurate to within 0.6% of the PS_{true}. This is while only using 50% of the search space as the number of evaluations to be completed before termination. This method of evaluating the GA is used to find the search space percent. It is this percentage that will be used to determine the number of evaluations the GA will perform for all problem sizes.

Using the performance analysis to determine the number of evaluations

Several explanations of GA were reviewed in this research. Based on these explanations of GA given specifically by the work such as that in Altiparmak et al. (2006) and Costa et al. (2013), the purpose of GA can be simplified to the exploration or traversal of a search space in an attempt to locate a solution or solutions.

The mechanism used to traverse the search space and the solution or solutions sought after can differ widely. The intent for the GA developed in this research is that it can find optimal or near-optimal solutions without traversing the entire search space. It is used as an alternative to using a brute force search of the entire search space. Therefore, for it to be effective, the GA should not require an exhaustive search of the search space. Noting this, it should be possible to find a number of evaluations (expressed as a percentage of the search space) such that the number of evaluations performed by the GA is lower than that of a brute force approach. This percentage must also allow the GA to generate a nondominated set that falls within a 5% error margin of the Pareto-optimal set (PS_{true}). Failure to find a search space percent that is less than 100 would mean that the GA is ineffective and a brute force approach can be used.

To find the percentage of the search space needed to generate solutions that are within the 5% error range various percentages were tested. Initially the percentages tested were determined by the binary search. The initial value of 100% was tested and subsequent test percentages were determined by dividing the previous test by 2. This process was repeated until the test found a (failing) value that generated an error percent greater than 5.

The penultimate test (the last percentage that gave an error within the threshold) would be the percentage of the search space used to determine the number of evaluations for all problems. During the tests, however, all values above 1% were found to generate error percentages under 5. Due to the time taken to perform these tests an adjustment was made to the process in order to increase the speed at which the failing value (an error percent greater than 5) could be found. The tests started with 1 as the initial search space percentage. This initial percentage was then altered by a factor of 10 as opposed to 2, in order to find an average error percent greater than 5. Table 18.3 shows the results of each test performed.

Table 18.3 Results for each test performed with different problem sizes.

Size	1.00000%			0.10000%			0.01000%		
	Test 1	Test 2	Test 3	Test 1	Test 2	Test 3	Test 1	Test 2	Test 3
5	0.05416	0.09211	0.06613	0.04403	0.06871	0.03116	0.01046	0.01046	0.03891
6	0.79239	0.6095	1.00999	0.93813	1.3385	0.12206	0.10105	0.16745	0.18923
7	0.75459	0.83883	0.59117	0.82251	0.70226	1.79555	1.55565	1.92438	1.75318
8	2.62686	2.33249	2.61077	2.40226	2.75333	1.1649	0.7975	0.96724	1.0217
9	2.27132	2.99413	2.61965	3.02466	3.34272	4.58538	5.38389	5.12704	5.38632
10	1.96775	4.12317	4.85451	4.34773	3.93745	1.44683	3.40863	2.86489	3.53469
11	0.83591	3.0331	3.4484	3.54296	3.82638	0.4769	2.45158	3.29168	3.19333

The results in Table 18.3 show that the search space percent value of 0.01 generates an average error percent greater than 5 for test numbers 1, 2, and 3 for the problem size of 9. The search space percentage of 0.1 is therefore chosen as the search space percentage as it is the smallest value that generates solutions within the 5% threshold. It is important to note that the results shown here operate with maximum and minimum thresholds. The reason for these thresholds are explained in detail in the following section.

Determining the maximum and minimum population and iteration size

There are limitations to the amount of resources that the GA can use. These limitations are dependent on the capabilities of the server that it will run on. Since the server has limited resources, the GA would need to limit the amount of resources it uses in a given instance. However, the GA must be given enough resources to be able to perform an acceptable level of optimization. Table 18.4 shows the actual results of the GA without the implementation of maximum and minimum thresholds.

Table 18.4 Actual results of the GA.

Problem sizes	Error percent with search space percent of 25	Error percent with search space percent of 12.5	Error percent with search space percent of 6.25	Error percent with search space percent of 3.125	Error percent with search space percent of 1.5625	Error percent with search space percent of 0.78125
5	2.296	100.000	11.095	100.000	2.951	100.000
6	11.042	24.459	2.834	8.484	2.327	3.622
7	0.590	2.695	2.633	7.715	1.149	2.407
8	1.148	3.965	0.529	2.048	1.731	4.151
9	0.755	2.501	0.750	3.517	0.099	1.278
10	0.264	0.750	0.157	0.644	0.512	1.616
11	0.110	0.319	0.138	0.336	0.066	0.210
12	0.601	0.839	0.173	0.162	0.065	0.116

Minimum number of evaluations

Table 18.4 shows that the algorithm was unable to perform within the 5% error threshold on a problem size of 7 and below for this percentage. As seen in the percentage error in the tests shown in Table 18.5 the percentage error in tests with problem sizes of 8–12 meets the 5% threshold but smaller problem sizes do not.

This is a somewhat expected result as it is understandable that at some percentage for a small problem size the number of evaluations would be too small for the GA to converge. A minimum value must therefore be found so that if the percentage used causes the number of evaluations to drop below this threshold it would be overwritten with the minimum value. This would prevent a small problem from causing the number of evaluations performed to be too small for the algorithm to converge.

A fixed number of evaluations was set for all test cases. To determine which value was used in each test case, the sequence of numbers 1, 2, and 5 was used. Each number in the sequence was then multiplied by 10 to generate the next three numbers in the test. The algorithm is then executed using each fixed number of evaluations for each problem size. The number of evaluations that allows all test cases to generate solutions within the 5% error was chosen as the minimum value for all problem sizes. This was because the larger problems would require more evaluations to perform within the 5% threshold. Therefore, if a minimum value tested allowed a larger problem to converge and fall within the 5% threshold we could expect that this would occur for smaller problems using the same minimum value. Table 18.5 shows the results of the tests for each test case.

Here the performance of the GA for each test for each problem size can be seen. At test 4 with a value of 500 evaluations, all test cases performed within the 5% error threshold. This is the smallest number that generates these results and is therefore chosen as the minimum number of evaluations for all problems.

Table 18.5 Test case results.

Problem size	50 Evaluations	100 Evaluations	200 Evaluations	500 Evaluations	1000 Evaluations	2000 Evaluations
5	1.20614	0.3227	0.2127	0.03432	0.00874	0
6	10.15782	6.96232	2.93066	0.34145	0.03303	0
7	2.68658	1.48363	1.31022	0.61716	0.16059	0.04282
8	4.88531	4.24066	3.81012	1.8721	1.04771	0.53819
9	6.06474	4.50656	4.33747	3.05553	1.5616	0.84273
10	7.30207	6.65453	5.52268	4.05968	2.67663	1.40023
11	6.03466	5.48641	4.67979	3.18825	2.0547	1.37604
12	6.83265	6.35646	5.36746	4.93164	4.59867	2.60922

Maximum number of evaluations

Determining the maximum number of evaluations that the GA would be allowed to perform would take an analysis of the number of farmers that would be using the system and the server's capabilities. The analysis can be described as a series of steps, namely:

1. determining the number of farmers that will use the system,
2. determining the average size of the problems that will be optimized,
3. calculating the time taken for the server to perform one chromosome evaluation,
4. calculating the total available computation time per week,
5. determining how much time each farmer in the system will have per week,
6. dividing the total time each farmer has per week by the chromosome evaluation time to calculate the maximum number of evaluations.

Each step is explained in greater detail in the sections that follow in the order listed in the steps shown above.

Determining the number of farmers that will use the system. A subset of 15 farmers in the industry is used in this research to provide initial information on the expected system load. An initial survey was conducted on these 15 farmers selected to gather information about how farming in the country is generally carried out. More importantly, the survey also collects information that indicates the number of times the GA will be used and the average size of the problem it would have to solve. The important data were collected, and the number of customers is shown in Fig. 18.4 for each farmer. This survey provided the information needed to perform the necessary steps in calculating the maximum number of evaluations.

Determining the average size of the problems that will be optimized. The average number of customers gives the average problem size expected for each optimization request. This is because the problem size is proportional to the number of customers. The average number of customers is calculated to be 24. The data shown in Fig. 18.4 support this average, as most of the problem sizes would be 10–25 customers.

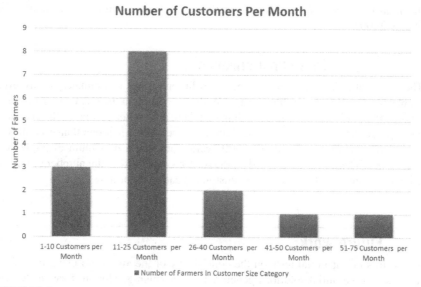

FIGURE 18.4

Data collected on customer size.

Calculating the time taken for the server to perform one chromosome evaluation.
Next, the total time it takes the server to perform one evaluation must be calculated.
This calculation must account for the increasing overhead incurred due to the size
of the chromosomes. To account for this, the average case chromosome or problem
size of 24 customers taken from our survey is used. The time taken to perform an
evaluation of a chromosome for 24 customers is 0.00072027 seconds for the server
we used.

***Calculating the total available computation time each farmer in the system will
have per week.*** A 24-hour day and 7-day week is used as the total amount of time
available for evaluations; this equates to 604,800 seconds of time for evaluations
per week. Given the 15 farmers that would be using this time each farmer would have
40,320 seconds of evaluation time a week. The total number of evaluations per farmer
per week would therefore be calculated as follows:

$$E_{max_i} = E_{weekly}/E_{worst}, \tag{18.6}$$

where E_{max_i} is the maximum evaluations per instance, E_{weekly} is the evaluation time
available in a week, and E_{worst} is the worst-case time for each evaluation.

This brings the number of evaluations for the GA using the current environment to
40,320/0.00072027 = 55,979,007 per farmer per week. This is also the total number
of evaluations that can be performed by a single instance of the GA. As discussed,
this number is going to be used to place a limit on the number of evaluations that the
GA can perform. Once a problem size requires a number of evaluations that exceeds

this limit the number of evaluations would be overwritten with the maximum value of 55,979,007.

18.6.4 Final review of performance

The GA is able to reduce the number of solutions it has to evaluate for any given problem to 0.1% of the problem's size. With this reduction the GA can guarantee an error percent of 5 or less. Understanding that there are limited resources the algorithm has a predefined minimum and maximum number of evaluations that it is allowed to perform. This is a minimum of 500 evaluations and a maximum of 55,979,007 evaluations. These limits are based on the server's capabilities, the number of farmers using the system, and the number of customers each farmer may have.

18.7 Future work

The results of our initial tests on the consistency of the algorithm using the subobjectives and the initial solutions generated are promising. Moving forward we will perform additional tests on the impact the subobjectives have on the algorithm. We will also compare the solutions generated by the algorithm to other solutions generated for the vehicle routing problem. We will also look at the resources other algorithms use to solve the vehicle routing problem and compare this to our solution to determine how well this algorithm performs with regard to resource utilization.

We also intend to increase the scalability of the entire system. This would primarily depend on the GA's ability to handle large problem sets. We intend to do this by modifying the algorithm so that it can distribute its workload across multiple servers that would be regulated by the algorithm itself. The GA would determine, based on problem size, the amount of resources it would need to generate a solution in the required time frame. It would then acquire these resources and execute the optimization.

Future work will also include the ability to provide personalized results according to the farmer's selection habits. More concretely, it would focus its optimization on solutions similar to those previously chosen by the farmer. The algorithm would use a feedback mechanism that would allow it to look at the solutions chosen from the solution sets it sent to the farmer in previous optimization requests. Using this, the algorithm would calculate the area on the Pareto-optimal front that it must focus on to generate a solution set focused on the farmer's preference.

References

Ahumada, O., Villalobos, J.R., 2009. Application of planning models in the agri-food supply chain: a review. European Journal of Operational Research 196 (1), 1–20.
Altiparmak, F., Gen, M., Lin, L., Paksoy, T., 2006. A genetic algorithm approach for multi-objective optimization of supply chain networks. Computers & Industrial Engineering 51 (1), 196–215.

Azad, N., Aazami, A., Papi, A., Jabbarzadeh, A., 2019. A two-phase genetic algorithm for incorporating environmental considerations with production, inventory and routing decisions in supply chain networks. In: Proceedings of the Genetic and Evolutionary Computation Conference Companion on – GECCO '19. ACM Press, Prague, Czech Republic, pp. 41–42. http://dl.acm.org/citation.cfm?doid=3319619.3326781.

Bandyopadhyay, S., Saha, S., 2013. Some single-and multiobjective optimization techniques. In: Unsupervised Classification. Springer, pp. 17–58.

Costa, E., Fabris, F., Loureiros, A.R., Ahonen, H., Varejão, F.M., 2013. Optimization metaheuristics for minimizing variance in a real-world statistical application. In: Proceedings of the 28th Annual ACM Symposium on Applied Computing – SAC '13. ACM Press, Coimbra, Portugal, p. 206. http://dl.acm.org/citation.cfm?doid=2480362.2480405.

de Keizer, M., Haijema, R., van der Vorst, J., Bloemhof-Ruwaard, J., 2012. Hybrid simulation and optimization approach to design and control fresh product networks. In: Proceedings of the Winter Simulation Conference. Winter Simulation Conference, p. 102.

Fonseca, C.M., Fleming, P.J., 1993. Genetic algorithms for multiobjective optimization: formulation, discussion and generalization. In: Icga, vol. 93. Citeseer, pp. 416–423.

Hoffmann, B., Chalmers, K., Urquhart, N., Guckert, M., 2019. Athos – a model driven approach to describe and solve optimisation problems: an application to the vehicle routing problem with time windows. In: Proceedings of the 4th ACM International Workshop on Real World Domain Specific Languages – RWDSL '19. ACM Press, Washington D.C., DC, USA, pp. 1–10. http://dl.acm.org/citation.cfm?doid=3300111.3300114.

Lehmann, R.J., Reiche, R., Schiefer, G., 2012. Future internet and the agri-food sector: state-of-the-art in literature and research. Computers and Electronics in Agriculture 89, 158–174.

Li, L., Chen, N., Wang, W., Baty, J., 2009. LocalBuy: a system for serving communities with local food. In: CHI'09 Extended Abstracts on Human Factors in Computing Systems. ACM, pp. 2823–2828.

Maharana, D., Choudhary, P., Kotecha, P., 2017. Optimization of bio-refineries using genetic algorithm. In: Proceedings of the 2017 International Conference on Intelligent Systems, Metaheuristics & Swarm Intelligence – ISMSI '17. ACM Press, Hong Kong, Hong Kong, pp. 26–30. http://dl.acm.org/citation.cfm?doid=3059336.3059360.

Mentzer, J.T., DeWitt, W., Keebler, J.S., Min, S., Nix, N.W., Smith, C.D., Zacharia, Z.G., 2001. Defining supply chain management. Journal of Business Logistics 22 (2), 1–25.

Rika, D., Sholomon, D., David, E.O., Netanyahu, N.S., 2019. A novel hybrid scheme using genetic algorithms and deep learning for the reconstruction of Portuguese tile panels. In: Proceedings of the Genetic and Evolutionary Computation Conference on – GECCO '19. ACM Press, Prague, Czech Republic, pp. 1319–1327. http://dl.acm.org/citation.cfm?doid=3321707.3321821.

Stevens, G.C., 1989. Integrating the supply chain. International Journal of Physical Distribution & Materials Management 19 (8), 3–8.

Multirobot coordination through bio-inspired strategies

Floriano De Rango, Nunzia Palmieri, Mauro Tropea

University of Calabria, DIMES Department, Cosenza, Italy

CONTENTS

19.1 Introduction

During the 2010s, the field of distributed robotics has been investigated actively, involving multiple, rather than single, robots. The field has grown dramatically, with a much wider variety of topics being addressed. Several new areas of applications of robotics, such as underwater and space exploration, hazardous environments, service

Nature-Inspired Computation and Swarm Intelligence. https://doi.org/10.1016/B978-0-12-819714-1.00030-0

robotics in both public and private domains, the entertainment field, and so forth, can benefit from the use of multirobot systems (MRSs). In these challenging application domains, multirobot systems can often deal with tasks that are difficult, if not impossible, to be accomplished by an individual robot. A team of robots may provide redundancy and contribute cooperatively to solve the assigned task, or it may perform the assigned task in a more reliable, faster, or cheaper way beyond what is possible with a single robot. However, the use of multiple robots poses new challenges; indeed the robots must communicate and coordinate in such a way that some predefined global objects can be achieved more efficiently.

An extensive amount of research has been carried out in the area of multirobot coordination mechanisms. Within these settings, a key challenge is to find ways in which the members of the team can coordinate their decision processes in order to increase the overall performance of the collective. Moreover, such decision processes could consider multiple objectives, possibly conflicting.

Swarm robotics is a new approach to the coordination of MRSs which consists of large numbers of mostly simple physical robots. It gets inspiration from swarm intelligence to model the behavior of the robots. Currently, swarm robotic algorithms are one of the most interesting research areas in the robotics field.

One of the most common approaches is to use algorithms based on biological inspiration, particularly social insects, in the development of similar behaviors in cooperative robot systems. Decentralized agents groups typically require complex mechanisms to accomplish coordinated tasks. In contrast, biological systems can achieve intelligent group behaviors with each agent performing simple sensing and actions. In these systems, each agent acts autonomously and interacts only with its neighbors, while the global system exhibits a coordinated and sophisticated behavior.

Bio-inspired metaheuristic algorithms have recently become the forefront of the current research as an efficient way to deal with many NP-hard combinatorial optimization problems and nonlinear optimization constrained problems in general (Yang, 2010). These algorithms are based on a particular successful mechanism of a biological phenomenon in nature in order to achieve the survival of the fittest in a dynamically changing environment. Examples of collective behavior in nature are numerous. They are based, mainly, on direct or indirect exchange of information about the environment between the members of the swarm. Although the rules governing the interactions at the local level are usually easy to describe, the result of such behavior is difficult to predict. However, through collaboration the swarms in nature are able to solve complex problems that are crucial for their survival. On the basis of these considerations, the chapter presents an application of the swarm intelligence-based approaches, which are strongly inspired by the biological behavior of social insects, for the coordination of a swarm of robots involved in a search and rescue mission in a hazardous environment. Many other bio-inspired techniques such as the genetic algorithm (GA) have been proposed in recent years to improve the coordination of nodes under resource constraints such as proposed in De Rango et al. (2009). However, in the case of robot coordination, more scalable and distributed techniques are necessary. For this reason approaches completely distributed without central control are used to coordinate the robots. Each of them utilizes only local information from its neighbors and then uses this information to make the best decisions from its point of view. The control law that each agent executes is simple, while the emerging global behavior is sophisticated and robust. Although the aim of this research is to develop effective coordination mechanisms for a team of mobile robots operating search and rescue in unknown and possibly hostile environments, the proposed approaches are generalized and they can be used for a wide range of applications with minor modifications.

19.2 Literature overview

Research related to this research topic includes the topics of multirobot exploration and multirobot coordination. This section provides a review of some of the most relevant works.

19.2.1 Multirobot exploration

Multirobot exploration has received much attention in the research community. The unknown area exploration should not lead to an overlap in robot movements and ideally, the robots should complete the exploration of the area within the minimum amount of the time.

Some exploration plans in the context of mapping are usually constructed without using environmental boundary information. However, most works combine different criteria in more complex utility functions. For example, Burgard et al. (2005) coordinated the robots in order to explore as much area as possible. A decision-theoretic approach trades off the utility and the cost of visiting targets. The cost of a target is defined as the length of the optimal path from a robot to it, whereas the utility of a target is defined as the area expected to be found when the robot arrives at it. On the other hand, some researchers focus on the exploration by using knowledge about environmental boundary information, as described in Wattanavekin et al. (2013). The authors assume that the robots already have the information of all obstacles. Therefore, when a robot encounters an obstacle, it can immediately grasp the obstacle. However, this is not practical in real-world applications in unknown areas. Other approaches, proposed by Gifford et al. (2010), coordinate the robots by means of dividing the environment into as many disjoint regions as available robots and assigning a different region to each robot.

Tree-cover algorithms, instead, use a precalculated spanning-tree to direct the exploration effort and distribute it among the agents. These algorithms require a priori knowledge of the environment. However, in real scenarios, especially in search and rescue missions, the considered area could have some uncertainty, thus building accurate maps may be problematic. Bio-inspired techniques have recently gained importance in computing due to the need for flexible, adaptable ways of solving engineering problems. Within the context of swarm robotics, most works on cooperative exploration are based on biological behavior and indirect stigmergic communication (rather than on local information, which can be applied to systems related to GPS, maps, wireless communications). This approach is typically inspired by the behavior of certain types of animals and insects, like ants, which use chemical substances known as pheromones to induce behavioral changes in other members of the same species. One of the well-known approaches is inspired by the collective behavior of insect colonies such as ants and fireflies (Dorigo and Stützle, 2003; Yang, 2010). These algorithms are based on decentralized local control, local communication, and the emergence of global behavior as the result of self-organization.

Ants and other social animals are known to produce pheromones and use them as a medium for sharing information to mark the paths that are used in the environment. Pheromone trails provide a type of distributed information that artificial agents may use to make decisions. Many works can be found in the literature using this kind of biology metaphor. Wagner et al. (1999) were the first who invested stigmergic multirobot coordination for covering/patrolling the environment. In their approach a group of robots is assumed able to deposit chemical odor traces and evaluate the strength of smell at every point they reach. Based on these assumptions, they used robots to model an unmapped environment as a graph and they proposed basic graph search algorithms for solving mainly robotic coverage problems. Kuyucu et al. (2015) used a guided probabilistic exploration of an unknown environment achieved via combining random movement with pheromone-based stigmergic guidance. Chen et al. (2013) proposed a fast two-stage ant colony optimization (ACO) algorithm which overcomes the inherent problems of traditional ACO algorithms. The basic idea is to split the heuristic search into two stages: the preprocess stage and the path planning stage. In the preprocess stage, the scent information is broadcast to the whole map and then ants do path planning under the direction of scent information. Ducatelle et al. (2011) uses a swarm of wheeled robots, called foot-bots, and a swarm of flying robots that can attach to the ceiling, called eye-bots, that serve as stigmergic markers for foot-bot navigation. However, in the exploration task, researchers use the concept of antipheromone so as to try to maximize the distance between the robots and to enforce a dispersion mechanism in different sites of the region of interest, with the aim to accomplish the mission as quickly as possible. Some examples of this approach can be found in Calvo et al. (2011) and Doi (2013) for surveillance missions, in Palmieri et al. (2019) for guiding of robots in search and rescue in a disaster site, and in Ranjbar-Sahraei et al. (2012) for multirobot coverage. Ravankar et al. (2016) use a hybrid communication framework that incorporates the repelling behavior of the antipheromone and attractive behavior of pheromone for efficient map exploration.

The use of physical substances for pheromone-based communication within robots is problematic and poorly understood. However, there is undergoing work in improving their use with promising results, and it is predicted that with improvements in sensing technology it may be possible that a robot could carry a life time supply of chemicals (Purnamadjaja and Russell, 2010). Other authors, like Payton et al. (2001), described techniques for coordinating the actions of large numbers of small-scale robots to achieve useful large-scale results in surveillance, reconnaissance, hazard detection, and path finding, using the notion of a "virtual pheromone," implemented using simple transceivers mounted to each robot. Unlike the chemical markers used by insect colonies for communication and coordination, our virtual pheromone is a symbolic message tied to the robots themselves rather than to fixed locations in the environment. Chemical trail following strategies have been implemented with real robots. For example, ethanol trails were deposited and followed by the robots: high concentrations of the pheromone yield high signal strength but the signal duration is short, while low pheromone concentrations yield low signal strength but a long signal duration (Fujisawa et al., 2008), but the use of decaying chemical trails by real robots can be problematic.

Other robotic implementations of insect-style pheromone trail following have instead used nonchemical substitutes for the trail chemicals. Other works that apply this similar approach were presented in Masár (2013). This latter work used a virtual pheromone system in which chemical signals are simulated with the graphics projected on the floor, and in which the robots decide their action depending on the color information of the graphics. Nevertheless, with recent developments in communication technology, electrical devices such as radio frequency identification devices (RFIDs) have gained much interest for such applications. Johansson and Saffiotti (2009) used RFIDs for mapping and exploring an unknown environment. In essence, most of the nature-inspired approaches use a combination of stochastic components or moves with some deterministic moves so as to form a multiagent system with evolving states. Such a swarming system evolves and potentially self-organizes into a self-organized state some emergent characteristics.

19.2.2 Multirobot coordination

Coordination in multirobot systems has been extensively studied in the scientific literature due its real-world applications including aggregation, pattern formation, cooperative mapping and transport, and foraging. All of these problems consist of multiple robots making decisions.

Decision making can be regarded as a cognitive process resulting in the selection of a course of actions among several alternative scenarios. Every decision making process produces a final choice. In MRSs, the decision making guided by planning can be centralized or decentralized in accordance with the group architecture of the robots (Yan et al., 2012). A solution is called centralized when a single element in the system is responsible for managing all the available resources. The strong point is that it can be used for the best known algorithms and usually this kind of approach has more information available than distributed or local algorithms. Studies belonging to the centralized architecture approach include Luna and Bekris (2011) and Yan et al. (2012).

On the other hand, there is no central control agent in distributed architectures, such that all the robots are equal with respect to control and are completely autonomous in the decision making process. Moreover, a decentralized architecture can better respond to unknown or changing environments, and usually has better reliability, flexibility, adaptability, and robustness (Yan et al., 2013). One of the most commonly used swarm-based approaches is the response threshold, where each robot has a stimulus associated with each task it has to execute. It continuously perceives the stimulus for each task; this stimulus reflects the urgency or importance of performing that task. When a robot perceives that a stimulus for a particular task exceeds its threshold, it begins completing the task. When the stimulus falls below this threshold (e.g., when the task is completed), the agent stops executing those behaviors. This response can be deterministic or probabilistic (Kalra and Martinoli, 2006). Some response threshold systems use such stimuli and the threshold value for calculating the probability of executing a task (de Lope et al., 2015; Palmieri et al., 2017).

In recent years, market-based approaches have become popular to coordinate MRSs. These methods have attempted to present a distributed solution for the task allocation problem (Trigui et al., 2014). Essentially, robots act as self-interested agents in pursuit of individual profit. They are paid in virtual money for tasks they complete and must pay in virtual money the value of the resources they consume. Tasks typically are distributed through auctions held by an auctioneer; this auctioneer is either a supervisor agent or one of the robots. Robots compete through bidding to win those tasks that they can complete inexpensively and thus maximize their profit. This price-driven redistribution simultaneously results in better team solutions. Zhao and Wang (2013) used this approach to collect and transport objects in an unknown environment.

Recently, bio-inspired algorithms inspired by a variety of biological systems have been proposed for self-organized robots. The self-organizing properties of animal swarms have been studied for better understanding the underlying concept of decentralized decision making in nature, but it also gives a new approach in applications to multiagent system engineering and robotics. ACO and its variants have been used as coordination techniques in coordinating robots. Hoff et al. (2010) presented two ant-inspired robot foraging algorithms which allow coordination between robots. This approach uses direct communication between the robots instead of environmental markers. They assume that the robots have limited sensing and communication capabilities and no explicit global positioning. De Rango and Palmieri (2012) used combined bio-inspired approaches based on ACO to guide the robots in an unknown mined area. Palmieri et al. (2015) used a hybrid approach that combines repellent and attractive pheromones to explore an area and recruit other robots, respectively.

Another well-known bio-inspired approach, called particle swarm optimization (PSO), takes inspiration from the behavior of birds (Kennedy and Eberhart, 1995). PSO-inspired methods and their extended versions have received much attention and have been applied for the coordination of mobile robots. Examples include guiding robots for targets searching in complex and noisy environments, as described by Derr and Manic (2009). Hereford and Siebold (2010) presented a version of PSO for finding targets in the environment. Modified versions of PSO are proposed to balance searching and selecting in a collective clean-up task (Li et al., 2013) for path planning in clutter environment (Das et al., 2016) and for mimicking natural selection emulated using the principles of social exclusion and inclusion (Couceiro et al., 2014).

Another nature-inspired algorithm, called bees algorithm (BA), which mimics the food foraging behavior of swarms of honey bees, and its modified versions have also been applied to robotic systems, demonstrating aggregation (Kernbach et al., 2009) and collective decision making (Jevtic et al., 2011; Contreras-Cruz et al., 2015). Other studies that take inspiration from the bees and ants have also been applied to robotic systems, such as task allocation (Momen, 2013), finding targets and avoiding obstacles (Banharnsakun et al., 2012), solving on line path planning (Liang and Lee, 2015; Garcia et al., 2009), and decision making to aggregate robots around a zone (De Rango and Palmieri, 2016; Arvin et al., 2014). A hybrid approach can be found in De Rango and Palmieri (2012).

Other approaches, called bacterial foraging optimization, take inspiration from the chemotactic behavior of bacteria such as *Escherichia coli*. Yang et al. (Yang et al., 2015) applied this method for a target search and trapping problem. An extensive review of research related to the bio-inspired techniques and the most behavior of the robots can be found in (Senanayake et al., 2016; Bayındır, 2016).

Other approaches use direct communication among the members of the swarm developing the protocol ad hoc. Direct communication refers to a process where robots exchange information directly between each other, often (but not necessarily) explicitly transmitting data to signal a particular status. Usually, according to the principle of local communication, information can be exchanged between nearby robots, which can then act upon received information modifying their behavior to improve the foraging performance. For example, ant-based routing is gaining more popularity because of its adaptive and dynamic nature and these algorithms consist in the continual acquisition of routing information through path sampling and discovery using small control packets called artificial ants. Some examples are AntHocNet, proposed by Di Caro et al. (2005), and Ant-Colony Based Routing Algorithm (ARA), described by De Rango and Tropea (2009a) and De Rango and Tropea (2009b).

On the basis of contribution emphasizing the importance of energy constraints in communication and node coordination such as proposed in De Rango et al. (2006, 2008), Fotino and De Rango (2011), and Guerriero et al. (2009), authors proposed bio-inspired routing strategies able to minimize the number of hops and the energy consumption (Derr and Manic, 2009) or to be able to combine more bio-inspired techniques (De Rango and Palmieri, 2016; De Rango et al., 2018; Palmieri et al., 2018).

19.3 Searching task

This chapter starts the discussion about one of the topics of the work that is the collective search task. The problem of collective search is a trade-off between searching thoroughly and covering as much area as possible. Solutions to the problem of collective search are currently of much interest in robotics, especially the study of distributed algorithms applied to this problem. The objective is to design ways in which, without central control, robots can use local information to perform the search and rescue operations (Countryman et al., 2015). The problem of coordinating a team of robots for exploration is a challenging problem, particularly in unstructured areas, as for example postdisaster and hazardous scenarios where direct communication is limited. Here, an algorithm is described for exploring the area inspired by the ant foraging model. The approach emphasizes the role played by individual robots and stresses some crucial aspects such as the lack of a governing hierarchy, self-organization of the robots, and indirect communication.

Ants in nature, indeed, have evolved over a long period of time and display remarkable behaviors that are highly suitable for addressing complex tasks. In social insects, pheromone communication serves a number of social functions, such as recognizing, aggregating, gathering food, mating, and alarm propagation, for the colony members. Swarm optimization algorithms, such as ACO, rely on pheromone trails to mediate (indirect) communication between the agents.

In this kind of coordination, the environment is used as a medium to transfer information among the robots: each robot deposits traces in the environment in order to send different types of signals, depending on what it wants to indirectly communicate. The accumulation of traces in the environment provides a shared memory, which allows memoryless simple robots to coordinate easily, while robots might not have any self-awareness of other agents. These algorithms are fully decentralized and rely on memoryless agents with very simple individual behaviors. Agents can only communicate through environment marking, as they only mark and move according to their local perceptions. In the robotics field, with the availability of various sensors, a range of environmental markers (such as chemicals, metals, heat sources, electronic tags) can be used as a way of encoding information in the environment (Kuyucu et al., 2012).

19.3.1 Ant-inspired techniques

Ant colonies provide some of the richest examples for the study of collective phenomena such as collective exploration. Exploration is a very important task in nature since it allows animals to discover resources, detect the presence of potential risks, forage for food, and scout for new home. Ant colonies operate without central control, coordinating their behavior through local interactions with each other. Ants perceive only local, mostly chemical and tactile cues. In a colony to monitor its environment, to detect both resources and threats, ants must move around so that if something happens, or a food source appears, some ant is likely to be near enough to find it (Gordon, 2010), (Countryman et al., 2015). Ant colonies, despite the simplicity of single ants, demonstrate surprisingly good results in global problem solving. Consequently, ideas borrowed from insects and especially from ant behavior are increasingly popular in robotics and distributed systems.

ACO was developed by Dorigo and Stützle (2003) inspired by the natural behavior of trail laying and following by ants. They live in colonies and their behavior is governed by the goal of colony survival rather than the survival of individuals. The behavior that provided the inspiration for ACO is the ant's foraging behavior, and in particular how ants can find shortest paths between food sources and their nest. When searching for food, ants initially explore the area surrounding their nest in a random manner. While moving, ants can leave and smell a chemical pheromone trail on the ground. When choosing their way, they tend to choose, in probability, paths marked by strong pheromone concentrations. As soon as an ant finds a food source, it evaluates the quantity and the quality of the food and carries some of it back to the nest. During the return trip, the quantity of pheromone that an ant leaves on the ground may depend on the quantity and quality of the food. The pheromone trails will guide other ants to the food source. The central component of an ACO algorithm is a parametrized probabilistic model, which is called the pheromone model (Dorigo and Blum, 2005). During the 2010s many variants of Dorigo's method have been proposed and applied in many robotics fields.

19.3.2 Ant-based team strategy for robot exploration

In this section, the exploration problem in the context of search and rescue operations is addressed, in which the mobile and autonomous robots must be able to decide the sequence of movements needed to explore the whole environment. The mainstream approaches for developing exploration strategies are mostly based on the idea of incrementally exploring environments by evaluating a number of candidate observation locations, in this specific case neighbor cells, according to a criterion and by selecting at each step the next best location. However, here the problem to build a map of the environment is not considered, since the main object consists of locating the largest number of targets in the minimum amount of time.

Differently from map building, search and rescue settings are strongly constrained by both time and battery limitations and generally require the amount of explored regions over the map quality. Since the robots should be required to be capable of various functionalities other than area exploration, it is desirable that both the integration to a swarm and the ability to explore are seamless and these actions should not consume a large amount of the robot's resources. Moreover, to be effective, a search strategy must attract robots towards unobserved areas so as to avoid the undesirable scenario where some areas are frequently revisited while others remain unexplored. Broadly speaking, the robots operate according to the following steps:

- (a) The robots perceive the surrounding cells using on-board sensors.
- (b) The robots compute the perceived information, in this case the concentration of pheromone, in neighbors cells.
- (c) The robots decide where to go next.
- (d) The robots move in their best local cell and start again from (a).

The basic intention behind the work described here is to design a motion policy which enables a group of robots, each equipped only with simple sensors, to efficiently explore potentially complex environments. As in biology, the basic idea

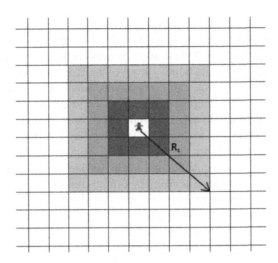

FIGURE 19.1

Example of pheromone diffusion.

pursued is to utilize the principle of pheromone-based coordination and to let each robot deposit pheromones on visited cells in order to inform, indirectly, the others about the already explored region. According to this approach, the robots need not communicate directly, but deposit pheromones on the borders of their territory for instructing other robots to not enter it. When the interior sensors detect pheromone, it should indicate to a robot that it is about to enter potentially explored territory, and therefore the robot could preferentially change direction. The objective is to develop a simple algorithm that can utilize pheromones to benefit from the existing physical properties of a real environment in achieving complex collective behavior within a large group of simple homogeneous robots. It does not need to keep a topology of the map in memory. Decision making is done probabilistically based on local pheromone information. It should be emphasized that the problems of sensors, leaving pheromone, or movements are not taken into account, since the main focus of the work is to design a self-adaptive decision making mechanism for performing the assigned task (Palmieri et al., 2019).

Broadly speaking, when the robots are exploring the area, they lay pheromone on the traversed cells and each robot uses the distribution of pheromone in its immediate vicinity to decide where to move. Like in nature, the pheromone trails change in both space and time. The pheromone deposited by a robot on a cell diffuses outwards cell-by-cell until a certain distance R_s such that $R_s \subset A \subset \mathbb{R}^2$ and the amount of the pheromone decreases as the distance from the robot increases (see Fig. 19.1). Mathematically, the pheromone diffusion is defined as follows. Assume that robot k at iteration t is located in a cell of coordinates $(x_k^t, y_k^t) \in A$. Then the amount of pheromone that the robot deposits at cell c of coordinates (x, y) is given by

$$\Delta \tau_c^{-r_{kc}} = \begin{cases} \Delta \tau_0 e^{\frac{-r_{kc}}{a_1}} - \frac{\epsilon}{a_2}, & \text{if } r_{kc} \leq R_S, \\ 0, & \text{otherwise,} \end{cases} \tag{19.1}$$

where r_{kc} is the distance between the robot and cell c, defined as

$$r_{kc} = \sqrt{(x_k^t - x)^2 + (y_k^t - y)^2}. \tag{19.2}$$

This means that pheromone spreads up to a certain distance, as in the real world, after which it is no longer perceivable by other robots. In addition, $\Delta \tau_0$ is the quantity of pheromone sprayed in the cell where robot k is placed and it is the maximum amount of pheromone, ϵ is a heuristic value (noise), and $\epsilon \in (0, 1)$. Furthermore, a_1 and a_2 are two constants to reduce or increase the effect of the noise and pheromone. It should be noted that if multiple robots can deposit pheromone in the environment at the same time, then the total amount of pheromone that can be sensed in a cell c depends on the contribution of many robots. Furthermore, the deposited pheromone concentration is not fixed as it evaporates with time. The rate of evaporation of pheromone is given by ρ $(0 \leq \rho \leq 1)$, and the total amount of pheromone evaporated in cell c at step t is given by the following function:

$$\xi_c^t = \rho \, \tau_c^t, \tag{19.3}$$

where τ_c^t is the total amount of pheromone on cell c at iteration t. Considering the evaporation of the pheromone and the diffusion according to the distance, the total amount of pheromone in cell c at iteration t is given by

$$\tau_c^t = \tau_c^{(t-1)} - \xi_c^{(t-1)} + \sum_{k=1}^{N^R} \Delta \tau_c^{k,t}. \tag{19.4}$$

19.3.3 Probabilistic decision making

At each time step, the algorithm selects the most appropriate cell for each robot, among a set of neighbor cells without the knowledge of the entire area. This happens because the robots have no global information about the environment. The aim is to avoid any overlapping and redundancy efforts; therefore, the robots must be highly dispersed in the area in order to complete the mission as quickly as possible, avoiding at the same time any wastage of the robot's resources such as energy. Each robot k, at each time step t, is placed on a particular cell c_k^t that is surrounded by a set of accessible neighbor cells $N(c_k^t)$. Essentially, each robot perceives the pheromone deposited into the nearby cells, and then it chooses which cell to move to at the next step.

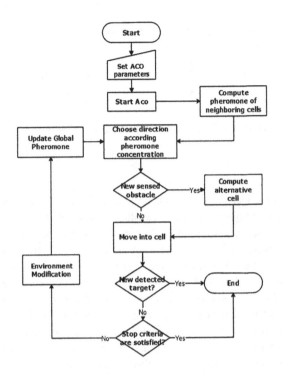

FIGURE 19.2

The flow chart of exploration tasks for a robot.

The probability at each step t for robot k of moving from cell c_k^t to cell $c \in N(c_k^t)$ can be calculated by

$$p(c|c_k^t) = \frac{(\tau_c^t)^\varphi \, (\eta_c^t)^\lambda}{\sum_{b \in N(c_k^t)} (\tau_b^t)^\varphi \, (\eta_b^t)^\lambda}, \quad \forall c \in N(c_k^t), \quad (19.5)$$

where $(\tau_c^t)^\varphi$ is the quantity of pheromone in cell c at iteration t and $(\eta_c^t)^\lambda$ is the heuristic variable to avoid the robots being trapped in a local minimum. In addition, φ and λ are two constant parameters which balance the weight to be given to pheromone values and heuristic values, respectively. Robot k moves into the cell that satisfies the following condition:

$$c = \arg\min[p(c|c_k^t)]. \quad (19.6)$$

In this way, the robots will prefer less frequently visited regions and more likely they will direct towards unexplored regions. Fig. 19.2 illustrates a simplified flowchart of the ACO-based strategy applied by each robot in *Forager State*.

19.3.4 **ATS-RE algorithm**

The exploration strategy is detailed in Algorithm 1, which provides the pseudocode for the pheromone-based control, which is executed periodically. At the first iteration of Algorithm 1, all cells are initialized with the same value of the pheromone trail, set to be zero, which represents that the cells have not yet been visited by any of the robots, so that the initial probabilities that a cell would be chosen is almost random. Then the robots move from a cell to another based on the cell transition rule in Eq. (19.5). Unexplored cells become more attractive to the robots in the subsequent iterations. Using this approach, the robots explore the area by following the flow of the minimum pheromone. Then the pheromone trails on the visited cells by ants are updated using Eq. (19.4). Algorithm 1 stops executing for a robot when it becomes a coordinator, it is recruited, or the mission is completed (that is, all cells have been visited at least once). See Fig. 19.3.

Algorithm 1: ATS-RE Algorithm

begin
    Step 1 : Initialization.
        Set t: {t is the time step}. Define φ, λ, a_1, a_2, ε, $\Delta\tau_0$, ρ, R_s
    Step 2 : Generation coordination system. For the whole swarm, set the
        initial locations in terms of coordinates in x and y directions.
    Step 3 : Procedure
    while *the stop criteria are not satisfied* **do**
        foreach *robot k in Forager State* **do**
            evaluate the current position c_k^t;
            evaluate neighboorhood $N(c_k^t)$;
            compute c according Eq. (4.6);
            if *(c.hasObstacle() or c.isOccupated() or c.isInaccessible())* **then**
                choose a random cell $c^* \in N(c_k^t)$;
                move robot k towards c^*;
            else
                move robot k towards c;
                deposit pheromone according to Eq. (4.1);
            end if
        end foreach
        foreach *cell $c \in A$* **do**
            update pheromone according Eq.(4.4);
        end foreach
        update t;
    end while
end

FIGURE 19.3

Algorithm 1: ATS-RE algorithm.

19.4 **Recruitment task**

The recruitment task aims to design a low-cost coordination mechanism that is able to form groups of robots at given sites where the targets are found. Once a robot detects a target, since it may not have sufficient resource capabilities to handle it, it acts as a strong attractor to the other robots to form a coalition that cooperatively works for the disarmament process of the target. The detection of a target may happen at any time during the exploration of the area, so the recruitment process takes place in real-time and possibly in different regions of the area.

For this purpose, wireless communication is used to share the information about the found targets, since direct communication may be beneficial when a fast reaction is expected and countermeasures must be taken. In this case, each robot is assumed to have transmitters and receivers, using which it can send packets to other robots within its wireless range R_t.

A key issue in this problem is how to avoid deadlock; that is, the situation where robots are waiting for a long time for the others to proceed the disarming process. These issues are particularly relevant in strictly collaborative tasks since the robots need to work collectively and adaptively for the disarmament of the hazardous targets, and each robot has only locally and partially information about the environment. The most common approach is in a greedy fashion in which a found target is instantaneously assigned to the robots without taking into account future events. Here, a flexible strategy is proposed in which the robots can react to future new events changing, eventually, the taken decisions. However, each robot must make individual decisions that could lead to retract itself from help requests. For example, for such kind of mission, it is possible to detect a target, while reaching another, or to receive another request, and thus changing decisions to move in a more convenient way from the robot's point of view. So at each time step, the robots will make the best selfish decision based on their positions and conditions, in response to the received help requests, trying at the same time to balance the two tasks. In order to tackle the problem, two communication mechanisms are proposed. The first approach considers a one-hop communication mechanism, meaning that the coordinator robots send the packets only to the direct neighborhood (robots within the communication range) and no forwarding of information can be done. Following this approach different bio-inspired algorithms have been proposed and compared.

The other approach is based on a multihop communication, which allows the spreading of the information among the team; an ant-based protocol has been designed and developed. It is worth mentioning that all proposed methods share the exploration algorithm, as described in Section 19.3. The main focus in this chapter is to describe these different recruitment bio-inspired algorithms considering the same exploration strategy. For details about the best performance of the described strategies, please refer to the references.

19.4.1 One-hop communication

This section treats the problem of recruiting the needed robots in target locations using only local spreading of the information about the detected targets. Essentially the information is sent using packets that contain mostly the coordinates of the detected targets. Therefore, the volume of information that is communicated among the robots is small, but it implies the robots still lack global knowledge of the environment. In this kind of approach, strongly inspired by the biological behavior of social insects, the decisions made by the robots are independent, and the other robots and the coordinators do not know the taken decisions; therefore, the coordinators robots will continue to send packets until the needed robots have actually arrived. At this purpose, bio-inspired techniques such as the firefly algorithm (FA), PSO, and distributed bee algorithm have been proposed as coordination mechanisms to form coalitions of robots and compared.

19.4.1.1 Firefly algorithm

FA is a nature-inspired stochastic global optimization method that was developed by Yang (2009). It tries to mimic the flashing behavior of a swarm of fireflies. A firefly in the search space communicates with the neighboring fireflies through its brightness, which influences the selection. Firefly swarms in nature exhibit social behavior using collective intelligence to perform their essential activities like species recognition, foraging, defensive mechanisms, and mating. A firefly has a special mode of communication with its light intensity that signals to the swarm about its information concerning its species, location, attractiveness, and so on. The two important properties of the firefly's flashing light are defined as follows:

- brightness of the firefly is proportional to its attractiveness;
- brightness and attractiveness of a pair of fireflies is inversely proportional to the distance between two.

These properties are responsible for visibility of fireflies, paving the way for communication with each other. The distance $r(i, j)$ between any two fireflies i and j, at positions x_i and x_j, respectively, can be defined as the Euclidean distance as follows:

$$r_{ij} = ||x_i - x_j|| = \sqrt{\sum_{d=1}^{D}(x_{i,d} - x_{j,d})^2}, \tag{19.7}$$

where $x_{i,d}$ is the dth component of the spatial coordinate x_i of the ith firefly and D is the number of dimensions. In the two-dimensional case, $r(i, j) : \mathbb{R}^2 \to \mathbb{R}$,

$$r_{ij} = \sqrt{(x_i - x_j)^2 + (y_i - y_j)^2}. \tag{19.8}$$

In FA, as the attractiveness function of a firefly j varies with distance, one should select any monotonically decreasing function of the distance to the chosen firefly defined as:

$$\beta = \beta_0 \, e^{-\gamma r_{ij}^2}, \tag{19.9}$$

where r_{ij} is the distance defined as in Eq. (19.7), β_0 is the initial attractiveness at the distance $r_{ij} = 0$, and γ is an absorption coefficient at the source which controls the decrease in light intensity. The movement of a firefly i which is attracted by a more attractive (i.e., brighter) firefly j is governed by the following evolution equation:

$$x_i^{t+1} = x_i^t + \beta_0 \, e^{-\gamma r_{ij}^2} (x_j^t - x_i^t) + \alpha(\sigma - \frac{1}{2}), \tag{19.10}$$

where the first term on the right-hand side is the current position of firefly i, the second term is used for modeling the attractiveness of the firefly as the light intensity seen by adjacent fireflies, and the third term is randomization, with α being the randomization parameter, determined by the problem of interest. Here, σ is a scaling factor that controls the distance of visibility, and in most cases we can use $\sigma \in [0, 1]$.

19.4.1.2 Firefly-based team strategy for robot recruitment

Concerning the considered problem, each coordinator robot k^* that has detected a target starts to behave like a firefly sending out help requests to its neighborhood $LN_{k^*}^t$. When a robot k receives this request and it decides to contribute in the disarming process, it stores the request in its list RR_k. If the list contains more requests, it must choose which target it will disarm. Using the relative position information of the found targets, the robot derives the distance between it and the coordinators and then uses this metric to choose the best target, which is usually the closer. The same information also allows to derive the next movement of the robots. The approach provides a flexible way to decide when it is necessary to reconsider decisions and how to choose among different targets.

It should be noted that the recruited robots do not respond to the received requests, since they can change their decision at any time, so the coordinator robots do not know which robots are arriving and continue to broadcast packets until the needed robots have arrived. This has some implications. First, not all recruited robots will go towards the target's locations, balancing the two tasks. Second, the order in which the requests are received is not as important as the allocation is not instantaneous. This allows an effective approach to reach solutions that the greedy strategy would miss. Third, the impact on communications is reduced, so that bandwidth used will increase slowly with the team size. Then the robots move towards the target's location according to a modified version of FA. The aim of this strategy is to increase the flexibility of the system that lets the robots be able to form groups effectively and efficiently in order to enhance the parallelism of the handling of the found targets, and at the same time move towards the targets location, avoiding overlapping regions and any redundancy (Fig. 19.4). Moreover, the algorithm allows to dynamically adjust the coordination task since it enables each robot to make the best choice from its own point of view.

19.4.1.3 Implementation of robot decision mechanism

The original version of FA is applied in the continuous space, and cannot be applied directly to tackle discrete problems, so the original algorithm has been modified properly. In the considered scenario, a robot can move in a two-dimensional discrete space and it can go just in the adjacent cells. This means that when a robot k, at iteration t, in the cell c_k^t with coordinates (x_k^t, y_k^t) receives a packet from a coordinator robot that has found a target, robot k will move in the next step $(t + 1)$ to a new position

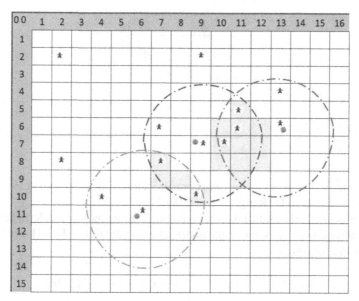

FIGURE 19.4

Example of an overlap region in which some robots are in the wireless ranges of different coordinator robots and thus they must decide towards which target to move.

(x_k^{t+1}, y_k^{t+1}), according to the FA attraction rules such as expressed below:

$$
\begin{cases}
x_k^{t+1} = x_k^t + \beta_0\, e^{-\gamma r_{kz}^2}(x_z - x_k^t) + \alpha(\sigma - \tfrac{1}{2}), \\
y_k^{t+1} = y_k^t + \beta_0\, e^{-\gamma r_{kz}^2}(y_z - y_k^t) + \alpha(\sigma - \tfrac{1}{2}),
\end{cases}
\tag{19.11}
$$

where x_z and y_z represent the coordinates of the selected target translated in terms of row and column of the matrix area and r_{kz} is the Euclidean distance between the target z and the recruited robot. It should be noted that a robot can receive more than one request. In the latter case, it will choose to move towards the brighter target within the minimum distance from the target as expressed in Eq. (19.9). A robot's movement is conditioned by the target's position and by a random component that it is useful to avoid the situation that more recruited robots go towards the same target if more targets have been found. This last condition enables the algorithm to potentially jump out of any local optimum (Fig. 19.4).

A key aspect occurs when a robot k moves too far from the target's position. Given a robot k located at step t in the cell of coordinates (x_k^t, y_k^t) and the target z with coordinates (x_z, y_z), the distance between robot k and the target z is the Euclidean distance r_{kz} as defined in Eq. (19.8). If $r_{kz} \geq (R_t + \Delta)\forall z \in RR_k$, this means that robot k moves too far from the target's locations and in this case, if it has not received

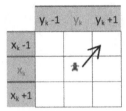

FIGURE 19.5

A possible selected cell.

other requests, it switches its role into Forager State. For details about potential robot states, please refer to De Rango et al. (2018).

In order to modify FA to a discrete version, the robot movements have been modeled by three kinds of possible value updates for coordinates $\{-1, 0, 1\}$, according to the following conditions:

$$
\begin{cases}
x_k^{t+1} = x_k^t + 1, & \text{if } [\beta_0 e^{-\gamma r_{kz}^2}(x_z - x_k^t) + \alpha(\sigma - \frac{1}{2}) > 0], \\
x_k^{t+1} = x_k^t - 1, & \text{if } [\beta_0 e^{-\gamma r_{kz}^2}(x_z - x_k^t) + \alpha(\sigma - \frac{1}{2}) < 0], \\
x_k^{t+1} = x_k^t, & \text{if } [\beta_0 e^{-\gamma r_{kz}^2}(x_z - x_k^t) + \alpha(\sigma - \frac{1}{2}) = 0],
\end{cases}
\tag{19.12}
$$

and

$$
\begin{cases}
y_k^{t+1} = y_k^t + 1, & \text{if } [\beta_0 e^{-\gamma r_{kz}^2}(y_z - y_k^t) + \alpha(\sigma - \frac{1}{2}) > 0], \\
y_k^{t+1} = y_k^t - 1, & \text{if } [\beta_0 e^{-\gamma r_{kz}^2}(y_z - y_k^t) + \alpha(\sigma - \frac{1}{2}) < 0], \\
y_k^{t+1} = y_k^t, & \text{if } [\beta_0 e^{-\gamma r_{kz}^2}(y_z - y_k^t) + \alpha(\sigma - \frac{1}{2}) = 0].
\end{cases}
\tag{19.13}
$$

A robot (e.g., robot k) that is in the cell with coordinates (x_k^t, y_k^t) as depicted in Fig. 19.5 can move, potentially, into eight possible cells according to the three possible values attributed to x_k and y_k. For example, if the result of Eqs. (19.12) and (19.13) is $(-1, 1)$, the robot will move into the cell $(x_k^t - 1, y_k^t + 1)$.

19.4.1.4 FTS-RR algorithm

The steps to be executed for FTS-RR are shown in Algorithm 2, see Fig. 19.6. Algorithm 2 is executed when one or more targets are found and some robots are recruited by others. If no targets are detected or all targets are removed or handled, the robots perform the exploration task according to Algorithm 1. More specifically, each recruited robot has the list of the requests in terms of target locations and evaluates the brightness of each of them, encoded as fireflies, taking into account their distances. At each step, the robots select the best from their list which has the maximum brightness. Next they move to the target's location according to firefly-based rules.

The proposed firefly-based approach is computationally simple. It requires only a few simple calculations (e.g., additions/subtractions) to update the positions of the robots. Moreover, the volume of information that is communicated among the robots is small, since only the position of the targets is sent. For this reason, FTS-RR has the benefit of scalability. In addition, the algorithm tries to form a coalition with the minimum size of involved robots, so the remaining robots are able, potentially, to conduct other search or disarmament tasks, allowing multiple actions at a time.

Algorithm 2: FTS-RR Algorithm

---

begin
   Step 1 : Initialization.
        Set t {t is the time step};
        Set the detected targets;
        Set the robots in Recruited State;
        Define the light absorption coefficient γ.
        Set the randomization parameter α;
        Set the random number σ;
        Set the attractiveness β_0;
   Step 2 : Generation coordination system.
        For the detected targets and the recruited robots, set the initial locations in terms of coordinates in x and y directions;
   Step 3 : Procedure.
   while *The stop criteria are not satisfied* **do**
      foreach *robot k in Recruited State* **do**
         set RR_k;
         evaluate the current position c_k^t;
         foreach *target $z \in RR_k$* **do**
            evaluate β according to Eq. (5.3);
            choose the best target z ;
         end foreach
         evaluate $N(c_k^t)$;
         compute the cell c_k^{t+1} according to Eqs.(5.6)-(5.7);
         if *(c_k^{t+1}.hasObstacle() or c_k^{t+1}.isOccupated() or c_k^{t+1}.isInaccessible())* **then**
            choose a random cell $c^* \in N(c_k^t)$;
            move robot k towards c^*;
         else
            move robot k towards c_k^{t+1};
         end if
      end foreach
      update t;
   end while
end

FIGURE 19.6

Algorithm 2: FTS-RR algorithm.

19.4.2 Particle swarm optimization

PSO is an optimization technique which uses a population of multiple agents (Kennedy and Eberhart, 1995). This technique is inspired by the movement of flocking birds and their interactions with their neighbors in the swarm. Each particle i moves in the search space and has a velocity v_i^t and a position vector x_i^t. A particle updates its velocity according to the best previous positions and the global best position achieved by its neighbors as follows:

$$v_i^{t+1} = \omega v_i^t + r_1 c_1 (g_{best} - x_i^t) + r_2 c_2 (p_{best} - x_i^t), \qquad (19.14)$$

where the individual best value is the best solution that has been achieved by each particle so far, called p_{best}. The overall best value is the best value (best position with the highest fitness function) that is found among the swarm, which is called g_{best}. Here, r_j ($j = 1, 2$) are uniformly generated random numbers between 0 and 1, while ω is the inertial weight and c_j ($j = 1, 2$) are the acceleration coefficients. In addition, Eq. (19.14) is used to calculate the new velocity v_i^{t+1} of a particle using its previous velocity v_i^t and the distances between its current position and its own best found position, that is, its own best experience p_{best} and the swarm's global best g_{best}. The new position of particle i is calculated by

$$x_i^{t+1} = x_i^t + v_i^{t+1}. \tag{19.15}$$

19.4.2.1 PSO for robot recruitment

Similarly as for FA, directly using this PSO-based decision strategy in the considered recruiting task would be problematic. Firstly on the two-dimensional map, there is only a limited number of possible directions for the robots to move and since we assumed that the robots can only move one cell at a time, the next position of the particles (robots) is limited to the neighbor cells. Moreover, in the recruiting phase, the object is to reach the target location (that is, g_{best}) and p_{best} is not taken into account. Therefore, a modified PSO version is proposed and this means that for each robot k at iteration t in a cell with coordinates (x_k^t, y_k^t), Eqs. (19.14) and (19.15) can be written as follows:

$$\begin{cases} v_{x_k}^{t+1} = \omega v_{x_k}^t + r_1 c_1 (x_z - x_k^t), \\ v_{y_k}^{t+1} = \omega v_{y_k}^t + r_1 c_1 (y_z - y_k^t) \end{cases} \tag{19.16}$$

and

$$\begin{cases} x_k^{t+1} = x_k^t + v_{x_k}^{t+1}, \\ y_k^{t+1} = y_k^t + v_{y_k}^{t+1}, \end{cases} \tag{19.17}$$

where (x_z, y_z) represent the coordinates of the detected target translated in terms of row and column of the matrix area. In order to modify the PSO to a discrete version, similar to case of the FA, the robot movements have been considered as three possible value updates for each coordinate $\{-1, 0, 1\}$, according to the following conditions:

$$\begin{cases} x_k^{t+1} = x_k^t + 1, & \text{if } [v_{x_k}^{t+1} > 0], \\ x_k^{t+1} = x_k^t - 1, & \text{if } [v_{x_k}^{t+1} < 0], \\ x_k^{t+1} = x_k^t, & \text{if } [v_{x_k}^{t+1} = 0] \end{cases} \tag{19.18}$$

and

$$
\begin{cases}
y_k^{t+1} = y_k^t + 1, & \text{if } [v_{y_k}^{t+1} > 0], \\
y_k^{t+1} = y_k^t - 1, & \text{if } [v_{y_k}^{t+1} < 0], \\
y_k^{t+1} = y_k^t, & \text{if } [v_{y_k}^{t+1} = 0].
\end{cases}
\tag{19.19}
$$

In this case, the PSO considers as metric the distance, thus when a robot receives more requests, it will choose to move toward the target at the minimum distance.

19.4.2.2 PSO-RR algorithm

In the described problem, the PSO algorithm is shown in Algorithm 3, see Fig. 19.7. Like FA, the steps are executed when the robots are recruited by others, but in the case when no targets are detected or all targets are handled, the robots continue to explore the area.

Algorithm 3: Particle Swarm based strategy algorithm

begin
  Step 1 : Initialization.
    Set t {t is the time step};
    Set the detected targets;
    Set the robots in Recruited State;
    Define the inertia weight ω;
    Set randomization parameter r_1;
    Set the acceleration coefficient c_1
  Step 2 : Generation coordination system.
    For the detected targets and the recruited robots, set the initial locations in terms of coordinates in x and y directions;
  Step 3 : Procedure.
  while *The stop criteria are not satisfied* **do**
    foreach *robot k in Recruited State* **do**
      set RR_k;
      evaluate the current position c_k^t;
      foreach *target $z \in RR_k$* **do**
        choose the best target z;
      end foreach
      evaluate $N(c_k^t)$;
      compute the cell c_k^{t+1} according Eqs.(5.12)-(5.13);
      if (c_k^{t+1}.hasObstacle() or c_k^{t+1}.isOccupated() or c_k^{t+1}.isInaccessible()) **then**
        choose a random cell $c^* \in N(c_k^t)$;
        move robot k towards c^*;
      else
        move robot k towards c_k^{t+1};
      end if
    end foreach
    update t;
  end while
end

FIGURE 19.7

Algorithm 3: PSO algorithm.

19.4.2.3 Artificial bee colony algorithm

Another evolutionary approach is the artificial bee colony (ABC) algorithm (Karaboga and Akay, 2009). This algorithm is inspired by the foraging behavior of honey bees when seeking a quality food source. In the ABC algorithm, there is a population of food positions and the artificial bees modify these food positions over time. The algorithm uses a set of computational agents called honeybees to find the optimal solution. The honey bees in ABC can be categorized into three groups: employed bees, onlooker bees, and scout bees. The employed bees exploit the food positions, while the onlooker bees are waiting for information from the employed bees about nectar amount of the food positions. The onlooker bees select food positions using the employed bee information and they exploit the selected food positions. Finally, the scout bees find new random food positions. Each solution in the search space consists of a set of optimization parameters which represent a food source position. The number of employed bees is equal to the number of food sources. The quality of a food source is called its "fitness value" and it is associated with its position.

In the algorithm, the employed bees will be responsible for investigating their food sources (using fitness values) and sharing the information to recruit the onlooker bees. The number of the employed bees or the onlooker bees is equal to the number of solutions in the population (SN). Each solution (food source) x_i ($i = 1, 2, \ldots, SN$) is a D-dimensional vector. The onlooker bees will make a decision to choose a food source based on this information. A food source with a higher quality will have a larger probability of being selected by onlooker bees. This process of a bee swarm seeking, advertising, and eventually selecting the best known food source is the process used to find the optimal solution. An onlooker bee chooses a food source depending on the probability value associated with that food source p_i calculated by the following expression:

$$p_i = \frac{fit_i}{\sum_{q=1}^{SN} fit_q},$$
(19.20)

where fit_i is the fitness value of the solution i evaluated by its employed bee, which is proportional to the nectar amount of the food source in position i, and SN is the number of food sources which is equal to the number of employed bees (BN). In this way, the employed bees exchange their information with the onlookers. In order to produce a candidate food position from the old one, the ABC uses the following expression:

$$x_{ij}^* = x_{ij} + \phi_{ij}(x_{ij} - x_{lj}),$$
(19.21)

where x_{ij}^* is the new feasible food source, which is selected by comparing the previous food source x_{ij} and the randomly selected food source, $l \in \{1,2,\ldots, SN\}$ and $j \in \{1,2,\ldots,D\}$ are randomly chosen indices, and ϕ_{ij} is a random number between $[-1, 1]$ which is used to adjust the old food source to become the new food source in the next iteration.

19.4.2.4 Artificial bee colony algorithm for robot recruitment

Similarly to the other two algorithms, the ABC algorithm has been modified to fit with our specific domain of interested as follows:

$$
\begin{cases}
x_k^{t+1} = x_k^t + \phi(x_k^t - x_z), \\
y_k^{t+1} = y_k^t + \phi(y_k^t - y_z),
\end{cases}
\tag{19.22}
$$

where $(x_z \; y_z)$ represent the coordinates of selected target translated in terms of row and column of the matrix area. Here, (x_k^t, y_k^t) is the current position of robot k and (x_k^{t+1}, y_k^{t+1}) is the new position of the recruited robot. In order to modify the ABC to a discrete version, like FA and PSO, the robot movements have been limited to three possible value updates for each coordinates, $\{-1, 0, 1\}$, according to the following conditions:

$$
\begin{cases}
x_k^{t+1} = x_k^t + 1, & \text{if } [\,\phi(x_k^t - x_z) > 0\,], \\
x_k^{t+1} = x_k^t - 1, & \text{if } [\,\phi(x_k^t - x_z) < 0\,], \\
x_k^{t+1} = x_k^t, & \text{if } [\,\phi(x_k^t - x_z) = 0\,]
\end{cases}
\tag{19.23}
$$

and

$$
\begin{cases}
y_k^{t+1} = y_k^t + 1, & \text{if } [\,\phi(y_k^t - y_z) > 0\,], \\
y_k^{t+1} = y_k^t - 1, & \text{if } [\,\phi(y_k^t - y_z) < 0\,], \\
y_k^{t+1} = y_k^t, & \text{if } [\,\phi(y_k^t - y_z) = 0\,].
\end{cases}
\tag{19.24}
$$

Essentially two cases could occur. The first is when a robot receives only one recruitment request; in this case, it will move towards the target location according to Eqs. (19.23) and (19.24). If a robot receives more than one request, it needs to decide which target it will move to. In this case, a concept according to the distributed bee algorithm presented in Jevtic et al. (2011) has been used.

Basically, when a robot k in cell c_k^t receives a packet from a coordinator in cell c_z^t, the cost of the target z for robot k at step t is calculated as the Euclidean distance between the robot and the target in the two-dimensional area, i.e.,

$$
r_{kz} = \sqrt{(x_k^t - x_z)^2 + (y_k^t - y_z)^2}, \quad \forall z \in RR_k.
\tag{19.25}
$$

Firstly, the concept of the *utility* of a target z for robot k, the reciprocal value of the distance, is introduced as

$$
\mu_z^k = \frac{1}{r_{kz}}.
\tag{19.26}
$$

Then, the probability that robot k chooses target z can be calculated by

$$p_z^k = \frac{\mu_z^k}{\sum_{b=1}^{RR_k} \mu_b^k}, \tag{19.27}$$

where $RR_k \subset F \subset T$. From Eq. (19.27), it is easy to show that

$$\sum_{z=1}^{RR_k} p_z^k = 1. \tag{19.28}$$

The underlying decision making mechanism adopts the roulette rule, also known as the wheel selection rule, that is, each target has been associated with a probability that it is chosen from a set of detected targets. Once all the probabilities are calculated according to Eq. (19.27), the robot will choose the target by spinning the wheel. Next the robot will move according to Eqs. (19.23) and (19.24). Such a coordination technique is well suited, like FA, to avoid that several robots approach the same target and spreads the robots over different target locations (Fig. 19.4).

19.4.2.5 ABS-RR algorithm

In the described problem, the algorithm for the bee-based strategy is shown in Algorithm 4 (see Fig. 19.8).

Like FTS-RR and PSO-RR, these steps are executed when the robots are recruited by others. In the case when no targets are detected or all the tasks about the targets are performed, the robots continue to explore the area until the mission ends.

It is worth pointing out that for all strategies, the decision mechanism is done at each step; this implies that if a recruited robot at step t chooses a target z, at the step $t + 1$ it takes again the decision and it could then choose another (better) target.

19.5 Influence of the parameters

Regarding the parameters of the proposed algorithms, a set of experiments have been performed in previous works in order to show and analyze the effectiveness of the proposal. The coefficients a_1 and a_2 in ATS-RE could be considered as noise in the diffusion of the pheromone. Noise is helpful since it can help to drive a robot to move through a region that has been covered to reach another region that, potentially, needs to be explored. Without noise, a robot would not move through these already explored cells, and could in fact become trapped. However, too much noise also has a negative impact because it marginalizes the effect of the pheromone. It is evident how too much or too little noise can negatively impact performance. A high value of a_1 means that the pheromone is more perceived and the impact of the distance decreases, while a high value of a_2 leads to a minor importance of the heuristic component. De Rango et al. (2018) show that balance of the two coefficients allows the swarm to

---

Algorithm 4: ABC-RR strategy

---

begin
  Step 1 : Initialization.
      Set t {t is the time step};
      Set the detected targets;
      Set the robots in Recruited State;
      Define randomization parameter ϕ
  Step 2 : Generation coordination system.
      For the detected targets and the recruited robots, set the initial locations in
      terms of coordinates in x and y directions;
  Step 3 : Procedure.
  while *The stop criteria are not satisfied* **do**
    foreach *robot k in Recruited State* **do**
      set RR_k;
      evaluate the current position c_k^t;
      foreach *target $z \in RR_k$* **do**
        evaluate p_z^k according to Eq. (5.21);
        choose the best target z according to the wheel-selection rule;
      end foreach
      evaluate $N(c_k^t)$;
      compute the cell c_k^{t+1} according to (5.17)-(5.18);
      if ($c_k^{t+1}.hasObstacle()$ or $c_k^{t+1}.isOccupated()$ or $c_k^{t+1}.isInaccessible()$) **then**
        choose a random cell $c^* \in N(c_k^t)$;
        move robot k towards c^*;
      else
        move robot k towards c_k^{t+1};
      end if
    end foreach
    update t;
  end while
end

FIGURE 19.8

Algorithm 4: ABC-RR strategy.

effectively execute exploration tasks and generally a good performance was observed when $a_1 = 0.5$ and $a_2 = 0.5$ (see Table 19.1).

Regarding the parameters of the FTS-RR algorithm, Palmieri and Marano (2016) analyzed how the values of the parameters influence the performance of the algorithm. Indeed, when the complexity of the task increases a high coefficient of attraction can negatively influence the performance. This happens because a high value of the attractiveness means that the weight of the attraction in Eq. (19.9) increases and it is possible that more robots, in an overlap region, go towards the same target, creating potentially an unnecessary redundancy, increasing the time to complete the task.

The randomization and light absorption coefficient parameters are important, especially when the complexity of a task increases in terms of number of targets and number of robots needed to handle a target. It is reasonable to expect that by increasing the number of robots the efficiency of the swarm improves and the values of parameter do not influence significantly the total performance.

For PSO and ABC techniques, we have used the values of previous studies by Clerc and Kennedy (2002) and Zhang et al. (2016), respectively (see Table 19.2).

Table 19.1 Parameters used in the exploration algorithm.

Parameter	Value
Sensing range R_s	4
ρ	0.2
$\Delta \tau_0$	2
φ	1
λ	1
η	0.9
a_1	0.5
a_2	0.5
ε	Uniform [0, 1]

Table 19.2 Parameters used in the coordination algorithms.

Parameter	Value
α	0.2
β_0	0.5
γ	$\frac{1}{L}$ (L=max{m,n})
σ	Uniform [0, 1]
ω	0.729
r_1	Uniform [0, 1]
c_1	2
ϕ	Uniform [−1, 1]

19.6 Conclusion

Coordination is one of the most challenging research issues in distributed MRSs, aiming to improve performance of a robotic system in accomplishing complex tasks. Social insect-inspired coordination techniques achieve these goals by applying simple but effective heuristics from which elegant solutions emerge. Simple yet effective heuristics that avoid complex, heavy computation and establish lightweight interactions are highly desirable for MRSs. Biologically inspired solutions for the challenging problem of multirobot coordination are gaining traction. In this work different techniques that are able to coordinate and control groups of autonomous robots have been presented. The main feature of these techniques is to drive the group of robots to make the best decisions in order to perform search and rescue missions in unknown hazardous area.

One of the key issues is to specify the rules of behavior and interactions at the level of an individual robot in order to minimize unnecessary movements, turning, and communication that can cause wastage of the resources of the systems. The work is based on a hybrid strategy which combines both indirect and direct communication mechanisms. It is studied how robots can accomplish the mission in a distributed and self-organized way through a stigmergic process in the exploration task, using simple information locally sent by the robots in the recruitment task. The system has unique features such as the minimal information exchange and local interactions between simple homogeneous robots, achieving complex collective behavior. Such solutions are in line with the general approaches used in swarm robotics, and support the desired system properties of robustness, adaptivity, and scalability.

The work and approaches presented in this work have paved the way for exploring new bio-inspired techniques for optimizing complex tasks for swarming robots that can be used with the same modifications in many fields related to swarm intelligence.

References

Arvin, F., Turgut, A.E., Bazyari, F., Arikan, K.B., Bellotto, N., Yue, S., 2014. Cue-based aggregation with a mobile robot swarm: a novel fuzzy-based method. Adaptive Behavior 22 (3), 189–206.

Banharnsakun, A., Achalakul, T., Batra, R.C., 2012. Target finding and obstacle avoidance algorithm for microrobot swarms. In: 2012 IEEE International Conference on Systems, Man, and Cybernetics (SMC), pp. 1610–1615.

Bayındır, L., 2016. A review of swarm robotics tasks. Neurocomputing 172, 292–321.

Burgard, W., Moors, M., Stachniss, C., Schneider, F.E., 2005. Coordinated multi-robot exploration. IEEE Transactions on Robotics 21 (3), 376–386.

Calvo, R., de Oliveira, J.R., Figueiredo, M., Romero, R., 2011. Inverse ACO applied for exploration and surveillance in unknown environments. In: COGNITIVE 2011, the Third International Conference on Advanced Cognitive Technologies and Applications.

Chen, X., Kong, Y., Fang, X., Wu, Q., 2013. A fast two-stage ACO algorithm for robotic path planning. Neural Computing and Applications 22 (2), 313–319.

Clerc, M., Kennedy, J., 2002. The particle swarm-explosion, stability, and convergence in a multidimensional complex space. IEEE Transactions on Evolutionary Computation 6 (1), 58–73.

Contreras-Cruz, M.A., Ayala-Ramirez, V., Hernandez-Belmonte, U.H., 2015. Mobile robot path planning using artificial bee colony and evolutionary programming. Applied Soft Computing 30, 319–328.

Couceiro, M.S., Vargas, P.A., Rocha, R.P., Ferreira, N.M., 2014. Benchmark of swarm robotics distributed techniques in a search task. Robotics and Autonomous Systems 62 (2), 200–213.

Countryman, S.M., Stumpe, M.C., Crow, S.P., Adler, F.R., Greene, M.J., Vonshak, M., Gordon, D.M., 2015. Collective search by ants in microgravity. Frontiers in Ecology and Evolution 3, 25.

Das, P.K., Behera, H.S., Das, S., Tripathy, H.K., Panigrahi, B.K., Pradhan, S., 2016. A hybrid improved PSO-DV algorithm for multi-robot path planning in a clutter environment. Neurocomputing 207, 735–753.

de Lope, J., Maravall, D., Quiñonez, Y., 2015. Self-organizing techniques to improve the decentralized multi-task distribution in multi-robot systems. Neurocomputing 163, 47–55.

De Rango, F., Palmieri, N., 2012. A swarm-based robot team coordination protocol for mine detection and unknown space discovery. In: 2012 8th International Wireless Communications and Mobile Computing Conference (IWCMC), pp. 703–708.

De Rango, F., Palmieri, N., 2016. Ant-based distributed protocol for coordination of a swarm of robots in demining mission. In: Unmanned Systems Technology XVIII, vol. 9837, p. 983706.

De Rango, F., Tropea, M., 2009a. Energy saving and load balancing in wireless ad hoc networks through ant-based routing. In: 2009 International Symposium on Performance Evaluation of Computer & Telecommunication Systems, vol. 41, pp. 117–124.

De Rango, F., Tropea, M., 2009b. Swarm intelligence based energy saving and load balancing in wireless ad hoc networks. In: Proceedings of the 2009 Workshop on Bio-Inspired Algorithms for Distributed Systems, pp. 77–84.

De Rango, F., Guerriero, F., Marano, S., Bruno, E., 2006. A multiobjective approach for energy consumption and link stability issues in ad hoc networks. IEEE Communications Letters 10 (1), 28–30.

De Rango, F., Lonetti, P., Marano, S., 2008. Mea-dsr: a multipath energy-aware routing protocol for wireless ad hoc networks. In: IFIP Annual Mediterranean Ad Hoc Networking Workshop, pp. 215–225.

De Rango, F., Tropea, M., Santamaria, A.F., Marano, S., 2009. Multicast QoS core-based tree routing protocol and genetic algorithm over an hap-satellite architecture. IEEE Transactions on Vehicular Technology 58 (8), 4447–4461.

De Rango, F., Palmieri, N., Yang, X.S., Marano, S., 2018. Swarm robotics in wireless distributed protocol design for coordinating robots involved in cooperative tasks. Soft Computing 22 (13), 4251–4266.

Derr, K., Manic, M., 2009. Multi-robot, multi-target particle swarm optimization search in noisy wireless environments. In: 2009 2nd Conference on Human System Interactions, pp. 81–86.

Di Caro, G., Ducatelle, F., Gambardella, L.M., 2005. AntHocNet: an adaptive nature-inspired algorithm for routing in mobile ad hoc networks. European Transactions on Telecommunications 16 (5), 443–455.

Doi, S., 2013. Proposal and evaluation of a pheromone-based algorithm for the patrolling problem in dynamic environments. In: 2013 IEEE Symposium on Swarm Intelligence (SIS), pp. 48–55.

Dorigo, M., Blum, C., 2005. Ant colony optimization theory: a survey. Theoretical Computer Science 344 (2-3), 243–278.

Dorigo, M., Stützle, T., 2003. The ant colony optimization metaheuristic: algorithms, applications, and advances. In: Handbook of Metaheuristics. Springer, pp. 250–285.

Ducatelle, F., Di Caro, G.A., Pinciroli, C., Gambardella, L.M., 2011. Self-organized cooperation between robotic swarms. Swarm Intelligence 5 (2), 73.

Fotino, M., De Rango, F., 2011. Energy Issues and Energy Aware Routing in Wireless Ad Hoc Networks. INTECH Open Access Publisher.

Fujisawa, R., Dobata, S., Kubota, D., Imamura, H., Matsuno, F., 2008. Dependency by concentration of pheromone trail for multiple robots. In: International Conference on Ant Colony Optimization and Swarm Intelligence, pp. 283–290.

Garcia, M.P., Montiel, O., Castillo, O., Sepúlveda, R., Melin, P., 2009. Path planning for autonomous mobile robot navigation with ant colony optimization and fuzzy cost function evaluation. Applied Soft Computing 9 (3), 1102–1110.

Gifford, C.M., Webb, R., Bley, J., Leung, D., Calnon, M., Makarewicz, J., Banz, B., Agah, A., 2010. A novel low-cost, limited-resource approach to autonomous multi-robot exploration and mapping. Robotics and Autonomous Systems 58 (2), 186–202.

Gordon, D.M., 2010. Ant Encounters: Interaction Networks and Colony Behavior. Princeton University Press, Princeton, NJ.

Guerriero, F., De Rango, F., Marano, S., Bruno, E., 2009. A biobjective optimization model for routing in mobile ad hoc networks. Applied Mathematical Modelling 33 (3), 1493–1512.

Hereford, J.M., Siebold, M.A., 2010. Bio-inspired search strategies for robot swarms. In: Swarm Robotics from Biology to Robotics. IntechOpen.

Hoff, N.R., Sagoff, A., Wood, R.J., Nagpal, R., 2010. Two foraging algorithms for robot swarms using only local communication. In: 2010 IEEE International Conference on Robotics and Biomimetics, pp. 123–130.

Jevtic, A., Gutiérrez, A., Andina, D., Jamshidi, M., 2011. Distributed bees algorithm for task allocation in swarm of robots. IEEE Systems Journal 6 (2), 296–304.

Johansson, R., Saffiotti, A., 2009. Navigating by stigmergy: a realization on an RFID floor for minimalistic robots. In: 2009 IEEE International Conference on Robotics and Automation, pp. 245–252.

Kalra, N., Martinoli, A., 2006. Comparative study of market-based and threshold-based task allocation. In: Distributed Autonomous Robotic Systems 7. Springer, pp. 91–101.

Karaboga, D., Akay, B., 2009. A comparative study of artificial bee colony algorithm. Applied Mathematics and Computation 214 (1), 108–132.

Kennedy, J., Eberhart, R., 1995. Particle swarm optimization (PSO). In: Proceedings of the IEEE International Conference on Neural Networks. Perth, Australia, pp. 1942–1948.

Kernbach, S., Thenius, R., Kernbach, O., Schmickl, T., 2009. Re-embodiment of honeybee aggregation behavior in an artificial micro-robotic system. Adaptive Behavior 17 (3), 237–259.

Kuyucu, T., Tanev, I., Shimohara, K., 2012. Evolutionary optimization of pheromone-based stigmergic communication. In: European Conference on the Applications of Evolutionary Computation, pp. 63–72.

Kuyucu, T., Tanev, I., Shimohara, K., 2015. Superadditive effect of multi-robot coordination in the exploration of unknown environments via stigmergy. Neurocomputing 148, 83–90.

Li, J., Chen, Z., Liu, Y., Cai, Y., Min, H., Li, Q., 2013. A modified particle swarm optimization algorithm for distributed search and collective cleanup. In: 2013 International Joint Conference on Awareness Science and Technology & Ubi-Media Computing (iCAST 2013 & UMEDIA 2013), pp. 137–143.

Liang, J.H., Lee, C.H., 2015. Efficient collision-free path-planning of multiple mobile robots system using efficient artificial bee colony algorithm. Advances in Engineering Software 79, 47–56.

Luna, R., Bekris, K.E., 2011. Efficient and complete centralized multi-robot path planning. In: 2011 IEEE/RSJ International Conference on Intelligent Robots and Systems, pp. 3268–3275.

Masár, M., 2013. A biologically inspired swarm robot coordination algorithm for exploration and surveillance. In: 2013 IEEE 17th International Conference on Intelligent Engineering Systems (INES), pp. 271–275.

Momen, S., 2013. Ant-inspired decentralized task allocation strategy in groups of mobile agents. Procedia Computer Science 20, 169–176.

Palmieri, N., Marano, S., 2016. Discrete firefly algorithm for recruiting task in a swarm of robots. In: Nature-Inspired Computation in Engineering. Springer, pp. 133–150.

Palmieri, N., De Rango, F., Yang, X.S., Marano, S., 2015. Multi-robot cooperative tasks using combined nature-inspired techniques. In: 2015 7th International Joint Conference on Computational Intelligence (IJCCI), vol. 1, pp. 74–82.

Palmieri, N., Yang, X.S., De Rango Floriano, F., 2017. Self-adaptive mechanism for coalitions formation in a robot network. In: Proceedings of the 21st International Symposium on Distributed Simulation and Real Time Applications, pp. 200–203.

Palmieri, N., Yang, X.S., De Rango, F., Santamaria, A.F., 2018. Self-adaptive decision-making mechanisms to balance the execution of multiple tasks for a multi-robots team. Neurocomputing 306, 17–36.

Palmieri, N., Yang, X.S., De Rango, F., Marano, S., 2019. Comparison of bio-inspired algorithms applied to the coordination of mobile robots considering the energy consumption. Neural Computing and Applications 31 (1), 263–286.

Payton, D., Daily, M., Estowski, R., Howard, M., Lee, C., 2001. Pheromone robotics. Autonomous Robots 11 (3), 319–324.

Purnamadjaja, A.H., Russell, R.A., 2010. Bi-directional pheromone communication between robots. Robotica 28 (1), 69–79.

Ranjbar-Sahraei, B., Weiss, G., Nakisaee, A., 2012. A multi-robot coverage approach based on stigmergic communication. In: German Conference on Multiagent System Technologies, pp. 126–138.

Ravankar, A., Ravankar, A.A., Kobayashi, Y., Emaru, T., 2016. On a bio-inspired hybrid pheromone signalling for efficient map exploration of multiple mobile service robots. Artificial Life and Robotics 21 (2), 221–231.

Senanayake, M., Senthooran, I., Barca, J.C., Chung, H., Kamruzzaman, J., Murshed, M., 2016. Search and tracking algorithms for swarms of robots: a survey. Robotics and Autonomous Systems 75, 422–434.

Trigui, S., Koubaa, A., Cheikhrouhou, O., Youssef, H., Bennaceur, H., Sriti, M.F., Javed, Y., 2014. A distributed market-based algorithm for the multi-robot assignment problem. Procedia Computer Science 32, 1108–1114.

Wagner, I.A., Lindenbaum, M., Bruckstein, A.M., 1999. Distributed covering by ant-robots using evaporating traces. IEEE Transactions on Robotics and Automation 15 (5), 918–933.

Wattanavekin, T., Ogata, T., Hara, T., Ota, J., 2013. Mobile robot exploration by using environmental boundary information. ISRN Robotics 2013.

Yan, Z., Jouandeau, N., Ali-Chérif, A., 2012. Multi-robot heuristic goods transportation. In: 2012 6th IEEE International Conference Intelligent Systems, pp. 409–414.

Yan, Z., Jouandeau, N., Cherif, A.A., 2013. A survey and analysis of multi-robot coordination. International Journal of Advanced Robotic Systems 10 (12), 399.

Yang, X.S., 2009. Firefly algorithms for multimodal optimization. In: International Symposium on Stochastic Algorithms, pp. 169–178.

Yang, X.S., 2010. Nature-Inspired Metaheuristic Algorithms. Luniver Press.

Yang, B., Ding, Y., Jin, Y., Hao, K., 2015. Self-organized swarm robot for target search and trapping inspired by bacterial chemotaxis. Robotics and Autonomous Systems 72, 83–92.

Zhang, B., Liu, T., Zhang, C., 2016. Artificial bee colony algorithm with strategy and parameter adaption for artificial bee colony algorithm with strategy and parameter adaption for global optimization. Neural Computing and Applications, 1–16.

Zhao, T., Wang, Y., 2013. Market based multi-robot coordination for a cooperative collecting and transportation problem. In: 2013 Proceedings of IEEE Southeastcon, pp. 1–6.

Optimization in probabilistic domains: an engineering approach

20

Panagiotis Tsirikoglou[a], Konstantinos Kyprianidis[c], Anestis I. Kalfas[d], Francesco Contino[b]

[a]*Limmat Scientific AG, Zurich, Switzerland*
[b]*Université Catholique de Louvain, Thermodynamics and Fluid Mechanics Group, Louvain, Belgium*
[c]*Mälardalens Högskola, Energy Engineering, Västerås, Sweden*
[d]*Aristotle University, Department of Mechanical Engineering, Thessaloniki, Greece*

CONTENTS

20.1 Introduction

Engineering design is facing a transition from the classic one-off designs to a more complete design pipeline that includes design space exploration and optimization processes. This transition is led by the continuous endeavor for added value in products and processes and the strong objective-oriented policy for sustainable growth, while the big advancements in algorithmic development, numerical modeling, and computational resources are key enablers. A characteristic example of the merits of such transition is provided on aircraft design, which constitutes the application field of focus of the current work. Thus, examining the design process and related research over the last years, optimization is identified in different stages of product or service development aiming to gain substantial competitive advantages. In the field of aircraft design, competitive advantages can potentially have the form of fuel savings, noise reduction, an increase in passengers safety and comfort, and improved fleet management and flight data harvesting. On top of the industry needs for more sophisticated products, policy in the form of directives and legislation of aircraft industry is pushed to comply with future strategic goals for sustainable development, environment protection, and human safety.

Essentially optimization is an automated decision making process that incorporates couples search and decision making strategies to an engineering optimization case. This case describes in detail the problem considered for optimization, and hence roughly contains the design variables, parameters, and outputs of the underlying numerical model and the objectives/constraints that need to be optimized. Therefore, optimization is an objective-oriented design process, where the derived designs are propagated and selected as "best" with respect to a number of well-defined objectives. Thus, the definition of "best" design(s) in an optimization problem cannot stand alone as the given solution to address every need, but it is rather complementary to the objectives defined in the optimization problem.

In parallel, it progressively became well understood that the nature of engineering models and their variables or parameters is rather uncertain than deterministic. The impact of those input uncertainties is significant, and often optimal designs obtained by deterministic optimization approaches deteriorate from the desired performance point or even fail to satisfy critical constraints of the real engineering system. To address this issue the design variables and parameters are now defined using a probabilistic distribution function and a relative or absolute range in order to effectively describe the uncertainty, while an uncertainty propagation technique quantifies their effect on a quantity of interest, such as objectives or constraints. The described process enables the probabilistic design optimization (PDO) to define new designs of desired qualities, but also designs that retain those qualities under the presence of the input variations. As expected the integration of uncertainty quantification to the optimization process comes at a price. Furthermore, the propagation of those uncertainties to the quantities of interest of the case considered for optimization dictates the sampling of a new space on top of the exploration of the design space from the optimizer, as mentioned before. Thus, it is understood that PDO intensifies the computational cost issues that already exist in deterministic optimization cases (Bellman, 1961).

Until now, motivation, advantages, and shortcomings of the deterministic and probabilistic optimization have been identified. Moreover, a great number of optimization frameworks in both domains are developed and benchmarked against artificial optimization problems, where optimal designs are known (Li et al., 2013; Quagliarella et al., 2019). Despite the fact that the same frameworks are used, integration to engineering design is not a trivial process. The necessary pipeline that facilitates the automated evaluation of the objectives and constraints is a challenging task, particularly when various disciplines are combined, e.g., thermal power–electrical coupled simulations (Sahoo et al., 2019).

Not different to the rest of the engineering fields, integration of a full PDO framework to aircraft design cases presents the same difficulties. A wide range of numerical models are available, varying from fast, low-fidelity models to time consuming, high-fidelity approaches that capture large amounts of physics, such as computational fluid dynamics (CFD) methods. Moreover, the definition of the optimization problems can strongly challenge the convergence of the optimization algorithm towards a global optimal design when a large number of design variables and uncertainties are considered and/or design space is heavily constrained. Therefore, feasibility of PDO applications to aircraft design is not satisfied, and great care has to be taken in order to increase the overall computational efficiency. Minimizing the number of calls of the expensive, original engineering model greatly contributes to the reduction of the computational demands. The main enablers of this minimization are mathematical models (Forrester and Keane, 2009), often known as surrogate models, that can mimic the response of the original expensive model for an initial limited training database of designs. Apart from surrogate evaluation, the capability of an optimizer to locate fast the global optimal design and the computational demands of the uncertainty propagation technique are crucial to the computational efficiency of the overall PDO framework. It is understood that merits in computational time can be obtained from development in several research fields. The longstanding experience and continuous efforts so far have increased the technology readiness level of the various techniques and led to several software solutions, such as DAKOTA (Adams et al., 2015), UQLab (Marelli and Sudret, 2014), and OpenMDAO (Gray et al., 2019).

In the current chapter, an overview of the computational pipeline of a PDO framework is first provided. Then, we focus our analysis on three aspects of the pipeline that relate to the integration of PDO applications in aircraft design and strongly affect their feasibility: problem definition, surrogate models, and global optimization schemes. In the last section, the current state of the art of the PDO applications to aircraft design are reviewed and discussed in relation with the aforementioned aspects of the PDO computational pipeline.

20.2 Probabilistic design optimization framework

The integration of PDO to engineering cases requires the development and implementation of a well-defined pipeline that automates and iterates through multiple searches and decision making. Fig. 20.1 illustrates an indicative implementation of such pipelines.

It should be highlighted that the flowchart illustrates the pipeline executed in one iteration of the PDO run. The total amount of iterations is directly related to the computational budget. A block-by-block description of the pipeline is first delivered.

- **Design space:** Initial input and all candidate designs obtained by the optimizer in each iteration are sampled here. The design space is based on the definition of all design variables and their ranges. Finally, any possible constraints on the input design variables segment the design space, and are thus described here.
- **Uncertain space:** For each design in design space, a set of new designs is defined here, in order to quantify the impact of uncertainties in objectives and constraints. The size of this set depends on the uncertainty propagation technique and the

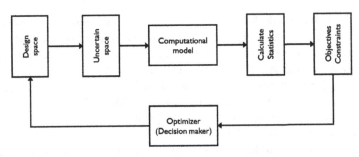

FIGURE 20.1

Computational pipeline of optimization in probabilistic domains. The pipeline describes one iteration of the optimization process. The overall optimization run constitutes a maximum number of iterations, defined by the computational budget.

dimensionality of the uncertain space. Finally, to perform this operation the probability distribution functions and ranges of all the uncertainties are employed.

- **Computational model:** The evaluation of the outputs of the engineering model is performed for the whole number of designs defined in design and uncertain space. The evaluation time is a key aspect that determines the overall computational budget and affects several choices regarding the structure of the PDO framework. To enable PDO for engineering models with relatively high evaluation time (a couple of minutes can be enough), the surrogate model totally or partially substitutes the original engineering model described above.

- **Calculate statistics:** The statistical measures of objectives and output constraints as defined in the optimization problem are calculated. The uncertainty propagation techniques sets the background methodology for this calculation.

- **Objective space:** The total set of objectives defined in the optimization problem form the objective space. Thus, all the calculated objectives and any possible output constraints are placed here.

- **Optimizer:** This block represents the selected optimization scheme. Two main operations are performed. Firstly, the candidate solutions are assessed and selected designs are propagated as the current optimal designs set. Moreover, the overall current best design, or set of designs (Pareto front) in the case of multiple objectives, is defined. Secondly, the search strategy is applied to current optimal designs set as described before, in order to further search the design space and provide new candidate solutions.

Please recall that the described actions are part of one iteration in the PDO framework. The number of iterations performed is dependent on the overall computational budget and the actual cost per iteration, i.e., how much original engineering model evaluation, nd how much time per evaluation are needed.

It is understood that the PDO framework consists of several methodologies that need to be effective and efficient to finally obtain a meaningful, optimal outcome in a realistic time frame. Moreover, its application to an engineering model requires a meticulous definition of the optimization problem, since the number of design variables, uncertainties, objectives, and constraints and their proper definition has a great impact on both the efficiency of the framework and the engineering impact of its optimal outcome.

Assuming that the optimization problem is defined properly and a robust engineering model is given, three methodologies of the PDO framework are important: the uncertainty propagation technique, the surrogate modeling, and the optimization scheme. Firstly, the propagation of the uncertainties is important for the accurate calculation of the statistical measures on the models' output, and the necessary amount of designs that are required to perform that operation. Moreover, the scale-up of the technique to a large number of uncertainties is of great importance for the integration of the PDO framework to more realistic engineering cases. A great deal of research is focused on developing such techniques (Abraham et al., 2017; Blatman and Sudret, 2011), aiming to accurately calculate the underlying partial differential equations of objectives and constraints using the lowest possible number of designs. Secondly, surrogate modeling can significantly extend the feasibility of PDO to more realistic, complex engineering cases. To achieve that, though, a proper surrogate management framework needs to be established in order to achieve good prediction accuracy and maintain it for cases that incorporate a large number of design variables and parameters. Thirdly, the selection of the optimization scheme affects the effectiveness of the search strategy and the quality of the decision making throughout the whole design process. Moreover, the class of the selected optimization scheme and its specific type determine the exploration and exploitation trade-off, and hence the amount of model evaluations needed until convergence to a local or global optimum solution.

In the current work we assume a robust engineering model and a state-of-the-art uncertainty propagation technique, and we focus on the following.

- **Problem definition:** The traits and effects of a proper problem definition are discussed.
- **Surrogate modeling:** The positioning of a surrogate model in a PDO framework and its managements is presented and thoroughly discussed.
- **Optimization scheme:** The selection of an optimization scheme is analyzed, based on the various trends of exploration/exploitation trade-off of the current state-of-the-art schemes.

20.2.1 Problem definition

Formulation of the deterministic optimization is the basis towards the proper definition of a PDO case. The conventional formulation of such a design problem uses a simple statement to link the objective(s) to the design variables, imposing input and output constraints:

$$\min_{x} \ J_m = f_m(x), \quad m = 1, 2, \ldots, M, \tag{20.1}$$

subject to

$$g_e(x) = 0, \quad e = 1, 2, \ldots, E, \quad h_i(x) \geq 0, \quad i = 1, 2, \ldots, I,$$

where the objective J_m is evaluated based on the outputs of the engineering model f_m seeking the global optimum design vector x^*. The search for this optimal design is subject to equality, g_e, and inequality, h_i, constraints that can affect the shape of the objective and design space, as described in the previous section. At this stage, the proper definition of design variables and its ranges and the selection of meaningful objective(s) are essential to lay the basis where PDO will be built upon. Last but not least, the proper definition of the constraints guarantees that the optimal outcome of the designs reflect the limitations of the actual technology.

Since a solid deterministic basis is formed, a problem statement for the final PDO case is needed. This additional layer constitutes the definition of the input uncertainties and the statistical formulations of the objectives and output constraints. These statistical measures quantify the effect of the uncertainties on the original objective J_m, hence becoming the new objectives of the optimization problem in the probabilistic domain. That said, the transformed new objectives need to address both the performance of the model and its sensitivity with respect to the defined input uncertainties. Two main approaches dominate the engineering design in probabilistic domains: robustness and reliability.

The term robustness is used to express and quantify the variance of each deterministic objective J_m, defined in Eq. (20.1), with respect to the input uncertainties. In this approach, each deterministic objective is transformed to its expectation and variance measures. Fig. 20.2 provides a visual example of this transformation, for a simple single-objective optimization problem that has one uncertain design variable.

To express that in an optimization case form, assume that the objectives, J_m, in Eq. (20.1) equal to one ($m = 1, J_1$), and hence we deal with a single-objective problem. As explained in the typical approach of robustness-driven design, the transition from deterministic to PDO doubles the number of objectives, since it is necessary to account here for both the expectation and the variance measures of the objective. The following equation describes the aforementioned rationale:

$$\min_{x} \ E[J_1], Var[J_1], \tag{20.2}$$

subject to

$$g_e(x) = 0, \quad e = 1, 2, \ldots, E,$$

$$h_i(x) \geq 0, \quad i = 1, 2, \ldots, I,$$

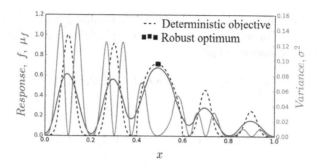

FIGURE 20.2

Transformation of a deterministic objective to an expectation measure and its variance under the presence of input uncertainties. The original deterministic objective is represented with the dashed line. As indicated, the new robust optimum is shifted to a more flat area, where variance is minimized.

where $E[J_1]$ and $Var[J_1]$ are the expectation and variance measures, respectively. Thus, the resulting optimization problem is biobjective. First, it should be identified if the relationship of the objectives is conflicting or not. For example, in Fig. 20.2 the resulting objectives in the probabilistic space are not conflicting, and hence one solution satisfies both. In the case of conflicting objectives, the extrapolation is straightforward, but note that higher-dimensional objective spaces provoke a few more issues in the computational pipeline of the PDO. The principal issue is that the presence of more than one conflicting objective changes the interpretation of "optimal" for the candidate designs through the course of optimization. In that case, the comparison of two candidate designs is not limited to the better/worse bipole, but it is rather described by the Pareto dominance relations (Deb et al., 2002). Thus, the optimal outcome is not a single (local or global) best design that could be captured for a given budget, but a set of designs in the form of an optimal Pareto front.

To avoid the increase in the number of the objectives and to utilize optimizers designed for single-objective decision making, collapsing of the various objectives to one weighted objective vector is followed. This collapsing is essentially the weighted summation of all the objective values in the objective vector. Despite the effectiveness of the method in reducing the size of the objective space, the optimal outcome consists of a single design, and hence the rest of design contained in the Pareto front is neglected. Repeated optimization runs with different assignment of weights in the collapsed objective vector can provide more designs that correspond to the true Pareto front. However, the accurate mapping between the weight values and the part of the Pareto front captured is not yet achieved.

As described, robust design optimization is the proper process to identify designs with good aspects and low variance at the same time. However, the operation of many engineering systems is characterized by various constraints, some of which prove critical. Therefore, reliability of the engineering systems under the presence of uncertainties becomes a primary design concern, and hence a different formulation of the optimization problem needs to be constructed. Assuming again a single-objective version ($m = 1, J_1$) of the classic deterministic optimization case (Eq. (20.1)) the reliability-based optimization problem is

$$\min_x E[J_1] \quad \text{or} \quad \min_x E[J_1] + w Var[J_1], \qquad (20.3)$$

subject to

$$P[g_e(x) = 0] \leq P_o, \quad e = 1, 2, \ldots, E,$$

$$P[h_i(x) \geq 0] \leq P_o, \quad i = 1, 2, \ldots, I,$$

where $E[J_1]$ and $Var[J_1]$ are the expectation and variance measures, respectively, w is an assigned weight value, and P represents the failure probability of the respective constraints, which has to be lower than a certain threshold noted as P_o. The main differentiating point is the use of the probability failure in the constraints of the underlying model. In this way, a minimum level of reliability under uncertainties, controlled by the P_o threshold, is maintained throughout the PDO process. The accurate calculation of the probability failure is a challenging task, since it requires the evaluation of more designs. Thus, continuous development of efficient schemes for the calculation of the probability of failure are detrimental for the feasibility for such schemes.

20.2.2 Surrogate model

Time is the universal constraint applied in any aspect of life and finally life itself. As expected, the available computational time is the main limitation in probabilistic design and optimization as well. There are several components of the computational pipeline (see Fig. 20.1) that control the overall budget requirements, such as the uncertainty propagation method, the exploration/exploitation trade-off of the optimizer, and the execution of the engineering model. The latter is critical to the feasibility of the probabilistic design and optimization process, particularly in the engineering field, where the calculation time per sample is relatively high or a high number of design variables and parameters are usually considered.

To quantify this effect, an example from simple aerodynamics is used, where the panel method (Drela, 1989) and Reynolds-averaged Navier–Stokes (RANS) equations are utilized to calculate lift and drag coefficients of an airfoil. The RANS method, which incorporates a larger amount of physics, is more than 10 times slower than the fast, but low-physics, panel method. Moreover, the absolute number regarding calculation time is 10 mins for the RANS calculation (on 24 cores) and 0.5 mins (single-core) for the panel method. Considering the absolute numbers, the RANS method is not prohibitive and someone can assume that the probabilistic design and optimization should be easily applied. However, if the evaluation of a full-factorial design of experiment (DoE) (Garud et al., 2017) is considered, the difference in computational time demands becomes significant as the number of design variables and

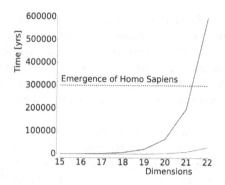

FIGURE 20.3

An illustrative example of the curse of dimensionality in engineering design, where an exponential increase in computational time is identified with respect to the number of design variables.

parameters is considered. Fig. 20.3 quantifies this difference for various numbers of design variables.

As can be seen, the increase in the number of variables, or search space in general, causes an exponential increase in the computational time needed to perform a full-factorial assessment of the model. This increase is not limited to the calculation of the full-factorial DoE, but it extends to operations such as optimization in deterministic and probabilistic domains, since searching strategies of the optimizers have to anticipate same-size or larger design spaces. In the case of PDO, the computational time demands are generally higher, due to the need of additional calculations to extract the statistics for each candidate design considered. Fig. 20.3 shows that a lower computational time per design delays the increase of the overall calculation time with respect to the design variables, thus making feasible probabilistic optimization for higher-dimensional design problems. Following the given example and considering the low complexity of the example given, it is apparent that the number of evaluations of the engineering model is crucial to the feasibility of the optimization process and the quality of the optimal outcome.

To enable the use of PDO in engineering design and extend its feasibility to more complex models or larger design spaces, approximations of the original engineering models are used. The aim of such approximation techniques, known as surrogate models, is to make sufficiently accurate predictions using the lowest number of original model evaluations possible. The significant computational advantages originating from their use led to the development of different kinds of surrogate models, such as kriging (Forrester and Keane, 2009) and its variants (Kleijnen, 2017), artificial neural networks (Cheng et al., 2016), polynomial regression (Forrester and Keane, 2009), and support vector regression (Smola and Schölkopf, 2004). The effectiveness and efficiency of the overall optimization scheme are controlled by two factors: the type of the surrogate models used and the structure of the scheme.

The first factor is related to the capabilities of the various models to derive quality predictions under certain conditions, e.g., linear or nonlinear original model responses. Chatterjee et al. (2019) performed a comparative assessment of different models indicating that anchored ANOVA, decomposition ANOVA, and polynomial chaos expansions (PCEs) (Wiener, 1938; Xiu and Karniadakis, 2002) are promising surrogates of the original engineering model, particularly for complex nonlinear responses.

The second factor is related to the structure of the overall optimization scheme, as regards the function and the management of the surrogate models within. Drawing experience from surrogate-based optimization schemes in deterministic domains, the management of surrogates is mainly divided in two cases. The first and simplest one is the a priori construction of the training database and the subsequent use of the surrogate models as global approximators over the whole design space. This management scheme is simple to implement, although the complete substitution of the original engineering model by the surrogate requires high prediction accuracy in the whole design space to derive optimal design close to reality. That is indeed feasible to achieve in lower-dimensional design spaces, and thus a lot of initial studies on surrogate-based optimization made use of this surrogate management structure. However, to obtain impactful designs, a high number of design variables and complex, high-physics models are needed, hence making the creation of a global approximator surrogate model a difficult task, both in terms of prediction accuracy and time demands. To tackle this rather strong limitation, adaptive formation of the training database of the surrogate model is suggested in the seminal study of Jones et al. (1998) and further developed in several studies (Shan and Wang, 2010). The basic principle behind all the adaptive management methods is the progressive build-up of the surrogate training database with designs that are of interest both for the course of the optimization run and the improvement of the prediction accuracy (Liu et al., 2018). That essentially translates to better prediction in design areas of interest.

While surrogate models as global approximators exist also in the probabilistic optimization field, adaptive formulation of the training database is a key aspect here. Moreover, the need to predict both the response of the original model and the behavior of its statistical measures adds a second level of prediction. To address this issue, one iteration of probabilistic optimization is split to the optimization and uncertainty loops, where different surrogate models are defined and used (Chaudhuri et al., 2019). The term "different" refers mainly to the designs that constitute their training databases and the type of data considered for prediction. Fig. 20.4 illustrates optimization and uncertainty loops in a typical iteration of a probabilistic optimization scheme (left) and the positioning of the surrogate models in such an iteration.

A common characteristic in some implementations of this multilevel surrogate modeling for probabilistic optimization is the combination of the design and uncertain space in one combined space (Arsenyev et al., 2015). This combined space allows the formulation of a training database that can describe the response of the original model with respect to all the parameters that induce a smaller (uncertainties) or larger (design variables) change to the inputs. The formulation of this design database al-

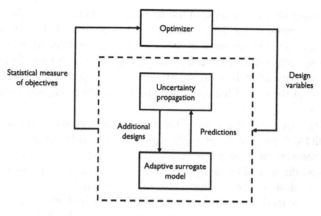

FIGURE 20.4

Positioning of an adaptive surrogate model in a sequential interaction of optimization and uncertainty loops. The flowchart is based on the work of Chaudhuri et al. (2019).

lows then the creation of the first-level surrogate and its sampling in order to obtain response data. The former, first-level surrogate is then coupled to the uncertainty propagation technique that needs to sample the surrogate model on a local scale in order to transform the input uncertainties to the statistical measure of preference. Based on these data, which link the design variables to the selected statistical measure, a training database is formulated, and the second-level surrogate model is fitted. The latter is then coupled to the search strategy of the optimizer, since it explicitly links the design variables to the objective considered for optimization.

A closer examination of the overall structure of the aforementioned scheme highlights again the approach regarding the formulation and treatment of the surrogate models. In particular, the training databases associated to the first-level surrogate model greatly affect the prediction accuracy of the surrogates in both levels. That said, one-off construction of that training database requires an a priori selection of designs that are well spread in the whole combined space (Arsenyev et al., 2015). Moreover, the size of the training database is increasing in order to guarantee a lower bound on the prediction accuracy of the first level surrogate, e.g., $(25\ 30) \cdot N_d$. Following the deterministic optimization paradigm, such techniques are limited to lower-dimensional spaces, and hence they are usually rendered insufficient for the particularly high-dimensional problems of PDO. Therefore, adaptive formulation allows for a smaller, initial training database, and the progressive definition of designs that are of interest for both the optimizer and the surrogate model (Liu et al., 2018).

Finally, the use of surrogate models within the context of probabilistic optimization raises the question of good prediction performance at both a local and global scale of the design space. Two aspects of the aforementioned optimization schemes are of interest here: combined space and adaptivity. Firstly, the combined space enables the definition of a DoE that samples sufficiently well the variations at both local and global scale. This advantage comes from the fact that the combined space has the shape of a hypercube, where some of its edges represent the range of the uncertainties and the rest the range of the design variables. As expected, a large number of uncertainties increase the hypervolume of the combined space, thus necessitating a large initial training database to achieve satisfactory prediction accuracy. To alleviate that, an initial sensitivity analysis on an uncertainty level is suggested. The analysis allows the reduction of the size of the combined space by adding only the most influential uncertainties.

Secondly, adaptive construction of the training database supports the good predictions at both local and global scale as the optimization run progresses. The reason behind that improvement is the gradual convergence of the optimization search strategy towards a specific area of the design space. This convergence reduces the Euclidean distance of the designs that are propagated to the training input database, thus allowing the surrogate model to predict more accurately a limited area of the design space. Despite the progressive improvement of the performance, the size of the initial training database and space filling properties of the designs have to be carefully defined to avoid any deterioration in the overall prediction performance of the defined surrogate models.

20.2.3 Optimization scheme

The effect of an optimization scheme is crucial to the quality of the optimal designs. That effect is reflected mainly in the ability of the optimizer to explore the design space and capture the optimal design, and the required budget to perform the aforementioned tasks. Those two effects are represented as the exploration/exploitation trade-off. In the current section, the exploration/exploitation trade-off is discussed for different classes of optimizers with a focus on global, nature-inspired optimization schemes. It should be highlighted that the goal is not to produce an explicit and detailed analysis of the existent optimization schemes. We rather aim to examine the overall developments in the optimization field with respect to their exploration/exploitation trade-off, hence projecting their potential to the PDO field.

The exploration/exploitation trade-off is considered as one of the key performance indicators of every optimization scheme. The interpretation of this indicator relies on the fact that every optimizer is essentially a decision maker, with an underlying search strategy that produces a series of candidate designs. This underlying searching strategy can be aggressive, thus producing designs that exploit the maximum improvement of the objectives under a narrow range of options. In contrast, other search strategies can be totally explorative, thus obtaining a variety of new candidate designs, though without considering the improvement of objective values. Those two aspects are conflicting in the search strategies that are developed so far, complying with the common perception that someone cannot explore more options if aggressive decision making is performed. Therefore, assuming that exploration and exploitation are objectives that need to be maximized in the current and future developed optimization schemes, their different combination should form a Pareto front. Fig. 20.5 demonstrates a qualitative interpretation of the exploration/exploitation trade-off in the form of a Pareto front as described above.

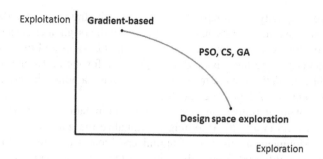

FIGURE 20.5

Qualitative Pareto front of exploration and exploitation as desired qualities of an optimization scheme. In the upper left fully exploitative gradient-based schemes are located, while in the lower right, extreme pure-design space exploration by space filling DoE is placed. Finally, all the gradient-free, nature-inspired optimizers lie in the middle part of the Pareto.

A brief analysis of the current developments in the optimization field is provided here, based on their location in the depicted Pareto. Starting from the upper left, the highly exploitative gradient-based schemes (Nocedal and Wright, 2006) are located. As the name states, gradient-based optimizers use gradient information to identify local areas of the design space that maximize the improvement of the defined objectives. Different mathematical approaches are employed in order to determine the search direction in each iteration (Nocedal and Wright, 2006) and calculate the gradients (Martins et al., 2003; Mader and Martins, 2012), thus modifying the overall search and computation efficiency. The main requirements for such optimization schemes are a starting design and the gradient information, while configuration of the algorithm is not an issue. Despite their significant computational efficiency and their sought mathematical background, gradient-based optimizers are strictly local search algorithms. Thus, their ability to locate the global optimum design is heavily dependent on the starting design, when complex engineer model responses are considered. To intensify the exploration and scale it up to a more global level, multistart, gradient-based optimization schemes have been developed (Chernukhin and Zingg, 2013). The definition of multiple new starting points allows the algorithm to perform the highly exploitative local search in several parts of the design space. Aiming to achieve the maximum of exploration of the design space, the multiple starting points are usually part of a space filling DoE, such as Latin hypercube sampling (Garud et al., 2017) or Sobol (Garud et al., 2017). This group of optimizers are located again in the upper left part of the Pareto front (see Fig. 20.5) but not in the extreme edge as the pure gradient-based optimizers.

Moving from the upper left to the middle and the lower right parts of the Pareto, the exploitation skills are more balanced to the exploration capabilities of the optimization schemes. In this area, the gradient-free class of optimizers is located. Gradient-free algorithms are a big family of stochastic optimizers that operate using only the value of the objectives to assess candidate designs and guide their search strategy. The simplicity and robustness of their search strategy allows them to perform under noncontinuous and complex responses of the original engineering model. The inputs of the algorithm consist of a set of initial designs and the configuration of their parameters. The latter is of great importance for the performance of the algorithm, while the definitions of the configuration set is a nontrivial task, due to the intrinsic randomness of their search strategies.

Several implementations of the gradient-free optimization paradigm exist in the literature (Boussaïd et al., 2013) aiming to capitalize the good search capabilities, while the development of this type of optimizers remains an animated topic of research. The main differentiation point between those implementations is the core research strategy often inspired by search patterns or processes that exist in nature (Kennedy and Eberhart, 1995; Michalewicz, 1995; Storn and Price, 1997; Yang and Deb, 2010; Yang, 2009). Among others, the original implementation of particle swarm optimization (PSO) (Kennedy and Eberhart, 1995) is a classic example of combination of exploration and exploitation skills. The optimizer exploits the food search mechanisms of birds in order to build an optimized search strategy. The particles (generalizing the term from bird) represent the candidate designs that the optimizer captures throughout the course of optimization. The critical part of this search mechanism is the definition of the velocity for each particle,

$$v_{i,d}^{t+1} = v_{i,d}^t + c_1 r_1^t (P_{i,d}^t - X_{i,d}^t) + c_2 r_2^t (P_{g,d}^t - X_{i,d}^t), \qquad (20.4)$$

where $X_{i,d}^t$ is the position of particle i in the design space during generation t and $P_{i,d}^t$ and $P_{g,d}^t$ are the local and global best position for the dimension d, respectively. Moreover, c_1 and c_2 are constants regulating two important terms in the updated velocity formula: cognitive and social. These two terms represent the two different mechanisms that a particle learns. The cognitive mechanism enables particle i to adjust its velocity towards the best position encountered until now, while the social mechanism allows the particle to drift its velocity towards the best solution captured by the whole swarm of particles. Finally, r_1^t and r_2^t represent random numbers uniformly distributed in $[0, 1]$. The randomization added to the search step enhances the exploration of the design space, while it helps to avoid the entrapment in local optima.

The numerous developments (Bonyadi and Michalewicz, 2017; Poli et al., 2007; Harrison et al., 2018) based on this initial simple idea are a strong indication of the capabilities of the PSO scheme, but also for the nature-inspired, global optimizers. Evidently, a quick examination of the research outcomes in the optimization development field demonstrates several new optimizers and variants of the most powerful ones (Tilahun et al., 2019; Al-Dabbagh et al., 2018; Jayabarathi et al., 2018). Despite the large production of optimization schemes, a relatively small fraction is used in expensive engineering design problems, due to their high computational demands and the lack of integration of novel optimizers to complex, i.e., multilevel, surrogate evaluation schemes. Using again the example of the Pareto front (see Fig. 20.5), efforts needs to be made in order to build upon the current developments and push the Pareto front to the direction where exploration and exploitation are both enhanced.

To this end, gradient-free and gradient-based schemes are coupled to optimization frameworks that aim to benefit from the advantages of both. The so-called hybridization employs a gradient-based optimizer and a switching criterion related to the convergence of the designs, in order to activate the gradient-based optimization scheme and increase the convergence speed (Bos, 1998). Despite the simplicity of the idea, it provided results that prove its benefits in real engineering problems (Chernukhin and Zingg, 2013; Vicini and Quagliarella, 1999; Bos, 1998). The merits of such optimization frameworks can be further extended by capitalizing the recent advancements in both optimization algorithmic development and surrogate modeling, aiming to create an optimization scheme with enhanced exploration/exploitation trade-off and maximized computational efficiency for PDO of expensive engineering cases, such as aircraft design.

20.3 Probabilistic optimization in aircraft design

In this section the advancements in aircraft design are examined from the PDO standpoint. Moreover, the different aspects of the computational pipeline as discussed above are linked to the different aspects or levels of aircraft design.

An aircraft as a finalized product is overly complicated, containing a very high number of components and several systems in place. Therefore, on a research level several engineering design frameworks are used to investigate single components, systems, and sets of systems at the aircraft level. To further analyze the overall research outcome, Fig. 20.6 illustrates the different levels of aircraft analysis and design.

Starting from the top, the aircraft level investigates the behavior and dynamics of the whole aircraft assuming that the output of the defined model encounters a number of phenomena originating from the different systems and components. The intractable complexity of the process necessitates the use of several assumptions in order to build the engineering model under consideration. The assumption making is essential to the reliability of the model and the engineering impact of the results. As expected, the fidelity of the computational approaches followed is low, due to limitations in the development of such large-scale engineering models and the computational resources.

The aircraft design as a whole, consists of several systems responsible for different functions, such as propulsion, electrical, landing, airframe, and more. The reduction in underlying complexity allows for an increase in the fidelity of the computational approaches. Despite the reduced levels, simplification through assumptions and lower-physics modeling is still necessary to render feasible studies of this scale. Finally, aiming to achieve more complete and thoroughly investigated designs, coupling of the various systems is becoming more and more necessary. Therefore, integrated systems design is proposed as one of the suitable approaches to increase the engineering impact of the designs produced.

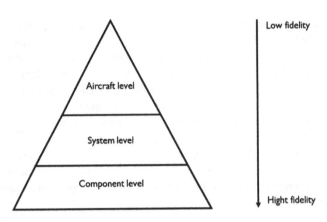

FIGURE 20.6

Segmentation of aircraft design: aircraft level, system level, and component level. As moving towards the integration of all components and systems in one entity, underlying models are simplified, i.e., levels of fidelity are decreasing.

Moving to the bottom of the design pyramid (see Fig. 20.6), components such as compressors, turbines, combustors, landing gear, electrical wiring, and many more are the basis of the whole design approach in aircraft vehicles. The relatively minimal complexity allows for the use of detailed designs and high-fidelity models such as CFD (Skinner and Zare-Behtash, 2018) approaches. Despite the increased level of details encountered in the design process, interactions with other components are largely neglected.

20.3.1 Applications on aircraft and system level

As described in the introduction section, engineering design in general is transitioning from a solely one-off design approach to a complete pipeline process that contains design space exploration, uncertainty quantification, and optimization in deterministic and probabilistic domains. In the current section, we examine the current status and outcomes of this design phase transformation regarding the first two top-level approaches in aircraft design, namely, the aircraft and the system level. Those approaches are treated together due to some similarities of the underlying models considered for optimization and the definition of the problem.

One-off design and assessment processes are the basis of aircraft development, since a nominal complete design is necessary to be obtained at least in the early design stage. Evidently, a lot of research studies are dedicated to the development of new design ideas (Drela, 2011a; Hall and Crichton, 2005) aiming to provide a complete view of a new candidate design. Building on this, new numerical schemes for design evaluation and optimization, and evolution of the computational framework are the enablers of research studies on the exploration and optimization of the initial one-off designs.

Aiming to demonstrate the described transformation, the example of the D8 transport aircraft configuration (Drela, 2011a) is discussed. The development of the D8 aircraft is supported within the N+3 initiative of NASA, aiming to shape future aircraft vehicles with significantly reduced fuel consumption. The initial development of the D8 aircraft is followed by further design examination, such as by Yutko et al. (2017), where boundary layer ingestion is introduced and the initial structure of the airframe is refined. Further than the one-off designs, Drela (2011b) demonstrated that deterministic optimization can produce more optimized variants of the initial aircraft, considering a number of different systems and assessing the various designs on a mission level analysis. A critical factor of the successful optimization run is the development of a set of models, namely, the Transport Aircraft System OPTimization (TASOPT) (Drela, 2010), which can assess different aircraft designs at feasible computational budgets. Finally, based on the strong basis of the available initial designs and efficient engineering models, probabilistic optimization of the D8 aircraft was performed by Ng and Willcox (2016). Considering the significant increase in computational costs, due to the need of a statistical estimator for every candidate design, an information reuse method is developed and coupled with the classic Monte Carlo (MC) method, achieving reductions of 90% compared with the original MC. As regards the optimal design outcome, a new aircraft configuration is captured that achieves 84% of predefined performance criteria, compared with an initial 22%.

Keeping aside the example of the complete design pipeline of the D8 aircraft, more probabilistic optimization studies on an aircraft design level are identified. Jaeger et al. (2013) introduced uncertainties in a short-range aircraft conceptual design optimization problem. Similar to the design of the D8 example, the availability of simple models for the aircraft design is one of the main enablers of this study (Birman and Druot, 2011). The modeling framework facilitates a range of models regarding the environmental impact, fuselage, wings and tail, propulsion system, and landing gear able to provide assessments of conceptual designs at the aircraft level. The performed optimization obtained robust aircraft designs that exhibit a 90%–95% chance of achieving the specified geometry and performance constraints. In a further attempt to integrate PDO to the conceptual design stage, Clark et al. (2019) used surrogate models as simplified, low-fidelity models to derive robust configurations of a generic fighter aircraft under mission uncertainties. The idea of using surrogate models in that design stage is particularly supported by a nondeterministic implementation of kriging (Bae et al., 2019), providing a reliable approximation tool for future use within such optimization frameworks. In all studies considered, the set of engineering models for the calculation of the quantities of interest are simplified, hence inducing some epistemic uncertainties. To assess the impact of those uncertainties, Molina-Cristóbal et al. (2014) introduced a novel epistemic uncertainty propagation technique coupled to a black-box modeling approach. The work targets the gas turbine system and its interactions with the airframe, thus focusing more one the integrated system design than the whole aircraft. Robust engine configurations that reduce the effect of the epistemic model uncertainties are obtained.

In previous sections, three aspects of the PDO pipeline were discussed: problem definition, evaluation of the engineering model, and optimization schemes. In the last part of this section a status of the applications as described above in relation to those three aspects is provided.

Firstly, probabilistic optimization coupled to the aircraft- and system-level design strongly relates to the problem definition. Despite the fact that the objective in such studies usually refers to some weight indicator on a mission level, i.e., maximum take-off weight, problem definition needs careful treatment regarding the constraint definitions. The latter are rather important due to the fact that constraints in the conceptual design phase represent critical technology limitations. Therefore, the complete definition of the probabilistic optimization problem significantly affects the engineering impact and feasibility of the underlying study.

Secondly, the evaluation of the engineering model is a relatively trivial task, since fast, simplified models are used at this level of design. Thus, the use of surrogate models, as global or adaptive approximators, valuable as it is considered, does not critically alter the feasibility boundaries of such applications.

Thirdly, the selection of the optimization scheme is slightly biased towards the gradient-based schemes. This preference stems from the strong mathematical background of those schemes, associated to their efficient handling of heavily constrained design spaces. Moreover, the relatively low evaluation cost of the engineering model enables the calculation of the necessary gradient information even with more computational time consuming techniques. This bias though does not exclude the use of derivative-free schemes. Particularly gradient-free optimizers designed for constrained spaces, such as COBYLA (Powell, 1994), come also into play when large search spaces (design or combined) with possible complex responses are formed. The low evaluation cost of the original engineering model again can support the increased computational demands of such optimization schemes. As expected, the availability of relatively fast engineering models simplifies the computational pipeline and allows for some flexibility in the choices of the optimization scheme. However, careful selection is necessary, since even low absolute evaluation time, e.g., several minutes, can result in computationally intractable optimization runs (see Fig. 20.3).

20.3.2 Applications at the component level

The literature analysis on the design optimization approaches applied to the aircraft and system levels indicates that PDO is gradually becoming part of the design, even in early stages. To complete the literature analysis, the status of the probabilistic design and optimization methods applied to the design of components and subcomponents of aircraft vehicles is examined.

Fundamental differences of component design with respect to aircraft and system design are identified in the response type and level of fidelity of the underlying engineering model and in the problem definition. The latter, in the case of component design, incorporates design specifications, variables, and parameters from a single technology domain, thus neglecting possible interactions. This does not mean of course that such optimization problems are simple to solve.

A vast number of research studies are available in the literature regarding the standalone design and deterministic optimization of aircraft components and subcomponents. The popularity of the subject stems from the large number of different components and subcomponents and the availability of different fidelity levels in the numerical approaches. Those two characteristics significantly increase the potential number of cases considered for design optimization, while they allow for a large variance in the complexity of the final case. Among others, airframe components such as wings and airfoils and gas turbine components such as turbine blades and compressors are thoroughly investigated and well documented (Skinner and Zare-Behtash, 2018; Du et al., 2017; Amrit et al., 2017; Chen and Agarwal, 2014). Due to this rich basis of design cases, penetration of deterministic optimization increases along with the advancements in different aspects of optimization pipeline, such as optimizers and surrogate modeling.

Based on the background described above, the animated research around uncertainty quantification and the development of more efficient uncertainty propagation techniques enables the realization of optimization in probabilistic domains. One of the first emerging and most popular applications is the PDO of an airfoil (Choi and Kwon, 2014; Wu et al., 2018; Rumpfkeil, 2012). The relatively low complexity of the overall optimization case and the continuous development of the lower- and higher-fidelity numerical approaches linked to the evaluation of the objectives allowed the fruitful production of research outcomes. Despite the generally lower evaluation times per design, the use of surrogate models is still necessary in order to increase the size of the combined space, thus creating a more comprehensive and detailed optimal design. Several studies employ a multilevel prediction scheme, as described in Section 20.2.2. As discussed, adaptivity and the definition of multiple levels in prediction improve the performance of the prediction scheme at a global and local scale. On that issue, Rumpfkeil et al. (2017) introduced a different technique, where clustering of the available training data is followed by the definition of multiple local surrogates. The different approximations made at the local level are probabilistically combined in an agglomerated final estimation. By extending these features to a multifidelity approach, good prediction accuracy is achieved for a limited size of the initial training database. Finally, the extensive study of this specific design problem led to the definition of benchmarks that use a specific airfoil geometry (Quagliarella et al., 2019), aiming to assess in a systematic and effective way novel PDO frameworks. Therefore, it is understood that the maturity level of methods related to uncertainty quantification is increasing.

In the current study, airfoils are considered as a partial design problem (sort of a subcomponent) of other components in the engine and the airframe systems. Due to the same reasons as the airfoil design problem, the wing, a component of the airframe system, is particularly well studied. Extending to PDO, wing-related applications are delivered (Liang et al., 2011), proving the readiness level of such design optimization problems as well. Due to this maturity, extensions of the former design problem to a multidisciplinary setting are identified. Jacome and Elham (2017) further optimize the wing geometry of a popular civil aircraft, using coupled aerodynamics-structural computation over a surrogate-assisted mission analysis. The successful optimization of the wing shape for given deterministic and probabilistic formulations of the objectives is performed in a cost-effective manner, while past flight data are used to define realistic uncertainty ranges. The latter is of great importance since a realistic problem definition significantly increases the engineering impact of the studies. In the same manner, but in a different component setting, Kamenik et al. (2018) use laser scans to model the manufacturing uncertainties of a high-pressure turbine blade. The defined uncertainties then serve as inputs in a probabilistic optimization scheme, aiming to derive robust shape designs. Despite the importance of realistic uncertainty ranges (see Section 20.2.1), a handful of research works combines the efforts of quantifying them with the optimization process.

Comparing to the application of probabilistic optimization at the aircraft and system levels, component-level design illustrates a significantly higher number of applications due to the higher number of possible cases, and their varying complexity. Following the same procedure as in the previous section, the application at the component level is discussed in relation with the three aspects of the computational pipeline of the probabilistic optimization, highlighted in the previous section: problem definition, evaluation of the engineering model, and optimization schemes.

Problem definition is a critical part in every optimization problem, both in deterministic and probabilistic domains. Different from the aircraft and system levels, the objectives here significantly vary, due to the focus on specific and not similar components. From the constraints standpoint, a varying level of complexity is also identified. In reality engineering search spaces (design and combined) are usually constrained, thus necessitating a more complex problem definition, if maximization of engineering impact is desired.

In a reverse trend, the engineering model is evaluated, through surrogate models are critical to the feasibility in most of the studies. This strong dependence stems from the generally increased amount of physics captured, pursuing an in-depth investigation of the isolated component. Kriging and its variants (Kleijnen, 2017) are key enablers of surrogate evaluation, while the research focuses on the construction of adaptive, multilevel surrogate prediction frameworks that can address the need of high prediction accuracy at both local and global levels.

Finally, a relatively strong bias towards the local gradient-based algorithms is identified. The main reason behind that choice is the generally increased computational demands for the evaluation of one design, e.g., turbine blade design cases. Therefore, the increased available budget per design limits the number of original model evaluations. Therefore, local, gradient-based optimizers are selected to further evolve the components considered for optimization. Moreover, adjoint formulations in shape optimization problems cheaply provide gradient information even in high-dimensional search spaces. To combine the necessary strong exploitation skills with more exploration, thus more alternate and possibly impactful designs, the use of hybrid scheme is suggested.

20.4 **Conclusions**

PDO applications in engineering, and particularly in aircraft design, are the results of an ongoing transition from one-off design processes to more complete design frameworks in the probabilistic domains. Longstanding research efforts focusing on a wide range of numerical techniques and algorithms support the development of such frameworks and their applications mainly through the maximization of the computational efficiency.

In this continuous pursuit of computational efficiency, two critical aspects of the PDO pipeline were identified and examined: the surrogate modeling and the selection of the optimization scheme. As regards the first one, global approximation using a priori training databases cannot be of use in the current status of engineering design problems, due to the higher-dimensional search spaces of the probabilistic optimization, i.e., for the same accuracy training databases of increasing size are needed. Moreover, PDO raised the additional demand of prediction accuracy at both global and local levels. A dominant solution captured was the multilevel, adaptive surrogate evaluation, creating effective prediction schemes at significantly reduced costs. However, the good performance of the surrogate at both local and global levels at the first iterations of the PDO process needs to be more investigated.

The maximization of the overall computational efficiency is strongly supported by an optimization scheme with strong exploration and exploitation skills. To establish a wide assessment, gradient-based and gradient-free optimizers were discussed within the context of the exploration/exploitation trade-off. The ongoing debate between the gradient-based and gradient-free algorithms, as the right way-to-go in engineering optimization, will be answered by optimization schemes that effectively combine the global exploration of the gradient-free algorithms and the local exploitation of the gradient-based schemes. To this end, building from the source optimization frameworks can facilitate good global exploration skills from nature-inspired algorithms and exploitation from the gradient-based ones.

Finally, the aircraft design field was examined with respect to the status of the PDO applications. Many interesting methods were identified, in relation to the surrogate evaluation and optimization scheme. However, one main characteristic was highlighted as a strong skill: knowledge of the PDO problem. The knowledge of the problem considered for optimization, which directly reflects the quality of the problem definition, has great impact on the derivation of meaningful results. Therefore, more studies that can define the ranges and types of uncertainties are encouraged, aiming to obtain a general picture regarding the fidelity of the technologies involved. Moreover, the continuous efforts to understand the aspects of different problems will enable safe generalizations and classifications, allowing the development of optimization frameworks operating on a more case-dependent basis.

References

Abraham, S., Raisee, M., Ghorbaniasl, G., Contino, F., Lacor, C., 2017. A robust and efficient step-wise regression method for building sparse polynomial chaos expansions. Journal of Computational Physics 332, 461–474.

Adams, B., Bauman, L., Bohnhoff, W., Dalbey, K., Ebeida, M., Eddy, J., Eldred, M., Hough, P., Hu, K., Jakeman, J., Stephens, J., Swiler, L., Vigil, D., Wildey, T., 2015. Dakota, a multilevel parallel object-oriented framework for design optimization, parameter estimation, uncertainty quantification, and sensitivity analysis: Version 6.0 user's manual. Sandia Technical Report SAND2014-4633.

Al-Dabbagh, R.D., Neri, F., Idris, N., Bab, M.S., 2018. Algorithmic design issues in adaptive differential evolution schemes: review and taxonomy. Swarm and Evolutionary Computation 43, 284–311.

Amrit, A., Du, X., Thelen, A.S., Leifsson, L.T., Koziel, S., 2017. Aerodynamic Design of the RAE 2822 in Transonic Viscous Flow: Single- and Multi-Point Optimization Studies.

Arsenyev, I., Duddeck, F., Fischersworring-Bunk, A., 2015. Surrogate-Based Robust Shape Optimization for Vane Clusters.

Bae, H., Clark, D.L., Forster, E.E., 2019. Nondeterministic kriging for engineering design exploration. AIAA Journal 57 (4), 1659–1670.

Bellman, R., 1961. Adaptive Control Processes: A Guided Tour. Princeton University Press.

Birman, J., Druot, T., 2011. Robustness assessment of a margin-setting process in preliminary aircraft sizing. In: 12e congrès annuel de la société Française de recherche opérationnelle et d'aide à la décision.

Blatman, G., Sudret, B., 2011. Adaptive sparse polynomial chaos expansion based on least angle regression. Journal of Computational Physics 230 (6), 2345–2367.

Bonyadi, M.R., Michalewicz, Z., 2017. Particle Swarm Optimization for Single Objective Continuous Space Problems: A Review.

Bos, A., 1998. Aircraft conceptual design by genetic/gradient-guided optimization. Engineering Applications of Artificial Intelligence 11 (3), 377–382.

Boussaïd, I., Lepagnot, J., Siarry, P., 2013. A survey on optimization metaheuristics. Information Sciences 237, 82–117.

Chatterjee, T., Chakraborty, S., Chowdhury, R., 2019. A critical review of surrogate assisted robust design optimization. Archives of Computational Methods in Engineering 26 (1), 245–274.

Chaudhuri, A., Marques, A.N., Lam, R., Willcox, K.E., 2019. Reusing Information for Multifidelity Active Learning in Reliability-Based Design Optimization.

Chen, X., Agarwal, R.K., 2014. Shape optimization of airfoils in transonic flow using a multi-objective genetic algorithm. Proceedings of the Institution of Mechanical Engineers, Part G: Journal of Aerospace Engineering 228 (9), 1654–1667.

Cheng, T., Wen, P., Li, Y., 2016. Research status of artificial neural network and its application assumption in aviation. In: 2016 12th International Conference on Computational Intelligence and Security (CIS), pp. 407–410.

Chernukhin, O., Zingg, D.W., 2013. Multimodality and global optimization in aerodynamic design. AIAA Journal 51 (6), 1342–1354.

Choi, S., Kwon, H.I., 2014. Robust Design Optimization Using a Trended Kriging Surrogate Model and Applications to Unsteady Flows.

Clark, D.L., Allison, D.L., Bae, H., Forster, E.E., 2019. Effectiveness-based design of an aircraft considering mission uncertainties. Journal of Aircraft 56 (5), 1961–1972.

Deb, K., Pratap, A., Agarwal, S., Meyarivan, T., 2002. A fast and elitist multiobjective genetic algorithm: Nsga-ii. IEEE Transactions on Evolutionary Computation 6 (2), 182–197.

Drela, M., 1989. Xfoil: an analysis and design system for low Reynolds number airfoils. In: Mueller, T. (Ed.), Low Reynolds Number Aerodynamics. In: Lecture Notes in Engineering, vol. 54. Springer, Berlin.

Drela, M., 2010. Tasopt 2.00–transport aircraft system optimization. MIT 3.

Drela, M., 2011a. Development of the D8 Transport Configuration.

Drela, M., 2011b. Simultaneous Optimization of the Airframe, Powerplant, and Operation of Transport Aircraft. Dept. of Aeronautics and Astronautics, Massachusetts Inst. of Technology.

Du, X., Leifsson, L., Koziel, S., Bekasiewicz, A., 2017. Airfoil design under uncertainty using non-intrusive polynomial chaos theory and utility functions, International Conference on Computational Science, ICCS 2017, Zurich, Switzerland, 12–14 June 2017. Procedia Computer Science 108, 1493–1499.

Forrester, A.I., Keane, A.J., 2009. Recent advances in surrogate-based optimization. Progress in Aerospace Sciences 45 (1), 50–79.

Garud, S.S., Karimi, I.A., Kraft, M., 2017. Design of computer experiments: a review. Computers & Chemical Engineering 106, 71–95.

Gray, J.S., Hwang, J.T., Martins, J.R.R.A., Moore, K.T., Naylor, B.A., 2019. OpenMDAO: an open-source framework for multidisciplinary design, analysis, and optimization. Structural and Multidisciplinary Optimization 59, 1075–1104.

Hall, Cesare A., Crichton, Daniel, 2005. Engine and instalations configuration for a silent aircraft. In: 17th ISABE Conference.

Harrison, K.R., Engelbrecht, A.P., Ombuki-Berman, B.M., 2018. Self-adaptive particle swarm optimization: a review and analysis of convergence. Swarm Intelligence 12 (3), 187–226.

Jacome, L.B., Elham, A., 2017. Wing aerostructural optimization under uncertain aircraft range and payload weight. Journal of Aircraft 54 (3), 1109–1120.

Jaeger, L., Gogu, C., Segonds, S., Bes, C., 2013. Aircraft multidisciplinary design optimization under both model and design variables uncertainty. Journal of Aircraft 50 (2), 528–538.

Jayabarathi, T., Raghunathan, T., Gandomi, A., 2018. The bat algorithm, variants and some practical engineering applications: a review. In: Nature-Inspired Algorithms and Applied Optimization. Springer, pp. 313–330.

Jones, D.R., Schonlau, M., Welch, W.J., 1998. Efficient global optimization of expensive black-box functions. Journal of Global Optimization 13 (4), 455–492.

Kamenik, J., Voutchkov, I., Toal, D.J.J., Keane, A.J., Högner, L., Meyer, M., Bates, R., 2018. Robust turbine blade optimization in the face of real geometric variations. Journal of Propulsion and Power 34 (6), 1479–1493.

Kennedy, J., Eberhart, R., 1995. Particle swarm optimization. In: Proceedings of the 1995 IEEE International Conference on Neural Networks, pp. 1942–1948.

Kleijnen, J.P., 2017. Regression and kriging metamodels with their experimental designs in simulation: a review. European Journal of Operational Research 256 (1), 1–16.

Li, X., Tang, K., Omidvar, M., Yang, Z., Qin, K., 2013. Benchmark functions for the CEC'2013 special session and competition on large-scale global optimization. Gene 7 (33).

Liang, Y., Cheng, X.Q., Li, Z.N., Xiang, J.W., 2011. Robust multi-objective wing design optimization via CFD approximation model. Engineering Applications of Computational Fluid Mechanics 5 (2), 286–300.

Liu, H., Ong, Y.S., Cai, J., 2018. A survey of adaptive sampling for global metamodeling in support of simulation-based complex engineering design. Structural and Multidisciplinary Optimization 57 (1), 393–416.

Mader, C.A., Martins, J.R.R.A., 2012. Derivatives for time-spectral computational fluid dynamics using an automatic differentiation adjoint. AIAA Journal 50 (12), 2809–2819.

Marelli, S., Sudret, B., 2014. UQLab: a framework for uncertainty quantification in MATLAB. In: Proceedings of the 2nd International Conference on Vulnerability, Risk Analysis and Management (ICVRAM2014). Liverpool, United Kingdom.

Martins, J.R.R.A., Sturdza, P., Alonso, J.J., 2003. The complex-step derivative approximation. ACM Transactions on Mathematical Software 29 (3), 245–262.

Michalewicz, Z., 1995. Genetic algorithms numerical optimization and constraints. In: Proceedings of the 6th International Conference on Genetic Algorithms, pp. 151–158.

Molina-Cristóbal, A., Nunez, M., Guenov, M.D., Laudan, T., Druot, T., 2014. Black-box model epistemic uncertainty at early design stage. An aircraft power-plant integration study. In: 29th Congress of the International Council of the Aeronautical Sciences.

Ng, L.W.T., Willcox, K.E., 2016. Monte Carlo information-reuse approach to aircraft conceptual design optimization under uncertainty. Journal of Aircraft 53 (2), 427–438.

Nocedal, J., Wright, S.J., 2006. Numerical Optimization. Springer.

Poli, R., Kennedy, J., Blackwell, T., 2007. Particle swarm optimization. Swarm Intelligence 1 (1), 33–57.

Powell, M.J.D., 1994. A direct search optimization method that models the objective and constraint functions by linear interpolation. In: Gomez, S., Hennart, J-P. (Eds.), Advances in Optimization and Numerical Analysis. Kluwer Academic, Dordrecht, pp. 51–67.

Quagliarella, D., Serani, A., Diez, M., Pisaroni, M., Leyland, P., Montagliani, L., Iemma, U., Gaul, N.J., Shin, J., Wunsch, D., Hirsch, C., Choi, K., Stern, F., 2019. Benchmarking Uncertainty Quantification Methods Using the NACA 2412 Airfoil with Geometrical and Operational Uncertainties.

Rumpfkeil, M., 2012. Optimization Under Uncertainty Using Gradients, Hessians, and Surrogate Models.

Rumpfkeil, M.P., Hanazaki, K., Beran, P.S., 2017. Construction of Multi-Fidelity Locally Optimized Surrogate Models for Uncertainty Quantification.

Sahoo, S., Zhao, X., Kyprianidis, K.G., Kalfas, A.I., 2019. Performance assessment of an integrated parallel hybrid-electric propulsion system aircraft. Paper No. GT2019-91459. American Society of Mechanical Engineers.

Shan, S., Wang, G.G., 2010. Survey of modeling and optimization strategies to solve high-dimensional design problems with computationally-expensive black-box functions. Structural and Multidisciplinary Optimization 41 (2), 219–241.

Skinner, S., Zare-Behtash, H., 2018. State–of–the–art in aerodynamic shape optimisation methods. Applied Soft Computing 62, 933–962.

Smola, A.J., Schölkopf, B., 2004. A tutorial on support vector regression. Statistics and Computing 14 (3), 199–222.

Storn, R., Price, K., 1997. Differential evolution–a simple and efficient heuristic for global optimization over continuous spaces. Journal of Global Optimization 11 (4), 341–359.

Tilahun, S.L., Ngnotchouye, J.M.T., Hamadneh, N.N., 2019. Continuous versions of firefly algorithm: a review. Artificial Intelligence Review 51 (3), 445–492.

Vicini, A., Quagliarella, D., 1999. Airfoil and wing design through hybrid optimization strategies. AIAA Journal 37 (5), 634–641.

Wiener, N., 1938. The homogeneous chaos. American Journal of Mathematics 60 (4), 897–936.

Wu, X., Zhang, W., Song, S., 2018. Robust aerodynamic shape design based on an adaptive stochastic optimization framework. Structural and Multidisciplinary Optimization 57 (2), 639–651.

Xiu, D., Karniadakis, G.E., 2002. The Wiener–Askey polynomial chaos for stochastic differential equations. SIAM Journal on Scientific Computing 24 (2), 619–644.

Yang, X.S., 2009. Firefly algorithms for multimodal optimization. Lecture Notes in Computer Science 5792, 169–178.

Yang, X.S., Deb, S., 2010. Engineering optimization by cuckoo search. International Journal of Mathematical Modelling and Numerical Optimisation 1 (4).

Yutko, B.M., Titchener, N., Courtin, C., Lieu, M., Wirsing, L., Tylko, J., Chambers, J.T., Roberts, T.W., Church, C.S., 2017. Conceptual Design of a D8 Commercial Aircraft.

Index

415

Printed in the United States
By Bookmasters